SPECIAL TOPICS IN EARTHQUAKE GEOTECHNICAL ENGINEERING

# GEOTECHNICAL, GEOLOGICAL AND EARTHQUAKE ENGINEERING

Volume 16



For further volumes:
http://www.springer.com/series/6011

# Special Topics in Earthquake Geotechnical Engineering

*by*

MOHAMED A. SAKR

*Egypt*

ATILLA ANSAL

*Turkey*

Springer

*Editors*
Mohamed A. Sakr
Faculty of Engineering
Tanta University
Tanta
Egypt
mamsakr@yahoo.com

Atilla Ansal
Kandilli Observatory and
Earthquake Research Institute
Boğaziçi University
Cengelkoy, Istanbul
Turkey
ansal@boun.edu.tr

ISSN 1573-6059
ISBN 978-94-024-0528-6 ISBN 978-94-007-2060-2 (eBook)
DOI 10.1007/978-94-007-2060-2
Springer Dordrecht Heidelberg London New York


Softcover re-print of the Hardcover 1st edition 2012


Printed on acid-free paper

Springer is part of Springer Science+Business Media (www.springer.com)

# Foreword

The Egyptian Society for Soil Mechanics and Geotechnical Engineering and the Technical Committee TC203 (previously TC4) on "Earthquake Geotechnical Engineering and Associated Problems" of the International Society for Soil Mechanics and Geotechnical Engineering has organised jointly the Satellite Conference on Earthquake Geotechnical Engineering held in Alexandria prior to the 17th International Conference on Soil Mechanics and Geotechnical Engineering October 2–3, 2009. The objectives were to generate an opportunity for the presentations and discussions on earthquake geotechnical engineering issues observed during the recent years.

A very prominent group of specialists in the field of earthquake geotechnical engineering delivered lectures of significant importance reflecting the recent developments in their field of interest during this conference. Later it was decided to compile all these presentations in one book so that it may be available to specialists and those that are involved with research and application in the field of geotechnical engineering.

The book is composed of 11 chapters. The first two chapters are concerned with geotechnical issues in the recent earthquakes one in Europe written by P. Monaco, G. Totani, G. Barla, A. Cavallaro, A. Costanzo, A. D'Onofrio, L. Evangelista, S. Foti, S. Grasso, G. Lanzo, C. Madiai, M. Maraschini, S. Marchetti, M. Maugeri, A. Pagliaroli, O. Pallara, A. Penna, A. Saccenti, F. Santucci de Magistris, G. Scasserra, F. Silvestri, A.L. Simonelli, G. Simoni, P. Tommasi, G. Vannucchi, and L. Verrucci titled as *Geotechnical Aspects of the L'Aquila Earthquake* and one in China written by I. Towhata and Y.J. Jiang titled as *Geotechnical Aspects Of 2008 Wenchuan Earthquake, China.*

The third chapter is about an experimental study on liquefaction written by T. Kokusho, F. Ito and Y. Nagao titled as *Effects of Fines and Aging on Liquefaction Strength and Cone Resistance of Sand investigated in Triaxial Apparatus*. Innovative miniature cone penetration and subsequent liquefaction tests were carried out in a modified triaxial apparatus on sand specimens containing fines. It has been found that one unique curve relating cone resistance $q_t$ and liquefaction strength $R_L$ can be established, despite the differences in relative density and fines content, the trend of which differs from the current liquefaction evaluation practice. In order to examine an aging effect on the relationship, sands containing fines added with a small amount of cement are tested to emulate a long geological period in a short time.

The fourth chapter was written by Ahmed Elgamal and Patrick Wilson titled as *Full Scale Testing and Simulation of Seismic Bridge Abutment-Backfill Interaction* where passive earth pressure at the abutments at high levels of seismic excitation that may provide resistance to excessive longitudinal bridge deck displacement was studied based on full scale tests and Finite Element (FE) simulations.

The fifth chapter is by Michael Pender, Liam Wotherspoon, Norazzlina M.Sa'Don, and Rolando Orense on the topic of "Macro Element for Pile Head Cyclic Lateral Loading". This chapter explores a macro element as an alternative which uses relatively simple formulae that are available for evaluating the lateral stiffness of long elastic piles embedded in elastic soil and an extension to handle nonlinear soil-pile interaction.

The sixth chapter is by M. Cubrinovski, Jennifer J. M. Haskell, and Brendon A. Bradley on the topic of "Analysis of piles in liquefying soils by the pseudo-static approach". This chapter addresses some of the key issues that arise in the application of the pseudo-static analysis to piles in liquefying soils, and makes progress towards the development of a clear modelling (analysis) strategy that will permit a consistent and reliable use of the simplified pseudo-static analysis.

The seventh chapter is written by Kyriazis Pitilakis and Vasiliki Terzi is titled as *Experimental and theoretical SFSI studies in a model structure in Euroseistest.* Within the framework of the EU research project (EUROSEISRISK Seismic hazard assessment, site effects and soil-structure interaction studies in an instrumented basin), a bridge pier model was constructed, instrumented and tested. The aim is the experimental investigation of the dynamic characteristics of the model, the study of the soil-structure-interaction effects, and in particular the wave field emanating from the oscillating structure to the surrounding ground.

The eighth chapter is written by Michele Maugeri, Glenda Abate, and Maria Rossella Massimino and titled as *Soil-Structure Interaction for Seismic Improvement of Noto Cathedral (Italy).* Noto Cathedral, one of the most famous example of Baroque architecture in Italy, was damaged by the December 13, 1990 earthquake ($M_L$ = 5.4). In particular, some columns of the right side of the main nave were severely fissured; due to the degeneration of these cracks, in 1996 the cathedral suffered a major partial collapse, which involved the columns of the right side of the main nave, the roof of the main nave and of the right nave and part of the dome. This chapter deals with pseudo-static and two 2-D FEM analyses of soil-structure interaction. These analyses were carried out for the maximum credible scenario earthquake bigger than the 1990 earthquake. Pseudo-static analysis as well as transient dynamic analysis by FEM modelling were performed.

The ninth chapter is by Susumu Yasuda and Kazuyori Fujioka and titled as *Study on the Method for the Seismic Design of Expressway Embankments.* In this chapter the behaviour of road embankments during past earthquakes in Japan are reviewed and methods for the seismic design of expressway embankments are discussed. There are several grades of roads in Japan. Though basic design methods for the embankments are similar, detailed methods are different. For example, the standard inclinations of slopes differ. The embankments for national expressways, which are of the highest grade, have been designed and constructed with special care.

The tenth chapter is by Kenji Ishihara titled as *Performances of Rockfill Dams and Deep-Seated Landslides during Earthquakes*. There has been increasingly high intensity of acceleration ever recorded during earthquakes in recent times. In consistence with this trend several characteristics have been unearthed of performances of rockfill dams and of large-scale landslides. With due considerations to these, damage features of high rockfill dams during recent earthquakes in Japan and China are briefly introduced herein, together with those previously reported. Large-scale landslides are mostly associated with development of deep-seated slip planes which pass through layers of cohesive soil deposits existing at great depths. Two cases of massive landslides that occurred in Japan during recent earthquakes are introduced and some interpretation given on the basis of the laboratory test results on volcanic pumice soils which are considered to have induced the landslides.

The eleventh chapter is by M.H. El-Naggar titled as *Bridging the Gap between Structural and Geotechnical Engineers In SSI For Performance-Based Design*. Performance-based design (PBD) involves designing structures to achieve specified performance targets under specified levels of seismic hazard, and requires the analysis of the entire soil-structure system. Dynamic soil-structure interaction (SSI) encompasses linear and nonlinear response features of the structure and foundation soil. Consequently, both structural and geotechnical expertise are needed. Most available computational tools, however, provide either elaborate models for the structure with simplified soil representation, or detailed geotechnical models with simplified structural idealization. This chapter discusses SSI pertinent to the analysis of soil-pile-structure interaction (SPSI) problems that are of direct concern to PBD of structures.

Mohamed Sakr and Atilla Ansal

# Contents

# List of Contributors

## Principal Contributers

**S. Aoi** Tsukuba

**J. J. Bommer** London

**D. M. Boore** Menlo Park

**K. W. Campbell** Oregon

**J. Clinton** Zurich

**I. G. Craifaleanu** Bucharest

**Z. Çağnan** Güzelyurt

**M. Çelebi** Menlo Park

**J. Douglas** Reykjavík

**T. van Eck** De Bilt

**P. Gülkan** Ankara

**L. Luzi** Milan

**B. Margaris** Thessaloniki

**R. Paolucci** Milan

**C. Péquegnat** Grenoble

**K. Pitilakis** Thessaloniki

**A. Roca** Barcelona

**E. Şafak** Istanbul

## Co-Contributors

**S. Akkar** Ankara

**G. Ameri** Milan

**D. Bindi** Potsdam

**I.S. Borcia** Bucharest

**C. Cauzzi** Zurich

**F.J. Chávez-García** México

**G. Cua** Zurich

**B. Dost** De Bilt

**D. Fäh** Zurich

**L. Frobert** Bruyères-le Châtel

**H. Fujiwara** Tsukuba

**D. Giardini** Zurich

**S. Godey** Bruyères-le Châtel

**X. Goula** Barcelona

**P. Guéguen** Grenoble

**F. Haslinger** Zurich

**R. Jacquot** Grenoble

**I. Kalogeras** Athens

**S. Koutrakis** Athens

**T. Kunugi** Tsukuba

**K. Makra** Thessaloniki

**M. Manakou** Thessaloniki

**M. Massa** Milan

**C. Michel** Zurich

**H. Nakamura** Tsukuba

**C.S. Oliveira** Lisbon

**M. Olivieri** Zurich

**F. Pacor** Milan

**C. Pappaioanou** Thessaloniki

**I.C. Praun** Bucharest

**R. Puglia** Milan

**D. Raptakis** Thessaloniki

**A. Savvaidis** Thessaloniki

**A. Skarlatoudis** Thessaloniki

**R. Sleeman** De Bilt

**T. Susagna** Barcelona

**N. Theodulidis** Thessaloniki

**C. Zülfikar** Istanbul

**P. Zweifel** Zurich

# Chapter 1
# Geotechnical Aspects of the L'Aquila Earthquake

**Paola Monaco, Gianfranco Totani, Giovanni Barla, Antonio Cavallaro, Antonio Costanzo, Anna d'Onofrio, Lorenza Evangelista, Sebastiano Foti, Salvatore Grasso, Giuseppe Lanzo, Claudia Madiai, Margherita Maraschini, Silvano Marchetti, Michele Maugeri, Alessandro Pagliaroli, Oronzo Pallara, Augusto Penna, Andrea Saccenti, Filippo Santucci de Magistris, Giuseppe Scasserra, Francesco Silvestri, Armando Lucio Simonelli, Giacomo Simoni, Paolo Tommasi, Giovanni Vannucchi, and Luca Verrucci**

**Abstract** On April 6, 2009 an earthquake ($M_L = 5.8$ and $M_W = 6.3$) stroke the city of L'Aquila with MCS Intensity I=IX and the surrounding villages with I as high as XI. The earthquake was generated by a normal fault with a maximum vertical dislocation of 25 cm and hypocentral depth of about 8.8 km. The deaths were about 300, the injured were about 1,500 and the damage was estimated as high as about 25 billion €. Both maximum horizontal and vertical components of the accelerations recorded in the epicentral area were close to 0.65 g. The paper summarises the activities in the field of earthquake geotechnical engineering aimed to the emergency and reconstruction issues. The ground motion recorded in the epicentral area is analysed; the geotechnical properties measured by in-situ and laboratory tests before and after the earthquake are summarised; site effects are preliminarily evaluated at accelerometric stations locations and damaged villages; the outstanding cases of ground failure are finally shown.

---

P. Monaco (✉) • G. Totani • S. Marchetti
Department of Structural, Water and Soil Engineering, University of L'Aquila, L'mAquila, Italy
e-mail: paola.monaco@univaq.it; gianfranco.totani@univaq.it; silvano@marchetti-dmt.it

G. Barla • S. Foti • M. Maraschini • O. Pallara
Department of Structural and Geotechnical Engineering, Politecnico di Torino, Torino, Italy
e-mail: giovanni.barla@polito.it; sebastiano.foti@polito.it; margherita.maraschini@polito.it; oronzo.pallara@polito.it

A. Cavallaro
CNR-IBAN, Catania, Italy
e-mail: acava@dica.unict.it

A. Costanzo
Department of Soil Protection, University of Calabria, Rende, Italy
e-mail: acostanzo@dds.unical.it

M.A. Sakr and A. Ansal (eds.), *Special Topics in Earthquake Geotechnical Engineering*, Geotechnical, Geological and Earthquake Engineering 16,
DOI 10.1007/978-94-007-2060-2_1, 

## 1.1 Introduction

The Abruzzo earthquake of April 6, 2009 caused considerable damage to structures over an area of approximately 600 km$^2$, including the urban centre of L'Aquila and several villages of the middle Aterno valley. Even for similar types of buildings, the distribution of damage within the affected area was irregular, due to both rupture directivity effects (Chioccarelli and Iervolino 2009) and site amplification phenomena.

In the first section of the paper, the ground motion in the epicentral area is described through the analysis of the records of the seismic stations located close to L'Aquila.

The second section presents a selection of results of in situ and laboratory tests carried out at various sites in the area of L'Aquila, including soil deposits of particular interest for the after-earthquake investigations. The data were collected from both previous tests by the geotechnical group of the University of L'Aquila and new investigations carried out after the earthquake by a task force of the Italian Geotechnical Society (AGI).

The third section shows the preliminary evaluation of site effects. In the middle Aterno valley, where experimental records were not available, the possible occurrence

A. d'Onofrio • L. Evangelista • A. Penna • F. Silvestri
Department of Hydraulic, Geotechnical and Environmental Engineering,
University Federico II, Naples, Italy
e-mail: donofrio@unina.it; lorenza.evangelista@unina.it; aupenna@unina.it;
francesco.silvestri@unina.it

S. Grasso • M. Maugeri
Department of Civil and Environmental Engineering, University of Catania, Catania, Italy
e-mail: sgrasso@dica.unict.it; mmaugeri@dica.unict.it

G. Lanzo • G. Scasserra
Department of Structural and Geotechnical Engineering, University La Sapienza, Rome, Italy
e-mail: giuseppe.lanzo@uniroma1.it; Giuseppe.scasserra@uniroma1.it

C. Madiai • G. Simoni • G. Vannucchi
Department of Civil and Environmental Engineering, University of Florence, Florence, Italy
e-mail: clau@dicea.unifi.it; gsimoni@dicea.unifi.it; giovan@dicea.unifi.it

A. Pagliaroli • P. Tommasi • L. Verrucci
CNR-IGAG, Rome, Italy
e-mail: alessandro.pagliaroli@uniroma1.it; paolo.tommasi@uniroma1.it; lucaverrucci@libero.it

A. Saccenti
ISMGEO Bergamo, Bergamo, Italy
e-mail: asaccenti@ismgeo.it

F. Santucci de Magistris
Department S.A.V.A, University of Molise, Campobasso, Italy
e-mail: filippo.santucci@unimol.it

A.L. Simonelli
Department of Engineering, University of Sannio, Benevento, Italy
e-mail: alsimone@unisannio.it

of site effects was back-figured from the variable building damage intensities. The damage level was compared among different villages, as well as within the same village, and related to the variable geological and morphological conditions, taking into account the different types of constructions.

The fourth section is devoted to ground failures in hard rock, in soft intensely fractured rocks and in coarse-grained materials. Also sinkholes and underground cavities are discussed.

In the fifth section, a case of liquefaction occurred in Vittorito village is analysed.

## 1.2 Ground Motion in the Epicentral Area

Figure 1.1 shows a map of the observed Mercalli-Cancani-Sieberg (MCS) intensity, which gives a picture of the non-uniform and asymmetric distribution of damage within the affected area. The main shock caused heavy damages in the centre of L'Aquila, where intensity value was reported varying between VIII and IX. Damages were even more significant in some villages located in the middle Aterno

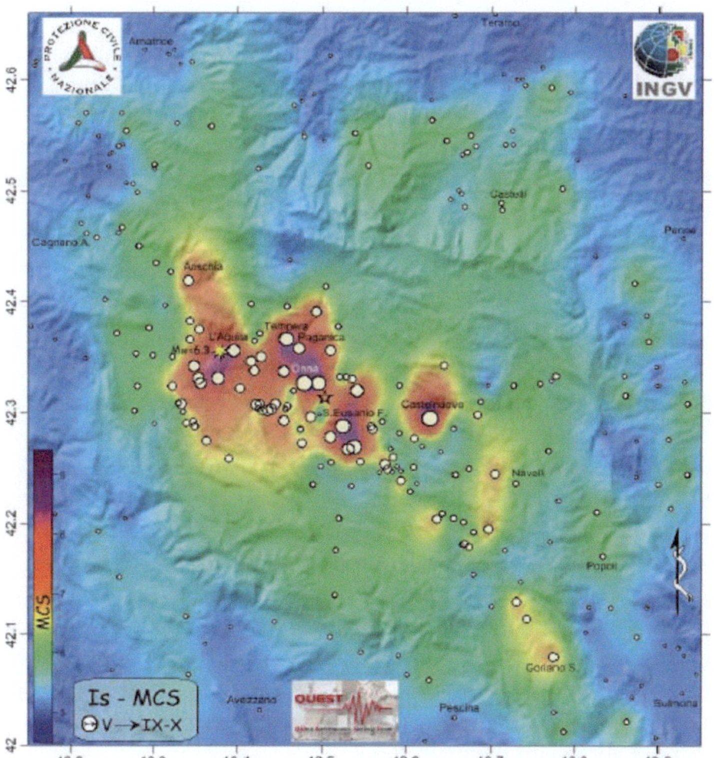


**Fig. 1.1** Map of the observed Mercalli-Cancani-Sieberg (MCS) intensity (Galli and Camassi 2009)

**Fig. 1.2** General view of the Aterno valley and bordering mountains showing the location of the strong-motion stations belonging to the DPC array (AQG, AQA, AQV, AQM, AQF, AQP); in the background the station AQK in the city of L'Aquila (vertical scale enlarged 2×)

valley, where intensities as high as IX-X were experienced in Castelnuovo and Onna. In total, 14 municipalities experienced a MCS intensity between VIII and IX, whereas those characterized by MCS intensity I ≥ VII were altogether 45 (Galli and Camassi 2009).

The Abruzzo earthquake is the first well-documented strong-motion earthquake instrumentally recorded in Italy in a near-fault area. The main event of April 6, 2009, was recorded by 56 digital strong motion stations which are part of the Italian Strong Motion Network (Rete Accelerometrica Nazionale, RAN), owned and maintained by the Department of Civil Protection (DPC). Fourteen stations are located in the Abruzzo region, while the remaining ones are distributed along the Apennines, mostly NW and SE of the source area. Five strong-motion stations were located within 10 km of the epicentre, on the hanging wall side of the normal fault; all of them recorded horizontal peak accelerations higher than 0.35 g. Four of these stations (AQG, AQA, AQV, AQM) belong to an array transversal to the upper Aterno valley, whereas one (AQK) is located in the centre of L'Aquila. A general view of the Aterno valley and the bordering mountains taken from Google Earth with the locations of the strong-motion stations is shown in Fig. 1.2. It can be seen in the figure that the transversal array also includes two other stations, namely AQF and AQP, which did not record the main shock.

A subsoil classification of the five stations that recorded the main shock is described by Di Capua et al. (2009), and is mainly based on the geological information

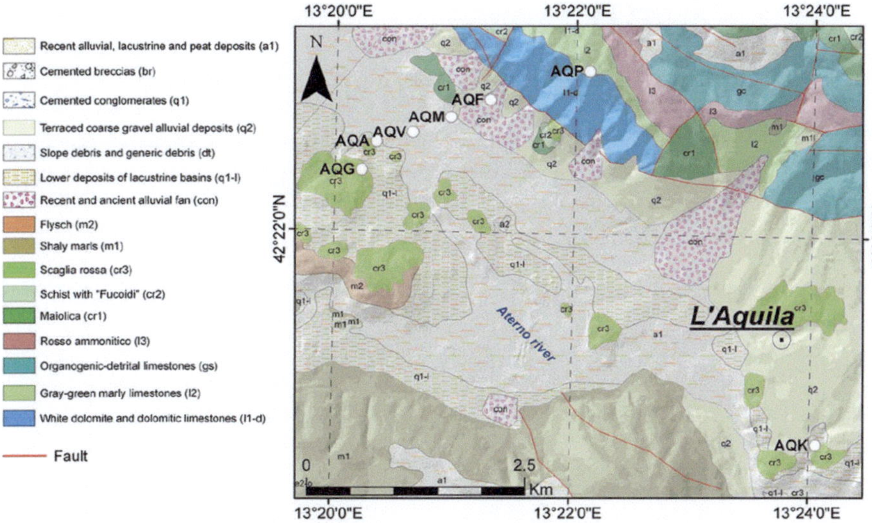


**Fig. 1.3** Geological map of Aterno Valley (Carta Geologica d'Italia, scale 1:100.000, Foglio 139 "L'Aquila"); white circles indicate strong-motion stations (Di Capua et al. 2009)

available. A geological map (scale 1:100.000) of L'Aquila and the eastern Aterno valley is shown in Fig. 1.3, together with the locations of the strong-motion stations. The city of L'Aquila is settled on cemented breccias having thickness of some tens of meters, and overlying lacustrine sediments resting on limestone. The Aterno valley is partly filled with Pleistocene lacustrine deposits formed by a complex sequence of pelitic and coarse-grained units overlying the limestone bedrock; these deposits are topped by quaternary alluvial deposits. The stations AQG and AQM are located on limestones, AQV and AQA on the recent alluvial deposits of the Aterno river, whereas AQK lays on cemented breccias. For both stations AQA and AQV, stratigraphic profiles are also available. Further, a shear wave velocity profile derived from cross-hole test is also available for station AQV (Fig. 1.4). Finally, it should be mentioned that AQK is installed in proximity of the entrance of the tunnel connecting a bus station with downtown L'Aquila, on the top of a retaining wall, and therefore its response might be affected by soil-structure interaction phenomena.

Table 1.1 lists the records available for all stations triggered by the main event and the aftershocks with local magnitude equal or greater than 4.0; for each seismic event considered, the magnitude, date and time of occurrence are reported. It is worth recalling that station AQM showed clipping during the main shock, i.e. recorded 1 g or more in the vertical and the horizontal directions. These recordings are still under investigation by DPC; therefore they will be not considered herein and are indicated with a question mark in the table. In the following, only records selected among those listed in Table 1.1 are examined and discussed, in order to highlight the possible occurrence of site effects.

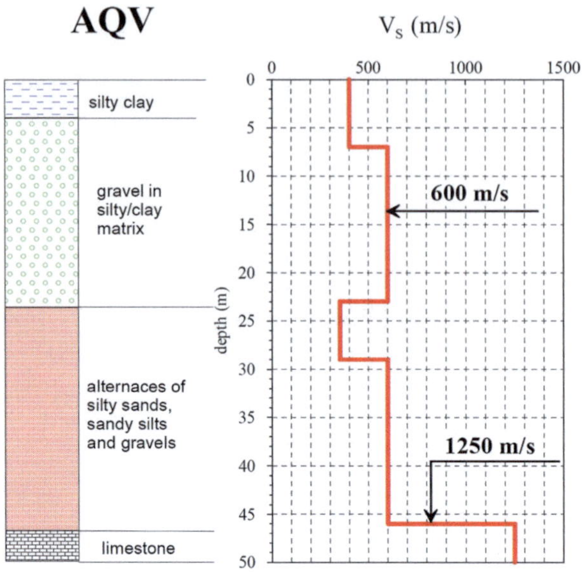


**Fig. 1.4** Available geotechnical information for the strong-motion station AQV

**Table 1.1** List of seismic events with $M_L \geq 4$ recorded by the RAN stations

| Event # | Date (d/m/y hh:mm UTM) | $M_L$ | AQV | AQG | AQA | AQK | AQF | AQP | AQM |
|---|---|---|---|---|---|---|---|---|---|
| 01 | 06-04-2009 01:32:39 | 5.8 | x | x | x | x | | | ? |
| 02 | 06-04-2009 02:37:04 | 4.6 | | x | | x | | | x |
| 03 | 06-04-2009 16:38:09 | 4.0 | x | | | x | | x | x |
| 04 | 06-04-2009 23:15:37 | 4.8 | x | | | x | | | |
| 05 | 07-04-2009 09:26:28 | 4.7 | x | x | | x | | x | x |
| 06 | 07-04-2009 17:47:37 | 5.3 | x | x | | x | x | x | x |
| 07 | 07-04-2009 21:34:29 | 4.2 | x | x | | x | | x | x |
| 08 | 08-04-2009 22:56:50 | 4.3 | x | | x | x | | x | x |
| 09 | 09-04-2009 00:52:59 | 5.1 | x | x | x | x | | x | x |
| 10 | 09-04-2009 03:14:52 | 4.2 | x | x | x | x | | | x |
| 11 | 09-04-2009 04:32:44 | 4.0 | x | x | x | x | | | x |
| 12 | 09-04-2009 19:38:16 | 4.9 | x | x | x | x | | x | |
| 13 | 13-04-2009 21:14:24 | 4.9 | x | x | | x | | | x |
| | | | 12 | 10 | 6 | 13 | 1 | 7 | 10 |
| | Total | | 59 | | | | | | |

The events simultaneously recorded at the stations AQV and AQG with AQG taken as reference station on rock, were considered first. The 5% damped acceleration elastic spectra for AQV and AQG are compared in Fig. 1.5 for both NS and EW components. In particular, the recordings of the main shock ($M_W$=6.3) and of the

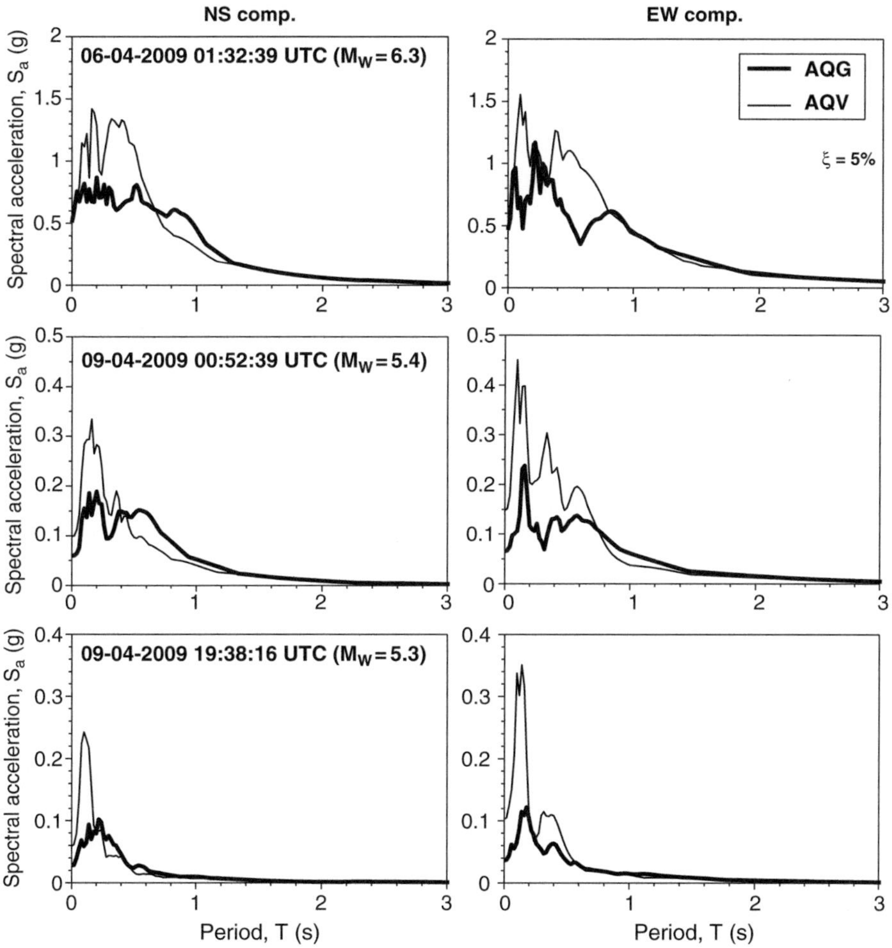


**Fig. 1.5** Comparison between 5% damped elastic acceleration response spectra from NS and EW components of accelerograms recorded at AQG and AQV strong motion stations

two higher magnitude aftershocks ($M_W$=5.4 and 5.3 respectively) are shown. Spectral acceleration values for AQV station are generally higher than those recorded at AQG, located on outcropping rock, especially in the period range 0.1–0.2 s and 0.3–0.5 s. This is clearly shown in Fig. 1.6 in which the spectral ratio AQV/AQG is plotted with reference to main shock and eight $M_L \geq 4$ aftershocks, recorded simultaneously at both stations. Moreover, by comparing the main shock spectral ratio with the corresponding average curve from the aftershocks, a clear evidence of soil non-linearity appears. In fact, main shock spectral ratios peaks are shifted towards higher periods (lower frequencies) with respect to aftershock peaks, due to soil stiffness reduction with increasing shaking intensity. Also, main shock

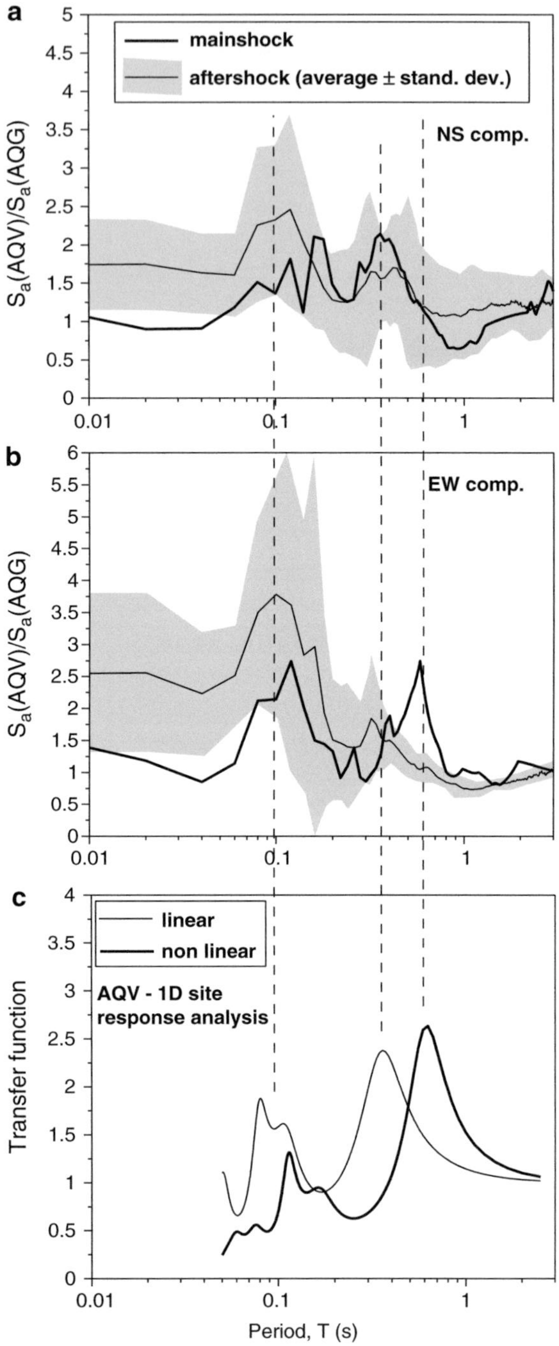


**Fig. 1.6** Spectral ratios AQV/AQG for the NS (**a**) and (**b**) EW components of the main shock and eight aftershocks (showed as average and average ± one standard deviation values); (**c**) linear and non-linear transfer functions from 1D site response analyses

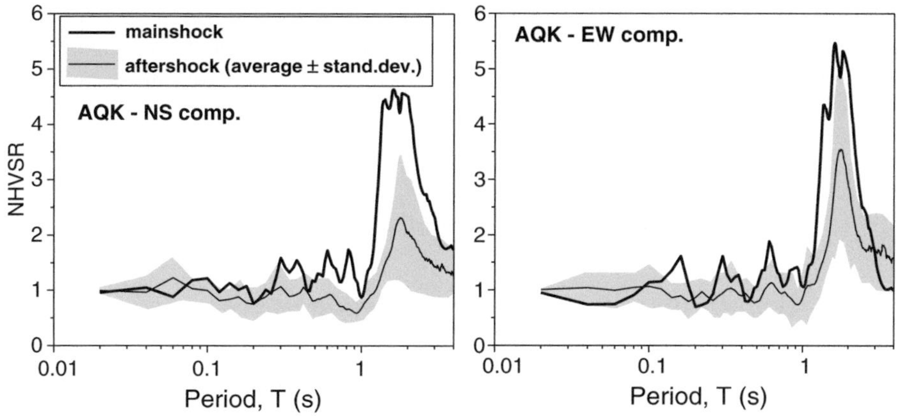


**Fig. 1.7** NHVSR computed at AQK station for main shock and 12 aftershocks (showed as average and average ± one standard deviation values)

amplitude of spectral ratios is lower than that recorded during aftershocks, due to increasing damping ratio with increasing acceleration level.

The experimental spectral ratios were confirmed by preliminary 1D frequency domain site response analyses carried out with the ProSHAKE computer code (EduPro Civil System 1998). The subsoil model was built according to the geotechnical data available for AQV station subsoil conditions (Fig. 1.4). Both linear and non-linear transfer functions between outcropping bedrock and surface were computed, showing a satisfactory agreement with the experimental spectral ratios (Fig. 1.6c). In the non-linear analysis, average normalised shear modulus and damping curves from literature were adopted for gravelly and silty soils; the AQV main shock recording was applied as input motion, after deconvolution to the bedrock.

The other station taken into account is AQK, located close to the city centre of L'Aquila, about 6 km from the six stations of the array. As this distance is comparable with that from the epicentre, array stations located on rock could not be used as reference sites. Preliminary information on likely site effects at AQK were therefore deduced by computing the Normalised Horizontal-to-Vertical Spectral Ratio (NHVSR), i.e. the ratio between the normalised 5% damped response spectrum of the horizontal component and the corresponding spectrum of the vertical component at the same station (Bouckovalas and Kouretzis 2001). The use of the response spectra was preferred rather than the Fourier spectra, mostly adopted in the literature, because it leads to smoother spectral ratios that can be more clearly interpreted. The NHVSR shown in Fig. 1.7 were computed for both NS and EW components of the records of the main shock and all of the 12 aftershocks listed in Table 1.1. A clear amplification of the horizontal components can be observed in the period range 1.5–2.0 s for all the events. Such evidence of low-frequency amplification at the AQK station is in agreement with previous results obtained from weak motion and ambient noise data, as well as from 2D numerical modelling of the basin underlying the city of L'Aquila (De Luca et al. 2005). Anyway, as AQK cannot be considered a free-field station, considering its location, further investigation is needed to explore the likely influence of soil-structure interaction phenomena on the NHVSR peaks.

## 1.3 Geotechnical Properties of Soils in the Aterno Valley

### *1.3.1 Geotechnical Properties from Routine In Situ and Laboratory Tests*

#### 1.3.1.1 Introduction

The collection and analysis of existing data from previous geotechnical investigations, generally carried out by *routine* in situ and laboratory testing techniques, is a primary step in planning after-earthquake site investigation programs (e.g. aimed at the characterization of reconstruction sites or at seismic microzonation), particularly in emergency scenarios which impose very strict time constraints for decisions.

Soon after the earthquake of April 6, 2009 a large amount of data from previous site investigations carried out in the area of L'Aquila was retrieved, organized in form of data base and made available to decision makers and investigators.

This section presents a selection of results of in situ and laboratory tests carried out at various sites in the area of L'Aquila (including sites/soil deposits of particular interest for the after-earthquake investigations), accumulated over the past years by the geotechnical group of the University of L'Aquila.

#### 1.3.1.2 Basic Geological Setting of the L'Aquila Basin

The area affected by the earthquake of April 6, 2009 is located within the central section of the Apennines, the mountain chain which traverses most of the length of the Italian peninsula.

The L'Aquila basin (middle Aterno River Valley) is a vast intra-Apennine tectonic basin elongated in NW-SE direction – parallel to many of the active normal faults – surrounded by the high peaks of the Gran Sasso and the Velino-Sirente mountains chains, mostly constituted by Meso-Cenozoic carbonate rocks and occasionally by marly-arenaceous rocks. The Aterno River is the principal hydrographic element of this area. A general view is shown in Fig. 1.8.

The current geological setting of the L'Aquila basin results from a complex sequence of depositional events, due to erosion and tectonics. The bottom of the valley was filled during the Quaternary with continental deposits of variable genesis and deposition age, resulting from lacustrine sedimentation followed by fluvial

**Fig. 1.8** General view of L'Aquila and its surroundings (view from SW, photos by G. Totani)

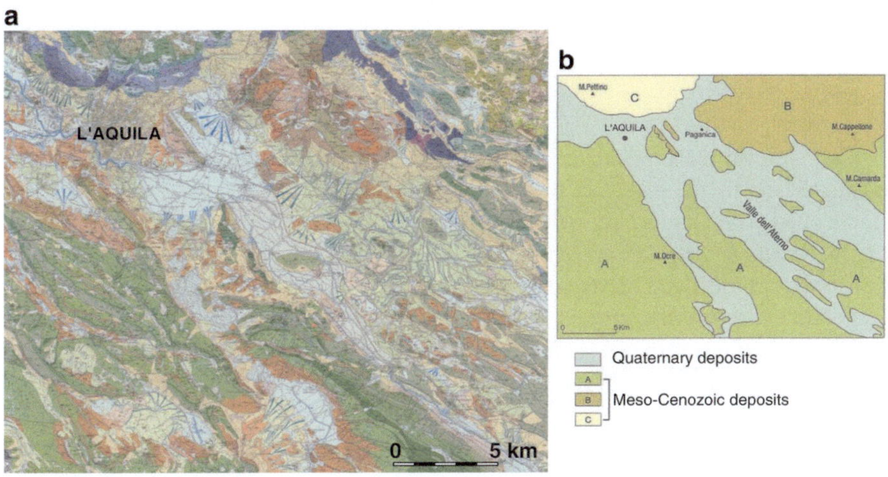


**Fig. 1.9** (**a**) Geological map of the area of L'Aquila. Principal formations: recent alluvial deposits (*light blue*), limestones (*green, brown*). (**b**) Schematic distribution of the Quaternary deposits in the L'Aquila basin and in the Aterno River Valley. (After Sheet No. 359 of the Geologic Map of Italy at the scale 1:50,000, APAT 2006)

sedimentation. The older Pleistocene lacustrine deposits, placed on the calcareous bedrock, form a complex depositional sequence of silt, sand and conglomerate units (Bosi and Bertini 1970; Bertini et al. 1989). The most recent Holocene deposits at the top, placed on the Meso-Cenozoic and Pleistocene deposits, are formed of alluvial soils (mostly sands and cobbles, less frequently sands and silts). The foot of the valley flanks and of the ridges located within the valley are covered by talus debris and locally by large debris alluvial fans. The relationship among the sedimentary styles is very complex, also due to the interplay of tectonics and climate changes. The maximum thickness of the Quaternary deposits is estimated to be about 400–500 m. The bedrock is constituted of limestone formations, generally outcropping along the sides of the valley and on ridges located within the Aterno River basin.

The geological setting of the L'Aquila basin is illustrated in Fig. 1.9a. The detail in Fig. 1.9b shows the schematic distribution of the Quaternary deposits which fill the bottom of the valley.

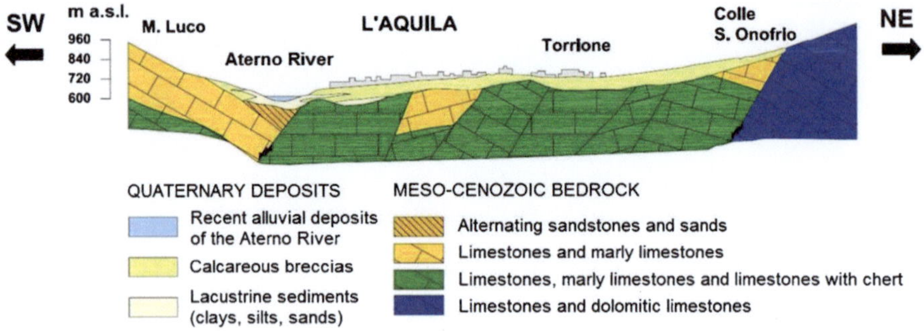


**Fig. 1.10** Geologic section in correspondence of the town centre of L'Aquila (Bertini et al. 1992a)

The area on the left side of the Aterno River basin (flow from NW to SE) is characterized by the presence of vast deposits of Pleistocene heterometric breccias (associated with Quaternary paleolandslides), known as "Megabreccias", placed on the lacustrine sequence overlying the bedrock. The "Megabreccias" are formed of limestone boulders and clasts ranging in size from some centimetres to some meters, embedded in sandy-silty matrix having a highly variable degree of cementation.

The old town centre of L'Aquila is located on a fluvial terrace on the left bank of the Aterno River. The highest elevation of the terrace is about 900 m a.s.l. in the NE part of the town and decreases down to 675 m a.s.l. in the SW direction. The terrace terminates in the Aterno River, which flows about 50 m below, at an elevation of 625 m a.s.l. The alluvial deposits forming the top of the terrace are composed of "Megabreccias", characterized by flat surfaces at the top and at the bottom and thickness of some tens of meters (Fig. 1.10).

#### 1.3.1.3 Results of Routine In Situ and Laboratory Tests in the Area of L'Aquila

This section presents a selection of results of site investigations carried out by *routine* in situ and laboratory testing techniques in the area affected by the April 6, 2009 earthquake. Most of the investigations were carried out in the past years, only a few of them were performed after the earthquake.

The soil deposit which has been more intensely investigated in this area, particularly in the West sector of the L'Aquila basin, is the lacustrine formation. Most of the results presented in this section relate to this deposit.

In detail, three different units may be distinguished within the lacustrine formation:

- an older unit at the bottom (placed on the bedrock) of highly variable composition, mostly gravels, sands and clays in variable combination;
- an intermediate unit, predominantly constituted by gravels and sands;

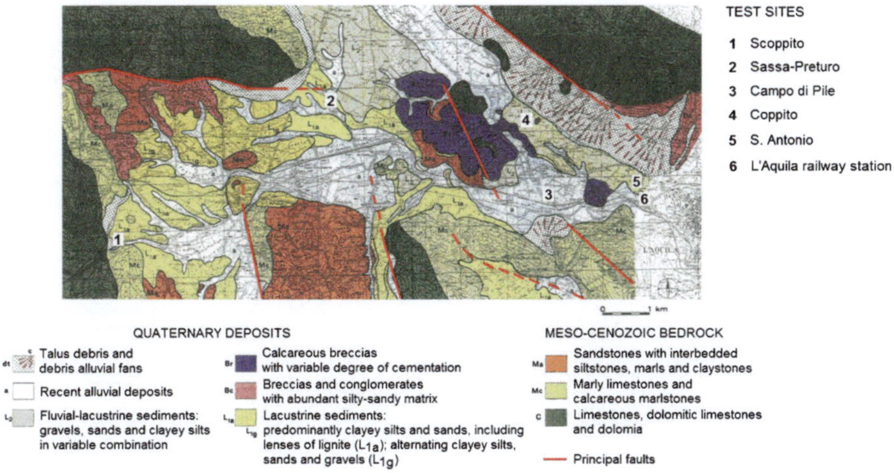


**Fig. 1.11** Schematic geological map of the West sector of the L'Aquila basin (Bertini et al. 1992b) and indicative location of six selected test sites

- a more recent unit at the top, predominantly constituted by sands and clays, having thickness of some tens of meters. Most of the results presented in this section refer to this top unit, which was encountered in this area within the commonly investigated depths.

The interest in characterizing these deposits (and, in general, the recent sediments of the Aterno River Valley) is also related to the higher concentration of structural damage caused by the earthquake on buildings founded on these soils, both in suburban areas of L'Aquila and in nearby villages (Onna and others), as mapped by after-earthquake field reconnaissance teams (e.g. GEER Working Group 2009).

The location of six selected test sites in the West sector of the L'Aquila basin (Scoppito, Sassa-Preturo, Campo di Pile, Coppito, S. Antonio, L'Aquila railway station) is shown in Fig. 1.11, superimposed to a schematic geological map of this area (Bertini et al. 1992b). Figures 1.12 and 1.13 show some examples of physical properties of the lacustrine deposits (grain size distribution, plasticity of fine-grained soils).

Several in situ flat dilatometer (DMT) tests have been carried over the past years in the area of L'Aquila. Some examples of DMT results are shown in Figs. 1.14–1.18.

In Figure 1.14 the results obtained by DMT in situ are compared to the results of laboratory tests (oedometer, triaxial UU) determined on samples taken at various depths at the same sites.

Figures 1.15–1.18 show the results of flat dilatometer tests carried out in the past at four sites located in the West sector of the L'Aquila basin (Scoppito, Campo di Pile, Coppito and L'Aquila railway station). The figures show the typical output of DMT results: the material index $I_D$ (indicating soil type), the constrained modulus $M$, the undrained shear strength $c_u$ and the horizontal stress index $K_D$ (related to OCR),

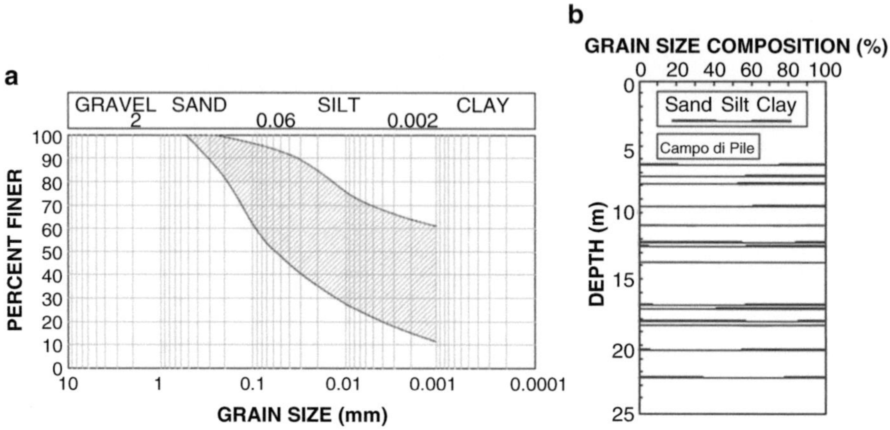


**Fig. 1.12** (**a**) Typical range of the grain size distribution of the lacustrine sediments in the West sector of the L'Aquila basin ($L_{1a}$ and $L_{1g}$ in Fig. 1.11). (**b**) Profile of grain size distribution at the site of Campo di Pile

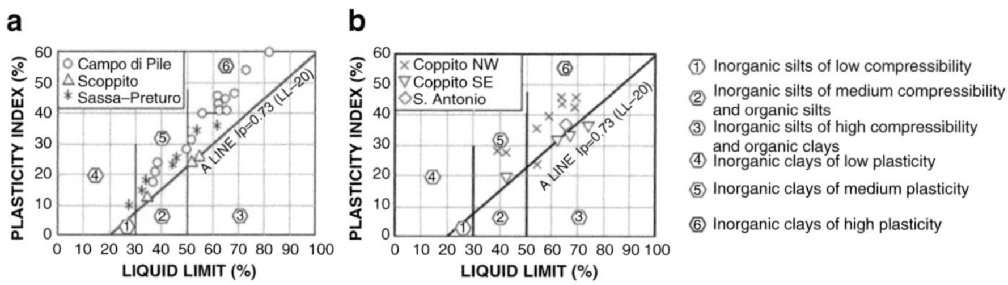


**Fig. 1.13** Plasticity properties of fine-grained soils at the sites of (**a**) Campo di Pile, Scoppito and Sassa-Preturo, and (**b**) Coppito and S. Antonio

obtained using common correlations (see TC16 2001). In addition to the usual DMT parameters, the profiles of the shear wave velocity $V_S$ roughly estimated from mechanical correlations with DMT data results using the correlation described later in this paper (see Fig. 1.28 ahead) are also shown in Figs. 1.15–1.18.

Figures 1.19 and 1.20 illustrate some results of site investigations carried out in the town centre of L'Aquila.

Figure 1.19 shows a typical stratigraphic profile obtained in the "Megabreccias" formation, which constitutes most of the top portion of the subsoil in the old town centre of L'Aquila. The profile of $V_S$ obtained by seismic dilatometer (SDMT) in a backfilled borehole at the same site (Piazza del Teatro, see Totani et al. 2009 – see next section) is also shown.

Figure 1.20 illustrates a stratigraphic condition encountered in the area of Via XX Settembre, most severely damaged by the April 6, 2009 earthquake. The stratigraphic logs in Fig. 1.20 was obtained from two boreholes executed within a short

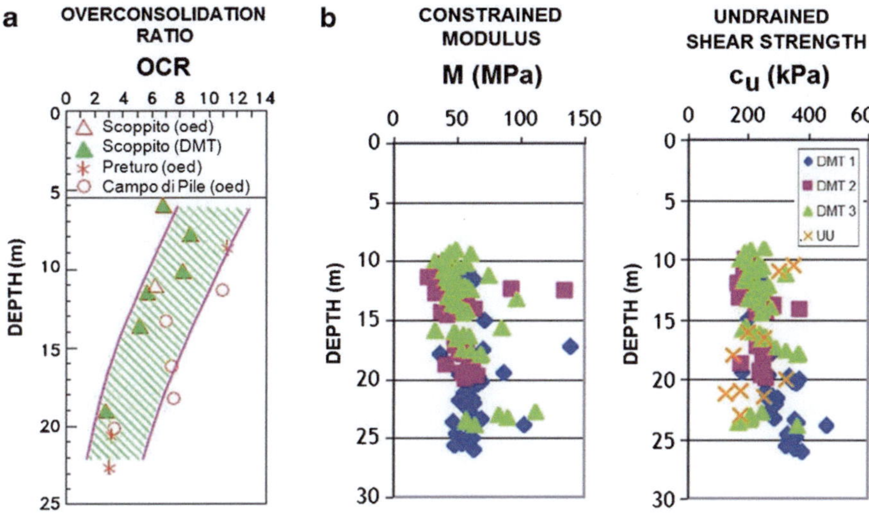


**Fig. 1.14** (**a**) Profiles of overconsolidation ratio OCR from laboratory oedometer tests and in situ DMT tests at the sites of Campo di Pile, Scoppito and Preturo. (**b**) Profiles of constrained modulus *M* and undrained shear strength $c_u$ from in situ DMT tests (compared to $c_u$ from laboratory UU triaxial tests) at the site of Campo di Pile

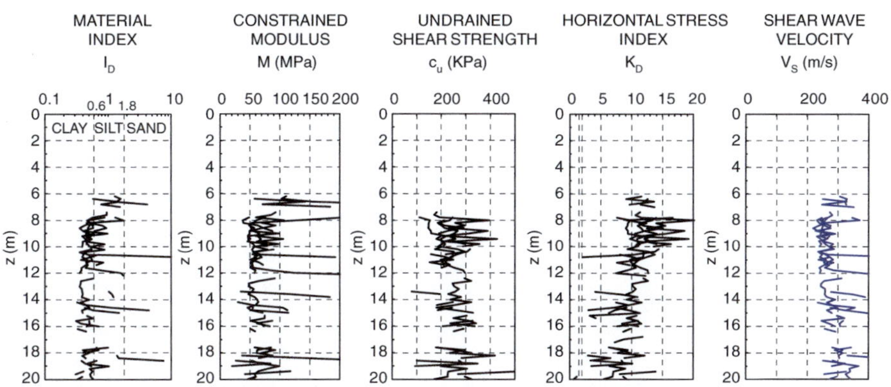


**Fig. 1.15** DMT profiles at the site of Campo di Pile – ex Seat-Philips plant, industrial zone (1987) and (on the *right*) profile of the shear wave velocity $V_S$ estimated from DMT data

distance, near a public building. The comparison of the two logs indicates that layers of fill material, having thickness of some meters, may be locally found in this area, possibly overlaying residual fine-grained soils at the top of the "Megabreccias" formation. It is supposed that most of these man-made fills originate from disposal of rubbles of masonry buildings destroyed in previous earthquakes that affected the town of L'Aquila, namely the 1703 earthquake.

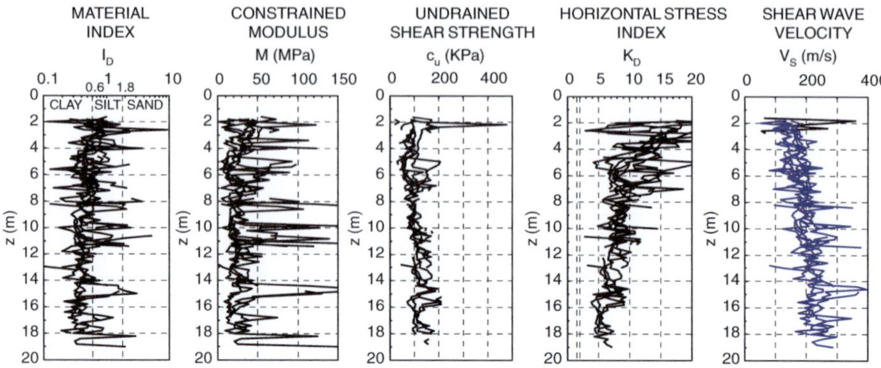


**Fig. 1.16** DMT profiles at the site of Scoppito – Sanofi-Aventis pharmaceutical plant (1991, 2000) and (on the *right*) profile of the shear wave velocity $V_S$ estimated from DMT data

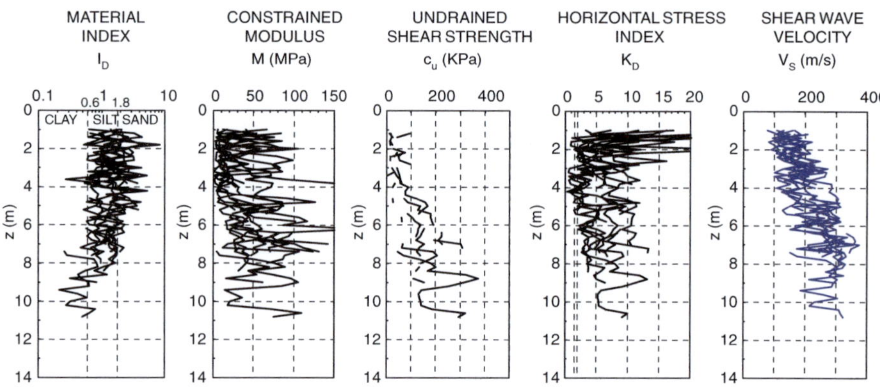


**Fig. 1.17** DMT profiles at the site of Coppito – San Salvatore Hospital (1993) and (on the *right*) profile of the shear wave velocity $V_S$ estimated from DMT data

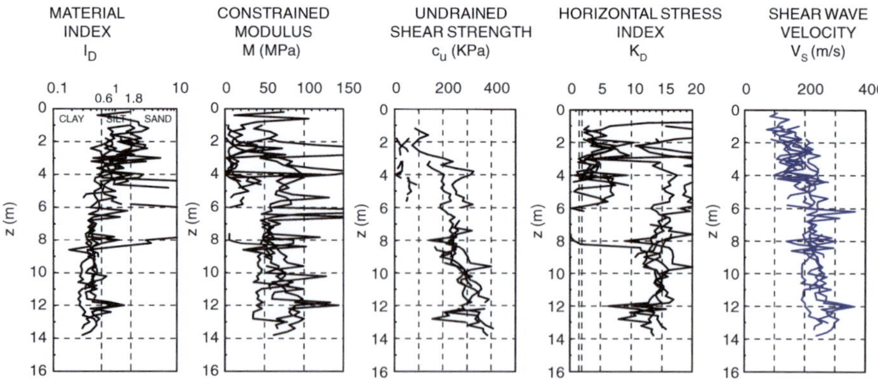


**Fig. 1.18** DMT profiles at the site of L'Aquila railway station – Nuovo Archivio di Stato, near Aterno River (1993) and (on the *right*) profile of the shear wave velocity $V_S$ estimated from DMT data

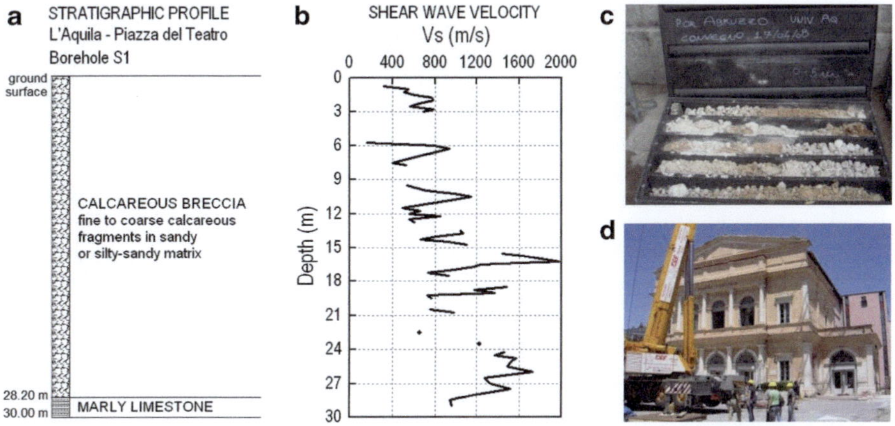


**Fig. 1.19** L'Aquila – Piazza del Teatro. (**a**) Typical stratigraphic profile of the subsoil ("Megabreccias") in the town centre of L'Aquila. (**b**) Profile of $V_S$ obtained by SDMT in a back-filled borehole (Totani et al. 2009). (**c**) Material obtained from the borehole. (**d**) The City Theatre after the April 6, 2009 earthquake

Results of site investigations carried out in the East sector of the L'Aquila basin, particularly in the bottom of the valley along the Aterno River (Fig. 1.21), are shown in Figs. 1.22–1.24.

Figure 1.22 shows an example of the coarse-grained deposits, mostly calcareous gravel, commonly encountered in this area, at elevations some meters higher than the bottom of the Aterno River valley.

Figure 1.23 shows the stratigraphic logs and the profiles of the cone penetration resistance obtained from boreholes and cone penetration tests (CPT) executed on the two banks of the Aterno River after the April 6, 2009 earthquake, in correspondence of the bridge on the road to Fossa, collapsed during the earthquake.

Figure 1.24 shows the superimposed profiles of the cone penetration resistance $q_c$ from several CPT tests carried out, some years ago, along the banks of the Aterno River from Onna to Fossa (see map in Fig. 1.21).

The results in Figs. 1.23 and 1.24 clearly indicate the poor geotechnical properties of the alluvial deposits in the flat area from Onna to Fossa, probably the poorest within the entire L'Aquila basin. This may contribute to explain the high concentration of damage caused by the earthquake in this zone.

### *1.3.2 Shear Wave Velocity from In Situ Tests*

Few weeks after the main shock, the Civil Protection Department (DPC) individuated about 20 sites to locate the first temporary houses for the homeless people (Fig. 1.25).

a

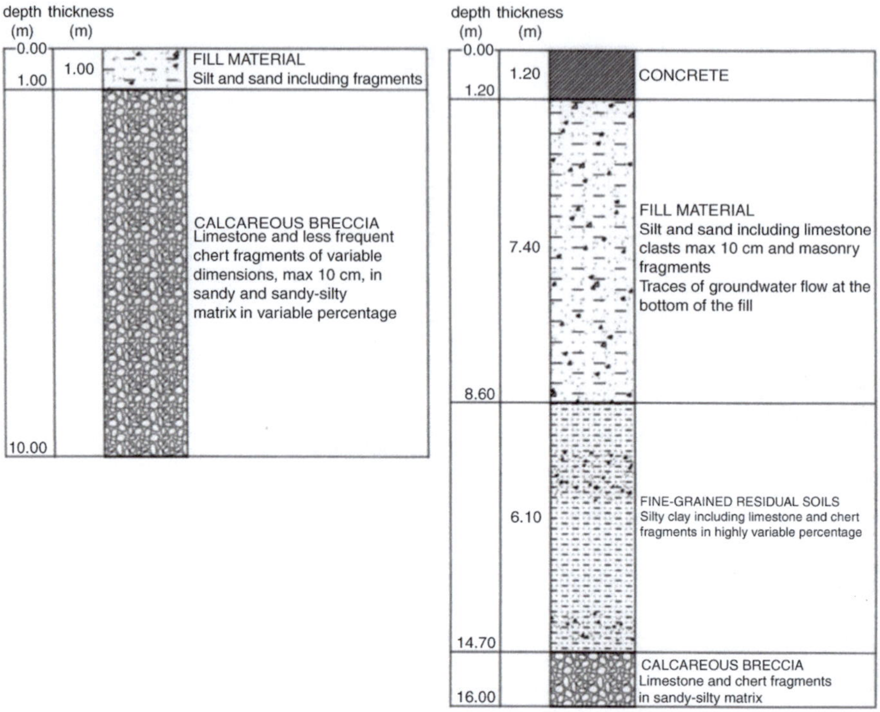


b

**Fig. 1.20** L'Aquila – Via XX Settembre. (**a**) Stratigraphic profiles from two boreholes executed within a short distance near a public building (on the *left* in the photo **b**)

**Fig. 1.21** Location of selected investigated sites in the East sector of the L'Aquila basin and along the Aterno River

These building stocks (C.A.S.E.) were conceived to be seismically isolated, so that one of the requirements to be satisfied for the location assessment was that the natural frequency of the subsoil should not be lower than 0.5 Hz. For an accurate yet quick dynamic subsoil characterization of such sites, surface wave tests were planned in all of them, and Down-Hole and laboratory tests only where the top of the seismic bedrock was not apparently shallow. A task force of the Italian Geotechnical Society (AGI) was committed to investigate on several of such sites by surface wave and seismic dilatometer tests.

In the following, first the seismic dilatometer test results are reported, then the test and interpretation procedures of surface wave tests are described, along with the results gathered in two deep and two shallow bedrock sites.

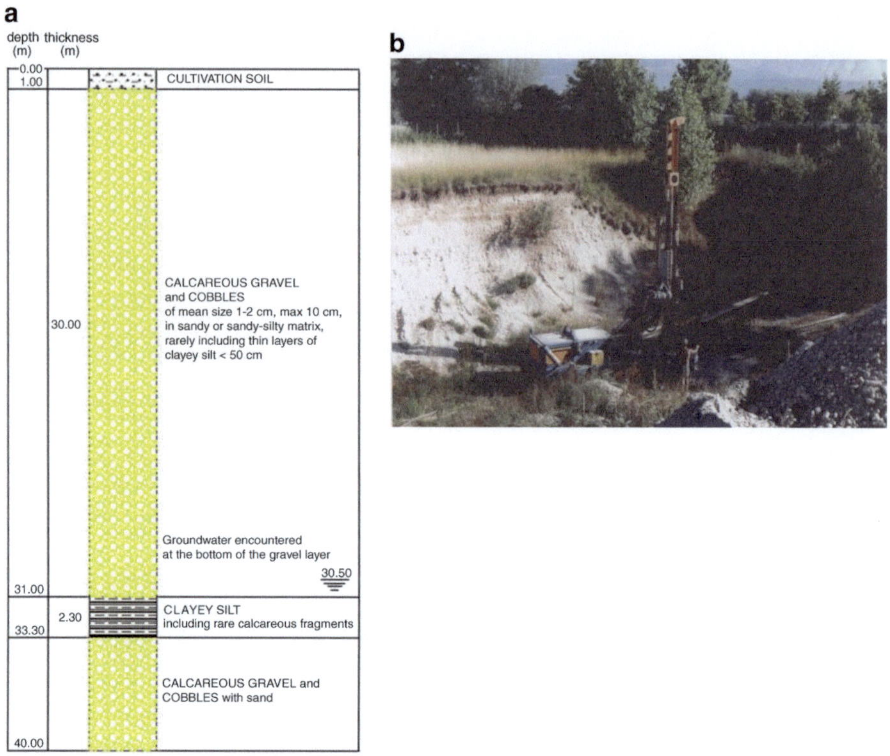


**Fig. 1.22** Example of typical coarse-grained deposits in the East sector of the L'Aquila basin (site of Monticchio)

#### 1.3.2.1 Seismic Dilatometer: Testing and Interpretation Procedures

The seismic dilatometer (SDMT) is the combination of the traditional "mechanical" flat dilatometer with a seismic module for measuring the shear wave velocity $V_S$.

The test is conceptually similar to the seismic cone (SCPT). First introduced by Hepton (1988), the SDMT was subsequently improved at Georgia Tech, Atlanta, USA (Martin and Mayne 1997, 1998; Mayne et al. 1999). A new SDMT system (Fig. 1.26) has been recently developed in Italy.

The seismic module (Fig. 1.26a) is a cylindrical element placed above the DMT blade, provided with two receivers spaced 0.5 m. The signal is amplified and digitized at depth.

The *true-interval* test configuration with two receivers avoids possible inaccuracy in the determination of the "zero time" at the hammer impact, sometimes observed in the *pseudo-interval* one-receiver configuration. Moreover, the couple of seismograms recorded by the two receivers at a given test depth corresponds to the same hammer blow and not to different blows in sequence, which are not necessarily identical. Hence the repeatability of $V_S$ measurements is considerably improved (observed $V_S$ repeatability ≈1–2%).

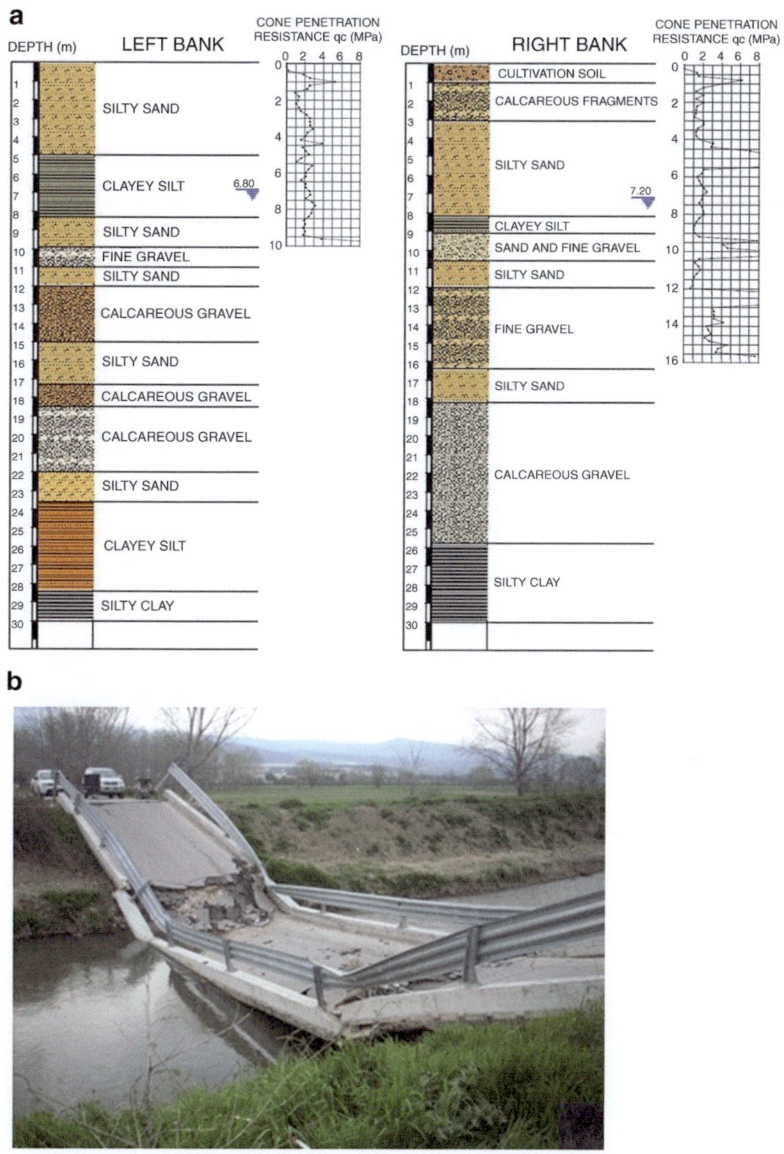


**Fig. 1.23** (**a**) Stratigraphic logs and profiles of the cone penetration resistance $q_c$ obtained from boreholes and CPT tests executed on the two banks of the Aterno River after the April 6, 2009 earthquake in correspondence of the collapsed bridge (**b**) on the road to Fossa

$V_S$ is obtained (Fig. 1.26b) as the ratio between the difference in distance between the source and the two receivers ($S_2 - S_1$) and the delay of the arrival of the impulse from the first to the second receiver ($\Delta t$). $V_S$ measurements are typically taken each 0.5 m of depth.

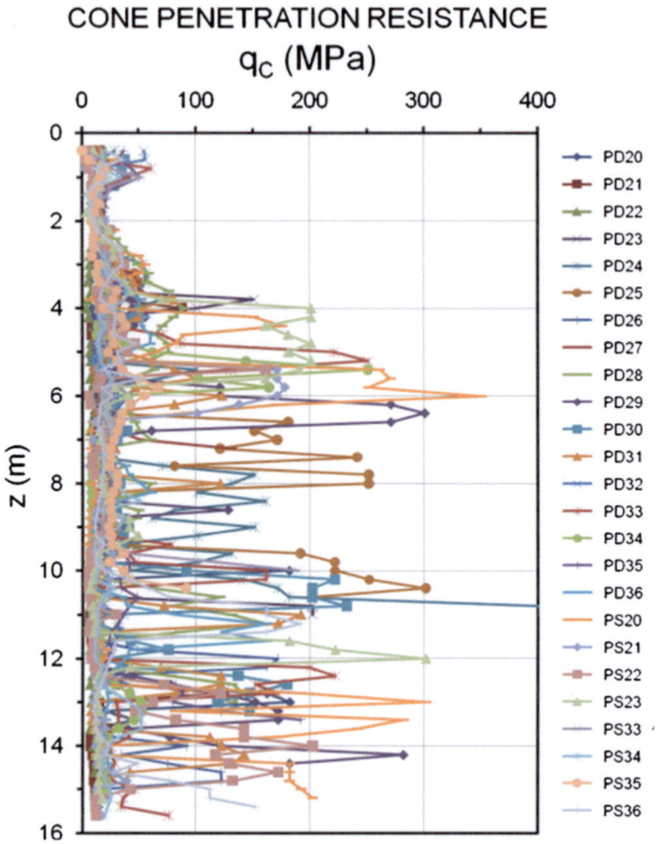


**Fig. 1.24** Profiles of the cone penetration resistance $q_c$ from several CPT tests carried out along the banks of the Aterno River from Onna to Fossa (see map in Fig. 1.21)

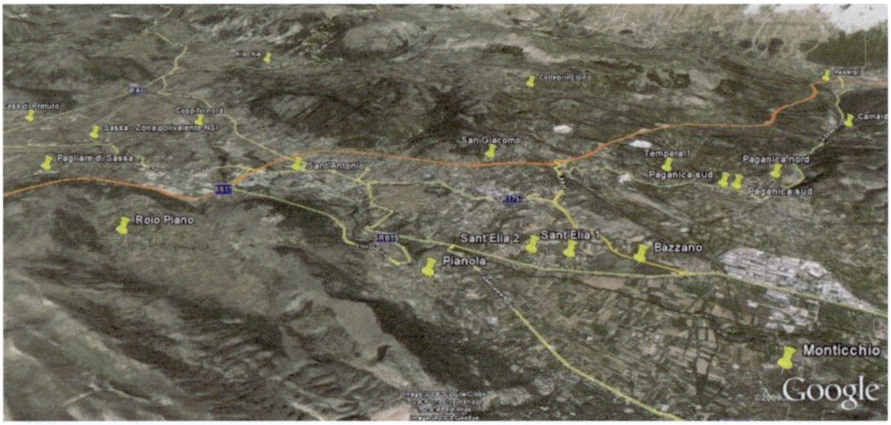


**Fig. 1.25** Location of the 20 temporary housing sites individuated by the DPC

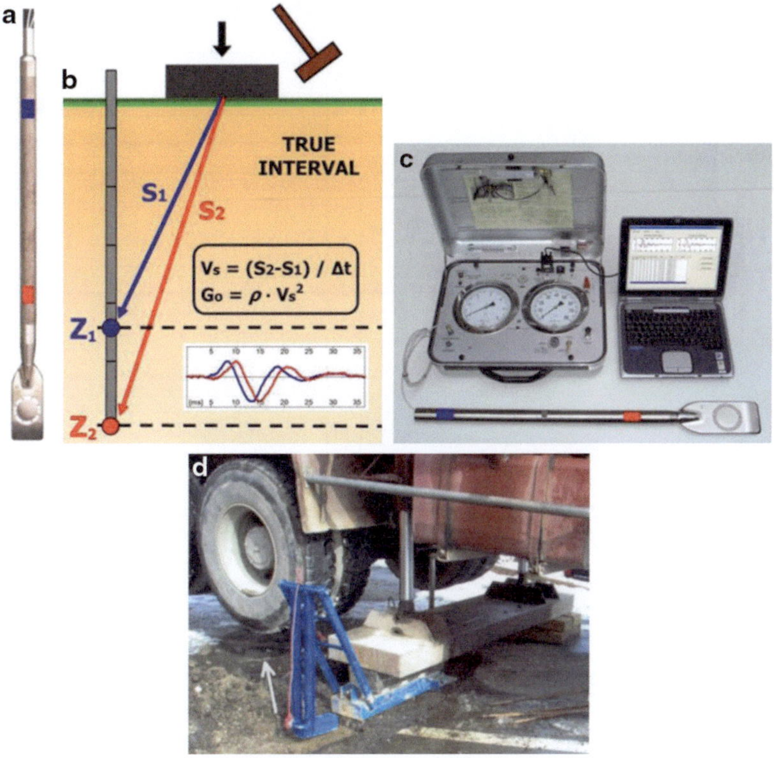


**Fig. 1.26** Seismic dilatometer. (**a**) DMT blade and seismic module. (**b**) Schematic layout of the seismic dilatometer test. (**c**) Seismic dilatometer equipment. (**d**) Shear wave source at the surface

The shear wave source at the surface (Fig. 1.26d) is a pendulum hammer (≈10 kg) which hits horizontally a steel rectangular base, pressed vertically against the soil (by the weight of the truck) and oriented with its long axis parallel to the axis of the receivers, so that they can offer the highest sensitivity to the generated shear wave.

The determination of the delay from SDMT seismograms, normally carried out using the cross-correlation algorithm, is generally well conditioned, being based on the waveform analysis of the two seismograms, rather than relying on the first arrival time or specific single points in the seismogram. Figure 1.27 shows an example of seismograms obtained by SDMT at various test depths at the site of Fucino; it is a good practice to plot side-by-side the seismograms as recorded and re-phased according to the calculated delay.

Several comparisons at various research sites (e.g. Fucino, AGI 1991; Bothkennar, Hepton 1988; Treporti, McGillivray and Mayne 2004; Zelazny Most, Młynarek et al. 2006) indicate quite good agreement between measurements of $V_S$ obtained by SDMT and by other in situ seismic tests (see Marchetti et al. 2008 for additional information).

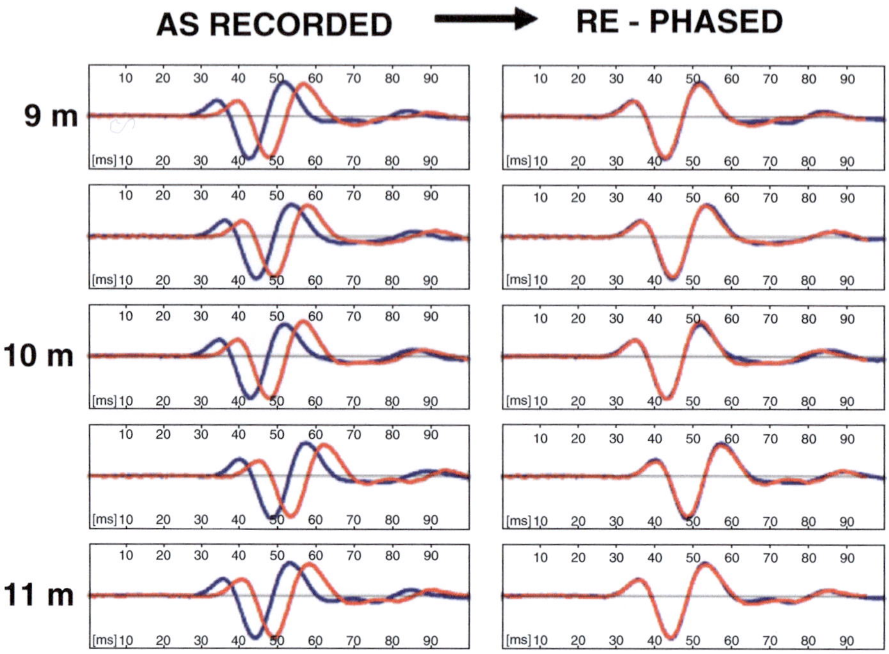


**Fig. 1.27** Example of seismograms obtained by SDMT at the site of Fucino (Italy)

The seismic dilatometer can therefore provide an estimate of the small-strain shear modulus $G_0$, which can be obtained by the value of $V_S$ using the theory of elasticity. Also, the SDMT can help to define or assess the decay of $G$ with shear strain γ, which can be empirically correlated to the operative modulus, $M_{DMT}$, as pointed out by Martin and Mayne (1997) and Mayne et al. (1999). More details on this subject may be found in Marchetti et al. (2008).

As a general rule, it is advisable to measure directly $V_S$, but estimates of $G_0$ (hence $V_S$) at sites where mechanical DMTs have been executed can be obtained from the parameters $I_D$, $K_D$, $M_{DMT}$ through the correlation in Fig. 1.28, obtained using the SDMT results at 34 different sites (Monaco et al. 2009). The examples in Figs. 1.15–1.18 show the shear wave velocity profiles resulting from the application of such procedure to four sites.

Considerable research work has been carried out in the last two decades on the possible use of the DMT/SDMT results to evaluate liquefiability (see e.g. Monaco and Marchetti 2007). A liquefaction case study (Vittorito), based on the use of both $K_D$ and $V_S$ obtained by SDMT, is presented in a subsequent section of this paper.

Practical advantages of the SDMT are the virtual independence of the results from the operator and the moderate cost. In fact SDMT obviously does not involve "undisturbed sampling", nor holes with pipes to be grouted, operations requiring, before testing, a few days pause for the cement to set up. A disadvantage of the

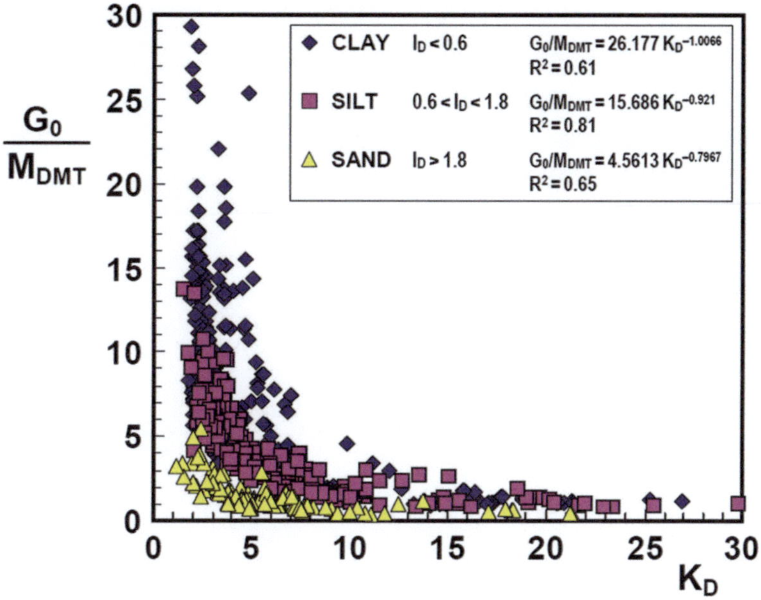


**Fig. 1.28** Ratio $G_0/M_{DMT}$ vs. $K_D$ (OCR) for various soil types (Monaco et al. 2009)

SDMT is the impossibility of penetrating very hard soils. However SDMT $V_S$ profiles (but not the other DMT parameters) can be obtained in non-penetrable soils by the following procedure (Totani et al. 2009):

- A borehole is drilled to the required test depth;
- The borehole is backfilled with sand;
- The SDMT blade is inserted and advanced into the backfilled borehole in the usual way (e.g. by use of a penetrometer rig) and $V_S$ measurements are carried out every 0.5 m of depth; no DMT measurements – meaningless in the backfill soil – are taken in this case.

Comparative tests at various penetrable sites, where both the usual method and the backfilling procedure were adopted, indicated that the values of $V_S$ obtained by penetrating the "virgin" soil are practically coincident with those measured in a backfilled borehole. The reliability of this latter procedure is therefore not affected by the borehole, since the shear wave path from the source to the dilatometer blade includes a short path in the backfill of very similar length for both upper and lower receivers.

#### 1.3.2.2 SDMT Results in the Aterno Valley

SDMTs were carried out at five locations (Fig. 1.29) in Spring 2009, i.e. in the first months following the 6 April earthquake. Figure 1.30a shows the SDMT results in

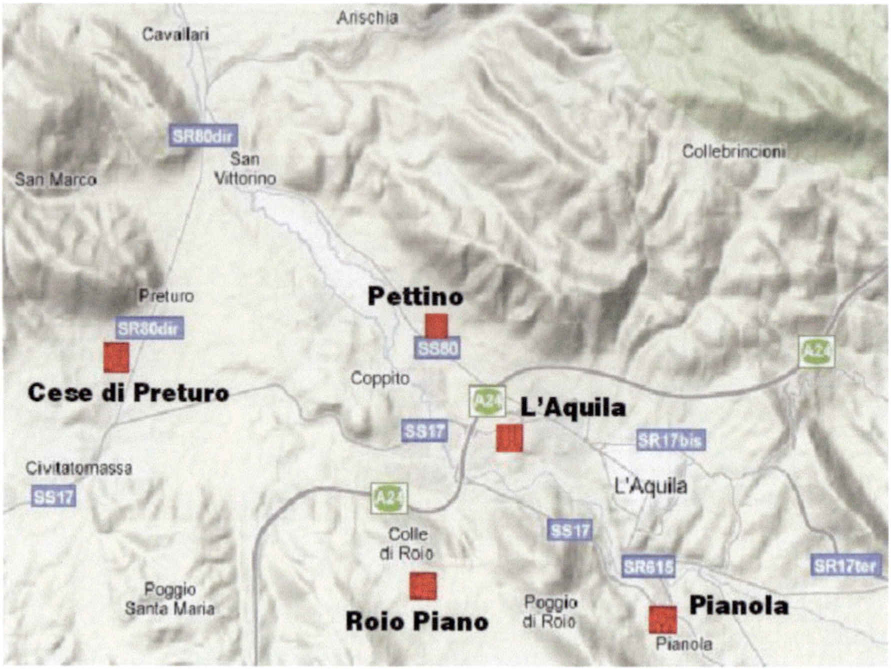


**Fig. 1.29** Map with the location of the five sites investigated by SDMT

terms of $V_S$ profile only (no DMT parameters) for two sites (L'Aquila and Pettino) where most of the subsoil was non-penetrable and the borehole backfilling procedure was followed. As a matter of fact, the measured $V_S$ results mostly higher than 400 m/s, reaching sometimes values as high as 2,000 m/s. The results in Fig. 1.30b were gathered in three sites of the C.A.S.E. project (Cese di Preturo, Pianola, Roio Piano) where the bedrock was expected to be deep; the measured $V_S$ values seldom trespass 400 m/s, even at the highest depths investigated.

The comparisons in Fig. 1.31 indicate a good agreement between $V_S$ measured by SDMT and that estimated from mechanical DMT using the correlation by Monaco et al. (2009) shown in Fig. 1.28. Comparisons between the $V_S$ profiles by SDMT and by other techniques (surface waves and down-hole tests) will be discussed later in this paper.

#### 1.3.2.3 Surface Waves: Testing and Interpretation Procedures

The Surface Wave Method (SWM) is a cost- and time-effective technique for evaluating the shear wave velocity profile. It is based on the dispersivity of surface waves in vertically heterogeneous media: the velocity of each harmonic component of the surface wave depends on the parameters of the medium affected by the wave propagation, and its penetration is proportional to the wavelength. Consequently, high frequency

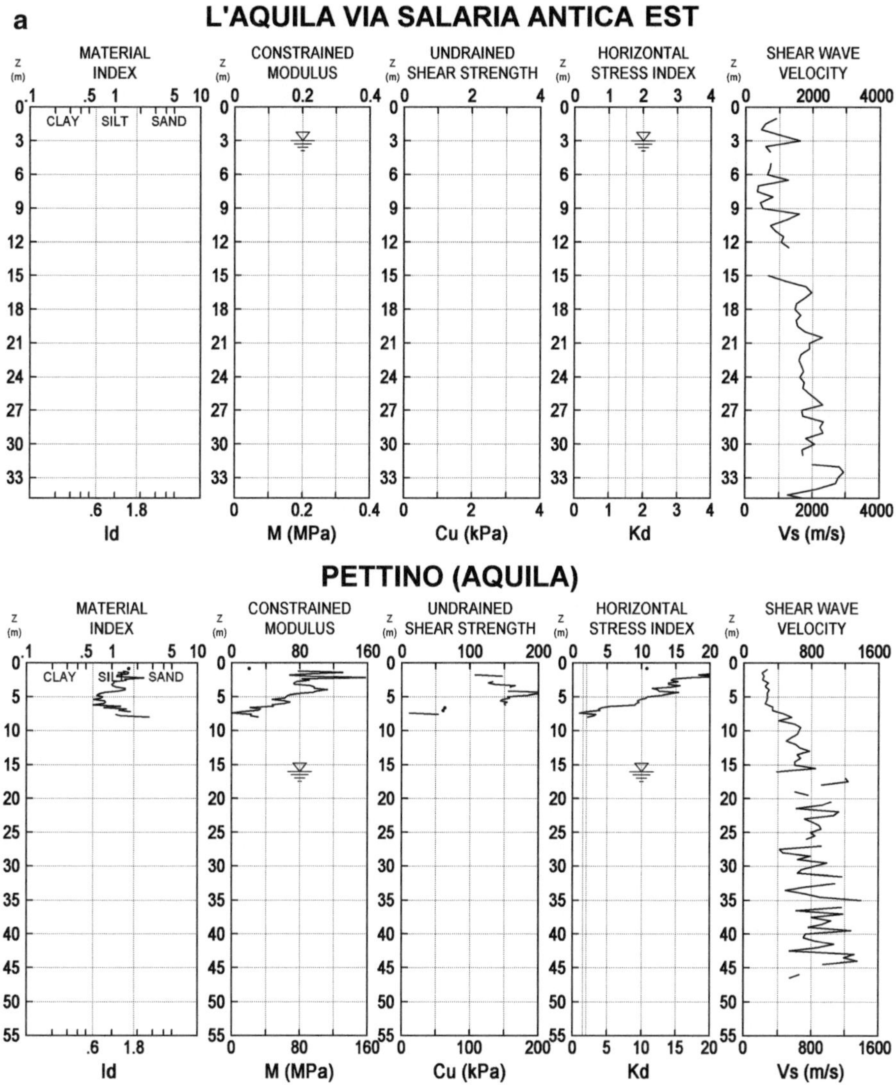


**Fig. 1.30** (**a**) SDMT results at the two sites investigated by the backfilling procedure

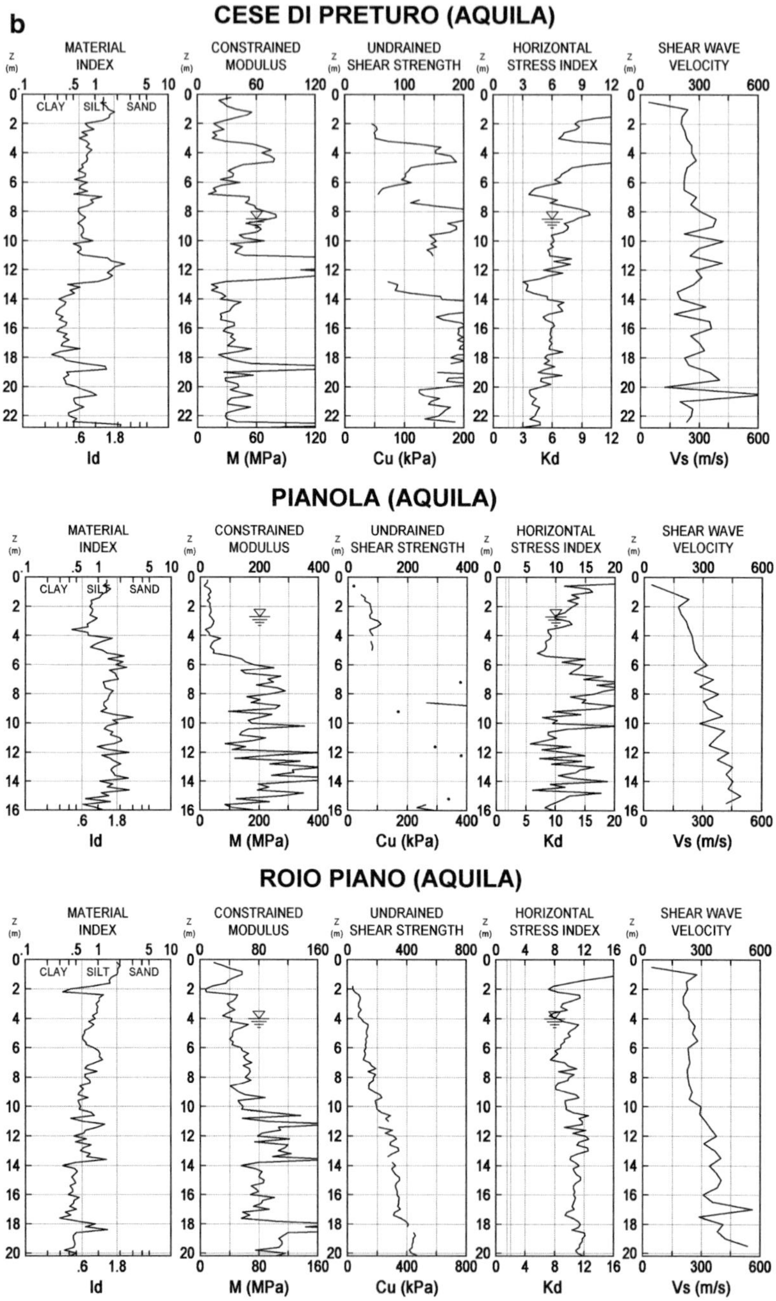


**Fig. 1.30** (**b**) SDMT results at the three sites investigated by the penetration procedure

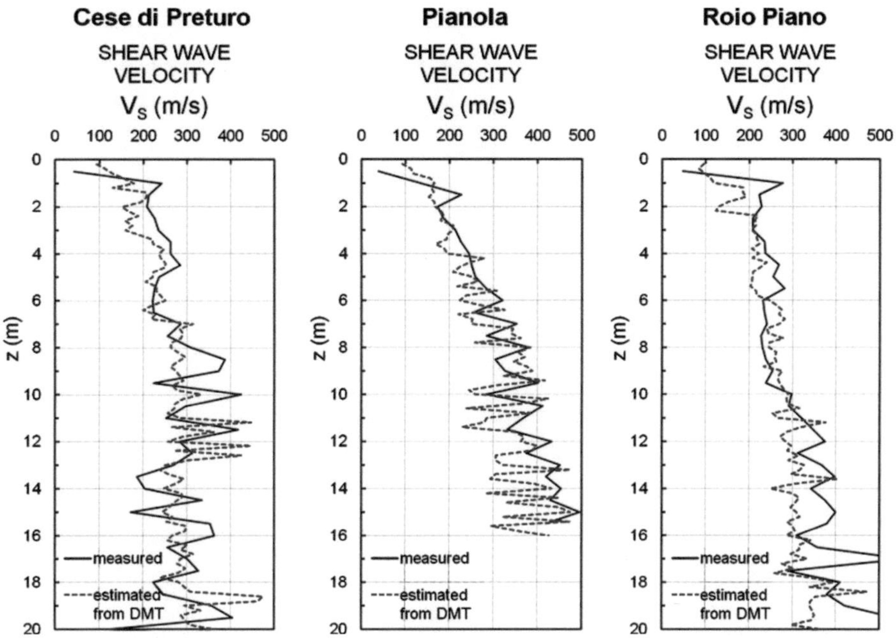


**Fig. 1.31** Comparison of profiles of shear wave velocity $V_S$ measured by seismic dilatometer SDMT and estimated from "mechanical" DMT data (correlation in Fig. 1.28) at three C.A.S.E. sites (Cese di Preturo, Pianola, Roio Piano)

surface wave components travel with a velocity slightly lower than the shear wave velocity of the shallow layers (Fig. 1.32a), whereas the velocity of the low frequency components is influenced also by deeper layers (Fig. 1.32b). An experimental 'dispersion curve' (velocity versus wavelength or frequency) can be extracted from field data using several processing techniques (e.g. Foti 2005; Socco and Strobbia 2004; Evangelista 2009). The shear wave velocity profile can be inferred by solving an inverse problem, based on the minimization of the distance between the theoretical dispersion curve and the experimental one.

In the Abruzzo sites (Fig. 1.25), multi-receiver surface wave (MASW) tests have been performed using both active and passive acquisition techniques. The multi--station active layout used harmonic or impulsive sources ($f>5$ Hz) and linear receiver arrays, whereas the passive layout ($f<10$ Hz) used two-dimensional receiver arrays. Dispersion curves were obtained as the amplitude peaks in the f-k (frequency-wave number) domain, where different branches of the dispersion curve can be identified. The inversion procedure was based on the Haskell-Thomson matrix method for the solution of the forward problem. For an elastic horizontally layered medium, the dispersion curve corresponds to the zeros in the velocity-frequency domain of the Haskell-Thomson matrix determinant, which depends on subsoil parameters (shear wave velocity, density, and Poisson's ratio). In Fig. 1.33 the shape of the absolute value of the Haskell-Thomson matrix determinant for a given profile is shown.

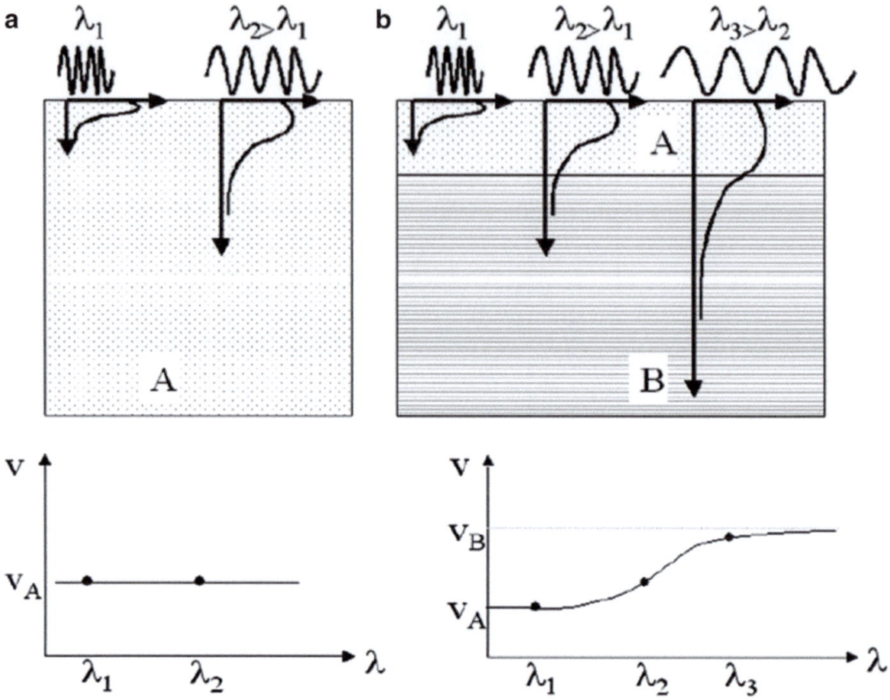


Fig. 1.32 Surface waves in (a) homogeneous and (b) layered subsoil (after Socco and Strobbia 2004)

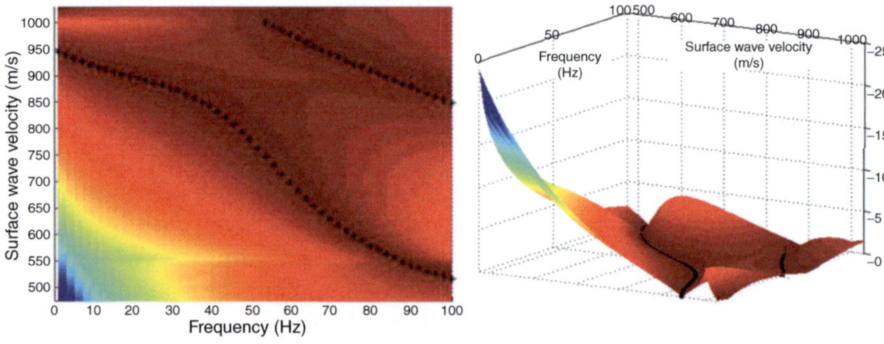


**Fig. 1.33** Determinant of the Haskell-Thomson matrix for a given profile: the coloured surface represents the value of the Haskell-Thomson matrix determinant in the velocity-frequency domain; black dots represent the zeros of the surface, i.e. the dispersion curve

The inversion consisted of the estimation of a layered shear wave velocity profile for realistic values of density and Poisson's ratio. A Monte Carlo algorithm has been used to perform the inversion of the experimental data. This algorithm evaluates the misfit of a set of soil profiles randomly generated between boundaries chosen by the user from the experimental dispersion curve. The used misfit function is the L1 norm of the vector of the Haskell-Thomson determinant, evaluated in correspondence of the experimental dispersion curve (Maraschini et al. 2008, 2009). This misfit function is suited for surface wave inversion, because it is cost-effective and allows taking into account all the modes of the experimental dispersion curve, with no assumptions needed on the mode each experimental point belongs to. The resolution of the results is so improved, and errors in the half space velocity estimation due to mode mis-identification are prevented. With the adoption of this stochastic approach, it is possible to appreciate the uncertainties associated to non-uniqueness of the solution, which is typical of any method based on the solution of an inverse problem. A discussion of the consequences of this uncertainty on seismic site response studies can be found in Foti et al. (2009).

#### 1.3.2.4 MASW Results at Sites with Deep Bedrock

At three test sites (Roio Piano, Pianola, Il Moro, see Fig. 1.25) active and passive tests have been performed to characterize both shallow and deep soil layers, respectively. In the active tests, the source is a hammer (5 kg) and the receivers are 48 geophones with natural frequency of 4.5 Hz. In the passive tests, the source is the ambient noise (microtremors); the direction of impinging waves is unknown, and it is therefore necessary to use 2D arrays of geophones in order to determine the direction of the dominant wave, and consequently its velocity. Circular arrays of 12 geophones having natural frequency of 2 Hz have been used. The data records have been analyzed in the f-k domain using a frequency beam-former approach (Zywicki 1999).

The integration of active and passive data allowed to retrieve a dispersion curve with a broad frequency range, and consequently, to increase the penetration depth without losing resolution in shallow layers.

The Roio Piano site is characterized by colluvial deposits and debris covers overlying sandy and silty layers; these latter lie on a stiff calcareous bedrock with variable depth (Tallini et al. 2009). As the expected bedrock position is deep, active and passive tests have been performed in order to increase the penetration depth. The length of the receiver array for active data was 72 m. Passive data were collected using a circular array (50 m diameter) of 12 geophones.

In Fig. 1.34 the dispersion curves are plotted as estimated from both active and passive data; it can be noted that the penetration depth is strongly increased by passive data, indeed more information are available for the low-frequency range. It is noteworthy the consistency of active and passive data in the common frequency region (5–10 Hz), in which there is an overlapping of the two datasets.

The dispersion curve is composed by three main branches, and it is inverted using the multi-modal Monte Carlo algorithm described above. In Fig. 1.35a the 20 best-fitting profiles are plotted using a colour scale such that the darkest colour corresponds to the best fitting profile, whereas in Fig. 1.35b the experimental dispersion curves are compared with the Haskell-Thomson matrix determinant of the best-fitting profile.

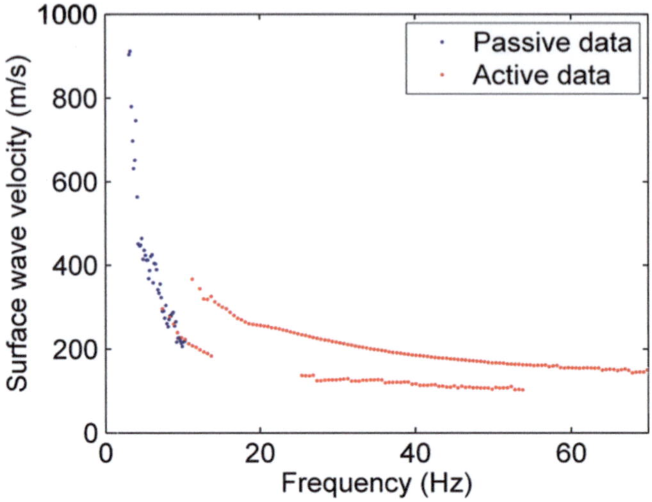


**Fig. 1.34** Active and passive data for Roio Piano site

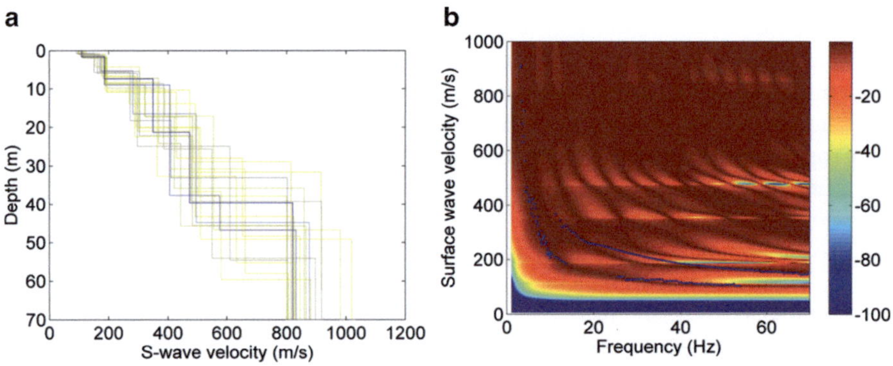


**Fig. 1.35** Roio Piano site: (**a**) Best fitting profiles from Monte Carlo inversion, the colours represent the normalized misfit, from blue (lower misfit) to yellow; (**b**) comparison of the Haskell-Thomson matrix determinant with the experimental data

It can be noted that the best-fitting profiles are in reasonable agreement each other, showing a gradual increase of the shear wave velocity with depth, and that all the branches of the experimental curve follow the minima of the misfit surface.

At this site, the active tests were repeated with the same experimental setup, moving the source position at the opposite end of the array, obtaining a double estimate of the high-frequency dispersion curve (shown in Figs. 1.34 and 1.35b). In Fig. 1.36 the best-fitting profiles associated to both source positions are compared with the stratigraphy and the profiles obtained with down-hole (DH) and the above described seismic dilatometer (SDMT) tests carried out nearby. It can be noted that

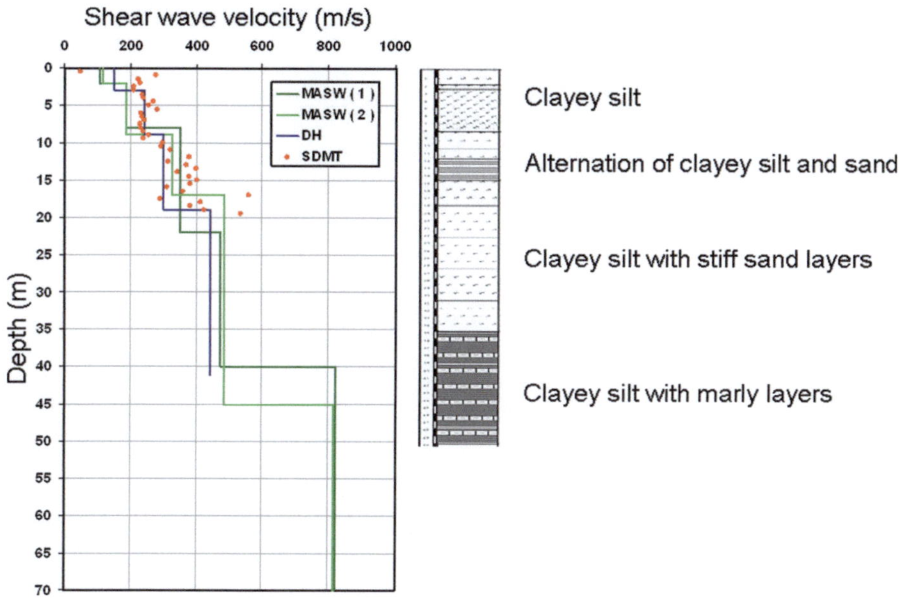


**Fig. 1.36** Roio Piano site: comparison of shear wave velocity profiles obtained by Down-Hole tests (DPC 2009) and SDMT tests

MASW, DH and SDMT test results are in very good agreement each other, with surface wave tests allowing to define a deeper velocity profile.

At the Pianola site, the geological survey anticipated the presence of sandy-silty layers overlying a coarse 'breccia' formation and sandstone-clay stone bedrock (Cavinato and Di Luzio 2009). The same testing and interpretation procedures as in the Roio Piano site were followed; again, the subsoil layering and the comparison with the DH and SDMT tests are presented in Fig. 1.37. Also for this site, the results of surface wave tests are in good agreement with the other test data, but they provide more information on deeper layers.

#### 1.3.2.5 MASW Results at Sites with Shallow Bedrock

In the three sites of Bazzano, San Giacomo and Sant'Antonio (see Fig. 1.25), the preliminary geological survey suggested the presence of shallow bedrock; therefore only high-frequency active MASW tests were performed. The source of surface waves was an electro-mechanical shaker controlled by a function generator, generating a harmonic force in the frequency range of 5–120 Hz. The receivers are 14 piezo-electric 1D accelerometers having a wide dynamic range (0.1 – 300 Hz), placed following a linear array having a total length of 29 m. The signal processing and inversion procedures were the same as those followed for the deep bedrock sites above described. Refer to Evangelista (2009) for detailed description of the tests in these three locations.

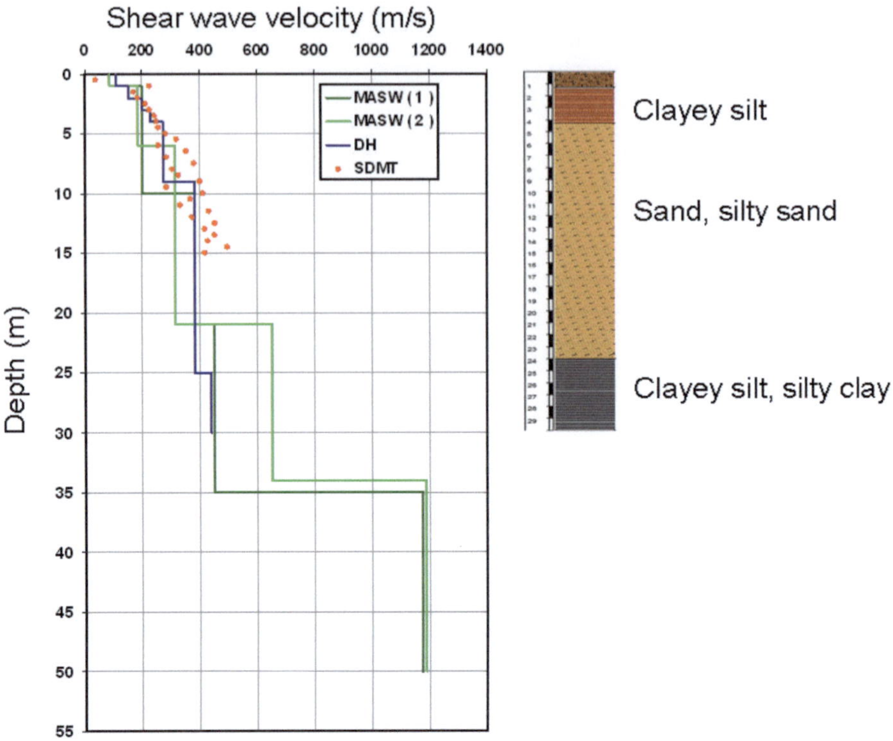


**Fig. 1.37** Pianola site: comparison of shear wave velocity profiles obtained by Down-Hole tests (DPC 2009) and SDMT tests

In the Bazzano area, the geological survey detected a sandstone formation locally covered by alluvial deposits (Cavuoto and Moscatelli 2009). The MASW reference vertical is located at the transition between the outcropping sandstone and the alluvial cover; as a matter of fact, the experimental shear wave velocity profile seems to identify shallow bedrock at 8 m (Fig. 1.38). In this case, however, no borehole was available in the vicinity to confirm this result.

The San Giacomo test site is located in a valley, developed along NE-SW within a complex Pliocene–Pleistocene 'megabreccia' formation, constituted by sandy and gravelly lumps in a finer matrix, overlain by alluvial Holocene deposits and made ground (Spadoni and Sposato 2009). The MASW test was performed in the middle of the valley, where a shallow cover of the Holocene sediments was expected, as it seems to be confirmed by the increasing trend of shear wave velocity profile from about 150 m/s until 500 m/s (Fig. 1.39).

In the upper part of the valley, around 250 m far from the MASW test site, a down-hole investigation showed, for the shallowest 5 m, the presence of man-made ground, resting on the Holocene alluvia (silty-clay in Fig. 1.40) and the 'megabreccia' (gravel and silty-sand in Fig. 1.40). Although the average velocity measured in the

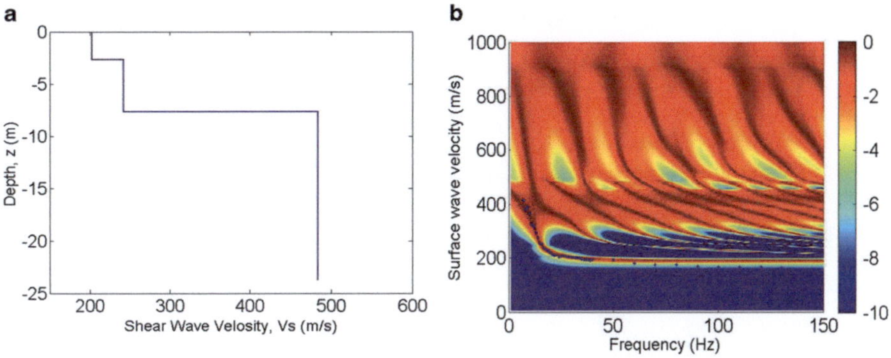


**Fig. 1.38** Bazzano site: (**a**) best fitting profiles from Monte Carlo inversion, (**b**) comparison of the Haskell-Thomson matrix determinant with the experimental data

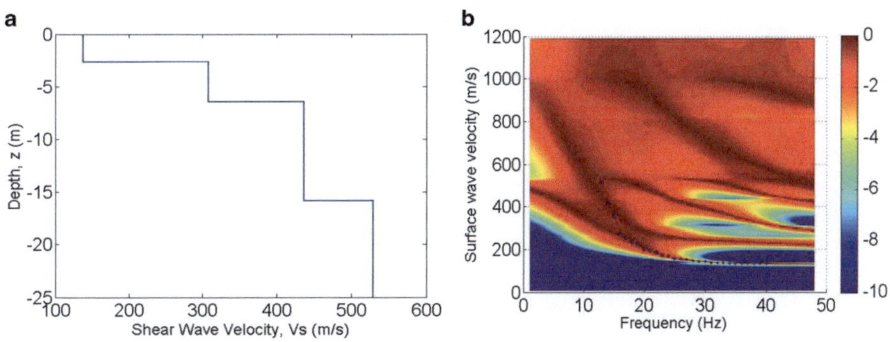


**Fig. 1.39** San Giacomo site: (**a**) best-fitting profile from Monte Carlo inversion, (**b**) comparison of the Haskell-Thomson matrix determinant with the experimental data

upper looser soils resulted comparable (about 300 m/s), from both MASW and DH tests it was not clear whether the coarse-grained megabreccia can be assumed as stiff bedrock.

### *1.3.3 Stiffness and Damping from Laboratory Tests*

A significant amount of advanced cyclic and dynamic laboratory tests were carried out on 18 undisturbed samples retrieved in 8 out of the 20 sites of the C.A.S.E. project. A network of soil dynamics laboratories was involved: ISMGEO (Bergamo), Politecnico di Torino, Universities of Catania, Florence, Naples and Rome 'La Sapienza'.

The testing programme consisted of standard classification tests, one-dimensional compression tests, cyclic-dynamic torsional shear tests (RC–TS) and double specimen direct simple shear tests (DSDSS). The cyclic/dynamic shear tests

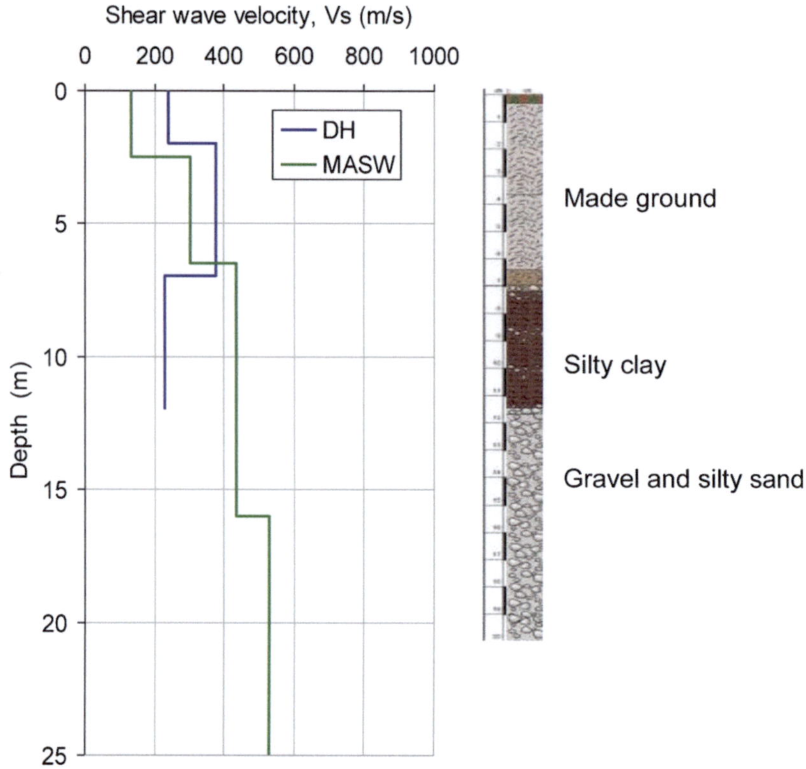


**Fig. 1.40** San Giacomo site: comparison of shear wave velocity profiles obtained by Down-Hole tests (DPC 2009)

were aimed to characterize the non-linear and dissipative behaviour of the medium- to fine-grained soils for site response analyses for both the C.A.S.E. temporary buildings and the ongoing microzonation studies on the whole damaged area. In the following, a synthetic description of the experimental data is given.

#### 1.3.3.1 Physical Properties

The values of the main index properties are summarised in Table 1.2, while the grain size distributions of the tested samples are shown in Fig. 1.41. Most of the tested samples are fine-grained soils, classifiable as silty clays to sandy silts, except for those retrieved at Tempera (S1–C3), Pianola (S1–C1) and Camarda (S1–C1) which are silty sands.

According to the particle size distributions, the tested soils can be divided into two main groups: those characterised by a CF lower than 30% (red curves in Fig. 1.41) and those with a CF higher than 30% (blue curves). Consistently with this main distinction, the data referring to the same groups are reported with the

**Table 1.2** Physical properties of the tested soils

| Sample | Depth (m) | $\gamma$ (kN/m$^3$) | $e_0$ | w (%) | $I_P$ (%) | Gravel (%) | Sand (%) | Silt (%) | Clay (%) | Laboratory | Tests |
|---|---|---|---|---|---|---|---|---|---|---|---|
| Cese S3-C1 | 4.0–4.8 | 19.13 | 0.752 | 26.1 | 36.7 | | 13 | 30 | 57 | Naples | RC-TS |
| Cese S3-C2 | 8.5–8.8 | 19.86 | 0.688 | 24.3 | 26.0 | | 3 | 47 | 50 | Naples | RC-TS |
| Cese S3-C3 | 17.5–18.0 | 18.66 | 0.880 | 31.7 | 37.4 | | | 60 | 40 | Naples | RC-TS |
| Monticchio S1-C1 | 15.0–15.3 | 17.98 | 1.058 | 37.5 | 27.1 | | 26 | 38 | 36 | Naples | RC-TS |
| Sassa S1-C1 | 7.5–8.1 | 17.40 | 0.980 | 36.0 | 25.1 | | 1 | 56 | 43 | ISMGEO | RC-TS |
| Sassa S1-C2 | 15.0–15.5 | 19.45 | 0.770 | 27.0 | 17.6 | | 5 | 71 | 24 | ISMGEO | RC-TS |
| Pagliare S2-C1 | 2.8–3.4 | 20.60 | 0.547 | 19.0 | 20.8 | | 20 | 44 | 36 | ISMGEO | RC-TS |
| Pagliare S2-C2 | 22.0–22.5 | 20.02 | 0.666 | 24.0 | 29.1 | | 1 | 53 | 46 | ISMGEO | RC-TS |
| Roio Piano S3-C1 | 4.0–4.5 | 20.57 | 0.616 | 22.0 | 16.0 | | 13 | 57 | 30 | Catania | RC |
| Roio Piano S3-C2 | 7.0–7.5 | 20.00 | 0.630 | 21.7 | 19.0 | | 20 | 51 | 29 | Rome | DSDSS |
| Roio Piano S3-C3 | 12.0–12.5 | 19.78 | 0.662 | 25.7 | 15.0 | | 16 | 79 | 5 | Florence | RC |
| Roio Piano S3-C4 | 15.0–15.4 | 17.50 | 0.684 | 10.5 | 16.9 | | 1 | 66 | 33 | Turin | RC-TS |
| Roio Piano S3-C7 | 33.0–33.4 | | | | 14 | | | 81 | 19 | Rome | DSDSS |
| Roio Piano S3-C8 | 49.6–50.0 | 15.68 | 1.440 | 54.0 | 32.9 | | 20 | 28 | 52 | Naples | RC-TS |
| Tempera S1-C2 | 5.5–6.0 | 17.37 | 1.167 | 39.4 | 28.0 | | 18 | 52 | 30 | Florence | RC |
| Tempera S1-C3 | 12.0–12.5 | 18.67 | 0.875 | 33.6 | 19.0 | 1 | 58 | 30 | 11 | Rome | DSDSS |
| Pianola S1-C1 | 6.0–6.5 | 19.72 | 0.616 | 24.8 | | | 60 | 33 | 7 | Rome | DSDSS |
| Camarda S1-C1 | 4.5–5.0 | 14.50 | 1.957 | 57.2 | 15.6 | 1 | 51 | 41 | 7 | Turin | RC-TS |

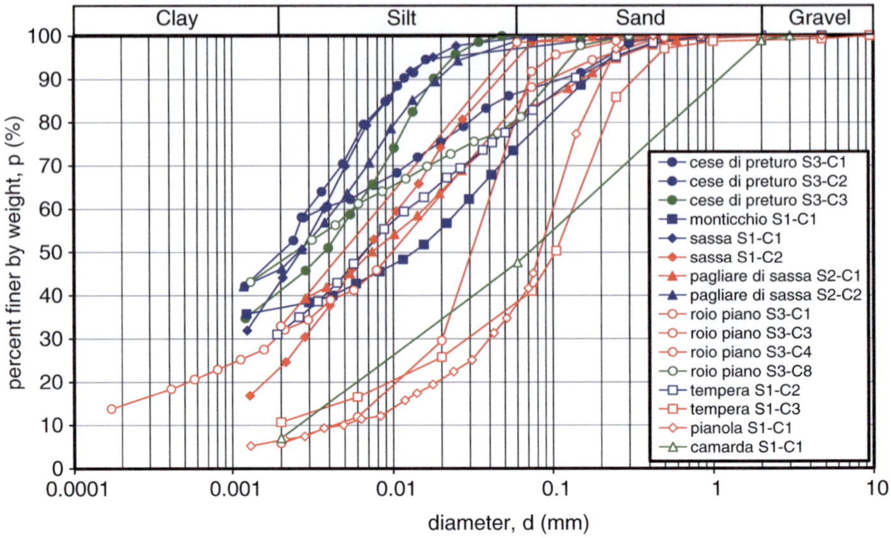


**Fig. 1.41** Particle size distribution of the tested soils

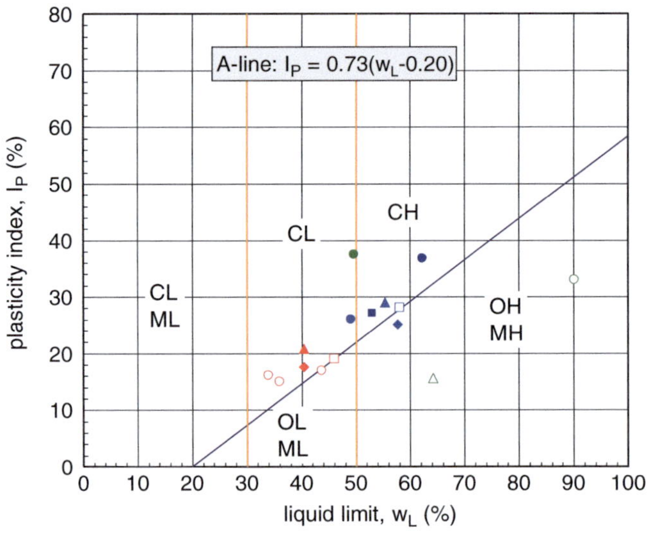


**Fig. 1.42** Plasticity chart of the tested soils

corresponding colour in terms of Atterberg limits in the plasticity chart in Fig. 1.42. Almost all the tested samples lie above the A-line and can be classified as inorganic clay of medium –high plasticity (USCS soil classification). On the basis of their plasticity properties, again it seems possible to recognise two main families in the Casagrande chart: all the red symbols pertain to soils characterised by a $I_P$ lower than 25% while all the blue ones are associated to $I_P$ higher than 25%.

It must be noted that evidences of organic matter has been found in some of the tested samples. This presence is associated with very high initial void index and water content (see Table 1.2). The corresponding determination of Atterberg limits appear to be outlier values in the Casagrande chart (green symbols in Fig. 1.42).

#### 1.3.3.2 Non-linear Stiffness and Damping

Cyclic and dynamic torsional shear tests have been carried out at the Universities of Catania, Florence, Naples, at the Politecnico di Torino, and at ISMGEO laboratory, while Double Sample Direct Simple Shear tests were carried out at the University of Rome. The equipment used for the torsional shear tests is a fixed-free Resonant Column/Torsional Shear device (RC-TS), in the different versions available at the various laboratories involved in the project. The Double Sample Direct Simple Shear tests were carried out using the device originally designed at the UCLA (Doroudian and Vucetic 1995) and updated by D'Elia et al. (2003).

The specimens were consolidated, either isotropically (RC-TS tests) or one-dimensionally (DSDSS tests) to the estimated in situ stress. At the end of the consolidation stage, cyclic and dynamic torsional shear tests or direct simple shear tests were performed with increasing strain levels, in order to investigate the behaviour of the soils from small to medium strains. As usual, the non-linear pre-failure behaviour has been interpreted with the linear equivalent model, characterised by the variation of the shear modulus, *G*, and the damping ratio, *D*, with the shear strain level, g.

The non-linear behaviour of the tested materials consistently reflected their differences in physical properties, as shown by the comparison of resonant column test data. Figure 1.43 shows the experimental results obtained for the different sets of samples in terms of normalised shear modulus, $G/G_0$ (Fig. 1.43a), and damping ratio, *D* (Fig. 1.43b), versus shear strain, g. The data are plotted with the same colour code used for the grain size distribution curves and for the plasticity chart.

Again, the curves define two quite different behaviours: the red curves (less fine - low plasticity soils) are characterised by a linear threshold strain not exceeding 0.005% and by a gentle decay trend, while the blue curves (finer – high plasticity soils) are characterised by higher values of the linear threshold (up to 0.02%) and a sharp evolution into the non-linear range. In the same plot the curves suggested by Vucetic and Dobry (1991) are reported for comparison.

The decay curve related to IP = 30% clearly separates the two main groups, confirming that the plasticity index is a key physical property to represent the non-linear soil behaviour. However, the literature curves do not overlap the overall trend of the experimental curves, which show a more pronounced decay when the shear strain becomes of the order of 0.02–0.07%; the outlier behaviour of the organic samples confirms the evidence of a complex soil behaviour for which the influence of fabric and structure cannot be caught by the 'standard' literature curves.

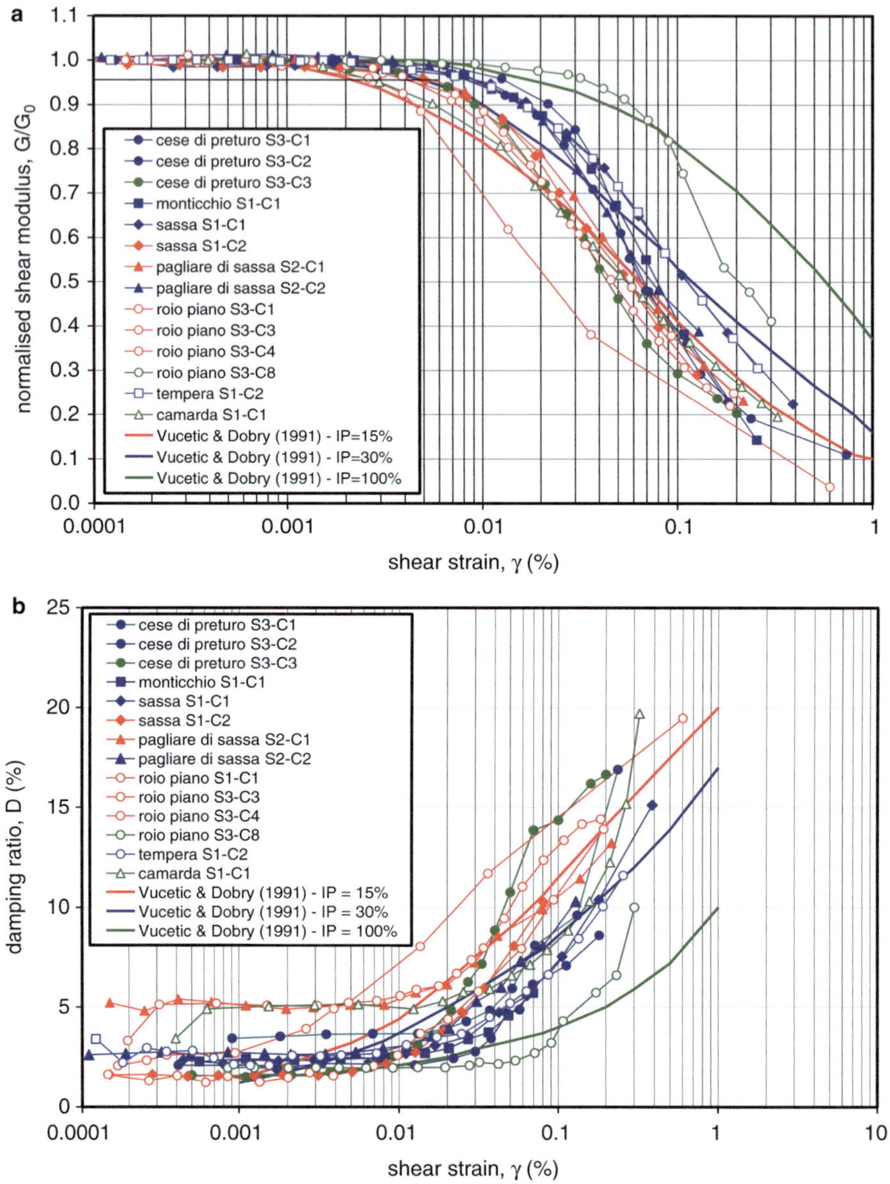


**Fig. 1.43** Normalised shear modulus (**a**) and damping ratio (**b**) vs. shear strain from RC tests

Similar comments pertain to the damping-strain curves (Fig. 1.43b), with the finer – high plasticity soils (blue group) showing less scattered trends with an initial damping, $D_0$, of the order of 2.5%, while that of the less fine – low plasticity samples varies between 1.5% and 5%. This latter is a quite unusual damping value for low plasticity soil and could be probably affected by some compliance problem in the measurement.

## 1.4 First Evaluation of Site Effects from Damage

### *1.4.1 Amplification Phenomena in the Middle Aterno Valley*

Within one week of the main shock, a multidisciplinary team coordinated by the GEER (Geoengineering Extreme Events Reconnaissance) Association reached the affected region (GEER Working Group 2009). The team was led by Jonathan Stewart, University of California at Los Angeles, and was formed with several researchers from U.S. and European universities and research institutions. Some of the Authors participated in the field reconnaissance on behalf of the Italian Geotechnical Association (Associazione Geotecnica Italiana, AGI).

Preliminary surveys in the towns and villages south-east of L'Aquila along the Aterno River valley showed significant differences in the damage distribution. As already said, the highest intensity (I = IX-X MCS) has been assigned to the villages of Onna and Castelnuovo, which are about 9 and 25 km far from the epicentre, respectively. Other villages such as San Gregorio, Poggio Picenze, Villa San Angelo were also heavily damaged. Other, less damaged villages are Monticchio, Fossa, Tussio, San Pio delle Camere, and Barisciano, among others. A rough indication of the damage distribution was documented for each village so that village-to-village comparisons could be attempted. In some villages, the coexistence of areas characterized by significant damages (with partial or total collapse of the buildings) with areas affected by little damage was also found. These differences suggested that site effects related to geologic conditions and different topographies may have played an important role in determining the damage distribution. Such effects may have been further amplified by the vulnerability of the buildings, when significantly different throughout a single village. Regardless of building vulnerability, a common description of damage intensity, shown in Table 1.3, was used.

In the following, evidence of site effects is illustrated by relating damage intensity with geological and morphological conditions in nearby villages at similar distance from the epicentre (village-to-village amplification), or within the same village (intra-village amplification). In particular the cases of Castelnuovo vs. San Pio delle Camere, Onna vs Monticchio and Poggio Picenze will be discussed herein in detail. A comprehensive inventory on the extent of damage to other villages of the middle Aterno valley and correlations with local geologic features can be found in the GEER report (GEER Working Group 2009).

**Table 1.3** Definition of damage categories (Adapted from Bray and Stewart 2000)

| Damage level | Description |
|---|---|
| D0 | No damage |
| D1 | Cracking of non-structural elements, such as dry walls, brick or stucco external cladding |
| D2 | Major damage to the non-structural elements, such as collapse of a whole masonry infill wall; minor damage to load bearing elements |
| D3 | Significant damage to load-bearing elements, but no collapse |
| D4 | Partial structural collapse (individual floor or portion of building) |
| D5 | Full collapse |

**Fig. 1.44** Google earth image showing the locations of Onna and Monticchio

### 1.4.1.1 Village-to-Village Amplification: Onna vs Monticchio

The hardest hit village near the city of L'Aquila was Onna (I=IX-X MCS), an old village on the floor of the Aterno valley, 9 km away from the epicentre at an average elevation of 580 ma.s.l. Monticchio (I=VI MCS), a village 1.3 km southwest of Onna (Fig. 1.44), is built on a gentle slope at the toe of the northern part of Cavalletto mountain at 600 m elevation approximately. Onna raises on Holocene calcareous alluvial and fluvial deposits of sand and gravel, and inter-bedded clay and silt, more than 5 m thick, overlying Pleistocene silty-clay deposits. Monticchio is founded on Mesozoic limestone and Pleistocene silts. Approximate geological cross-sections of the two villages, based on the geological map 1:25,000, are depicted in Fig. 1.45.

The village of Onna is composed mainly of 2–3 stories unreinforced masonry structures, with a minority of retrofitted structures. This village has a small number of newer reinforced concrete residential structures. Unreinforced masonry structures in Onna suffered a collapse (D5) rate of about 80% (Fig. 1.46a). Reinforced concrete structures suffered minor or no damage (Fig. 1.46b). The town was previously destroyed by the historical earthquake of 1461 (Rovida et al. 2009). During

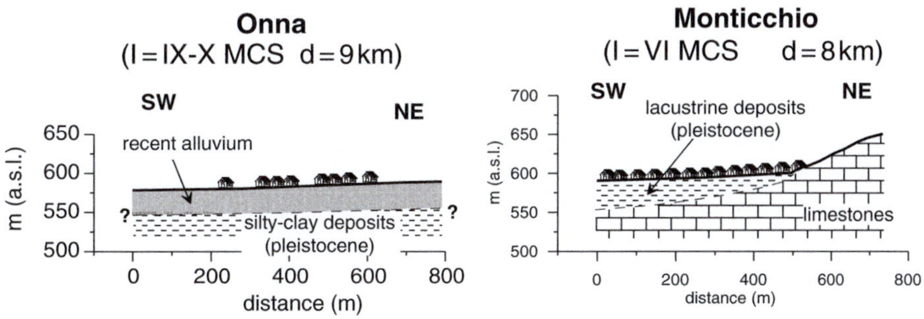


**Fig. 1.45** Approximate geological cross-sections of Onna and Monticchio (courtesy of Dr G. Di Capua, INGV)

**Fig. 1.46** Onna: (**a**) D5 damage level on masonry building structures; (**b**) D0 damage on an r.c. building (except for the tilting of the chimney)

that event, the village was reported to have suffered X MCS, and an eighteenth century chronicler Anton Ludovico Antinori reported that 'Nella Villa di Onda ne tampoco resto casa impiedi' (*'In the Onda village no house remained standing'*).

Monticchio consists mostly of 2–3 stories masonry buildings and to a lesser extent RC buildings. The intensity degree attributed was VI MCS. A D0-D1 damage was detected on both structure types throughout the village (Fig. 1.47).

The apparent elevated shaking intensity at Onna, as compared with the surrounding villages built on bedrock or stiffer alluvial debris, makes it a good candidate for speculation about site amplification effects on the valley fill compared with the surrounding bedrock.

#### 1.4.1.2 Village-to-Village Amplification: Castelnuovo vs San Pio Delle Camere

The villages of Castelnuovo (I=IX-X MCS) and San Pio delle Camere (I=V-VI MCS) are located about 25 km southeast of the April 6 epicentre, only 2 km apart (Fig. 1.48).

**Fig. 1.47** Monticchio : (**a**) D0 damage level on different building structures; (**b**) D0 damage on monumental masonry buildings; (**c**) D1 damage (fall of the cornice) on a masonry building

Castelnuovo (810–860 m a.s.l.) is settled on an elliptical hill mainly formed, according to recent geotechnical investigation, of fluvio-lacustrine deposits (lower to medium Pleistocene) belonging to the San Nicandro formation. This latter formation is eroded and filled with Holocene alluvial sediments downhill, at elevations below 800 m. San Pio delle Camere is located 2 km southeast of the village of Castelnuovo. It is a hill slope village built on breccias overlaying carbonate bedrock. Approximate geological cross-sections of the two villages, based on the geological map 1:25.000, are depicted in Fig. 1.49.

Castelnuovo consists of 2–3 stories masonry buildings of poor quality, some of them retrofitted. The centre of the village is located on the top of the hill and was nearly completely destroyed: a D5 damage level has been observed on 80% of masonry buildings, the remaining 20% being classified D4. A lower damage level has been observed on the buildings at the toe of the hill (Fig. 1.50). Note that, just like Onna, the town was already destroyed (I = X MCS) by the historical earthquake of 1461 (Rovida et al. 2009). In San Pio delle Camere, the housing stock of the village is similar to Castelnuovo. This village had no observable significant damage to any of the structures. Several fine cracks were observed in the exterior walls of some of the two-story and three-story residences, as shown in Fig. 1.51.

Fig. 1.48 Google earth image showing the locations of Castelnuovo and San Pio delle Camere

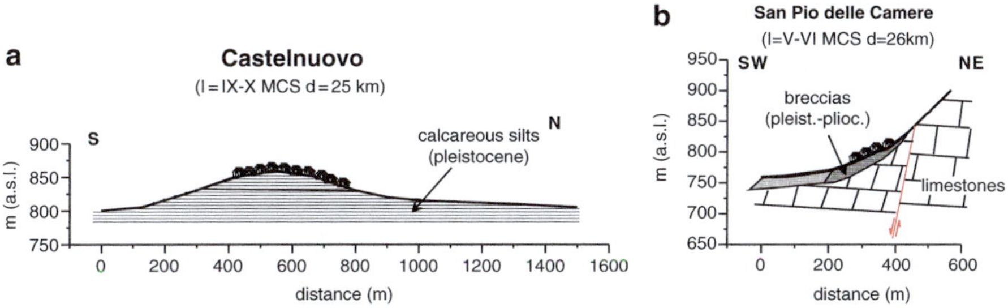


Fig. 1.49 Schematic geological cross-sections of Castelnuovo and San Pio delle Camere (courtesy of Dr G. Di Capua, INGV)

Fig. 1.50 Castelnuovo: (**a**) collapse (D5) of old masonry buildings on the hilltop; (**b**) limited damage (D1-D2) to masonry buildings at the toe of the hill

**Fig. 1.51** Typical unreinforced masonry structures in the village of San Pio delle Camere suffered no damage (D0), or slight cracking (D1). Structures in this village are similar to those in Castelnuovo

Based on the response of the structures, the shaking intensity at Castelnuovo was significantly greater than at San Pio delle Camere. It is also noteworthy that the damage to the top of the village of Castelnuovo was considerably worse than at the base. Accordingly, in Castelnuovo some factors related to topographic amplification may have contributed to the strong shaking at the highest elevations of the village. However, there is no indication of damaging topographic amplification at San Pio delle Camere.

#### 1.4.1.3 Intra-Village Amplification: Poggio Picenze

This village (695–760 m a.s.l.) lies along a slope located at the left (north) side of the river Aterno valley, 17 km away from the instrumental epicentre. An approximate geological cross-section SW-NE is shown in Fig. 1.52. The western side is settled on a coarse-grained Pleistocene formation; most of the historical centre is founded over the carbonate silt formation of San Nicandro, locally covered by layers of the Pleistocene gravel. This latter formation outcrops even at the toe of the hill. Outcrops of San Nicandro silts formation and Pleistocene conglomerate are depicted in Fig. 1.53.

The building heritage is constituted by a comparable number of masonry and r.c. houses, 2–3 floors high. Both irregular rubble and stone course masonry houses are widespread in the town; the former being more heavily damaged (Fig. 1.54a). In the

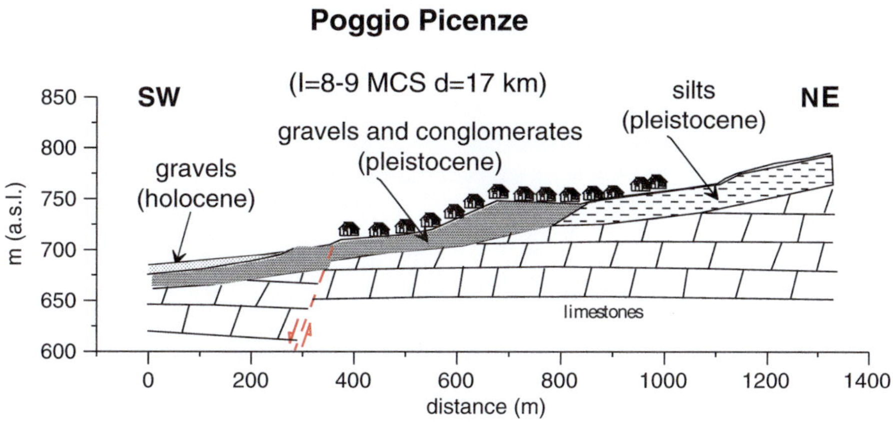


**Fig. 1.52** Schematic geological cross-sections of Poggio Picenze (courtesy of Dr G. Di Capua, INGV)

**Fig. 1.53** Outcrops of the (**a**) San Nicandro carbonate silts (note the conglomerate cover) and of the (**b**) Pleistocene conglomerate

historical centre, the old buildings including the monumental church (Fig. 1.54b) were heavily damaged (D3-D4 on the average), but the r.c. buildings lying nearby were almost unaffected by the shaking (Fig. 1.54c). Minor damage, even for monumental buildings, was detected in the western and downhill parts of the town, where the foundation soil is the coarse-grained Pleistocene formation (Fig. 1.54d).

## 1.5 Ground Failures

In this section, only the ground failures phenomena, interpreted as subsoil deformations and collapse following seismic shaking, will be described. Other source-related failure phenomena, like the surface fault rupture observed near Paganica, are described in greater detail in the report by GEER Working Group (2009).

**Fig. 1.54** Poggio Picenze: (**a**) variable damage (D3-D4) on masonry buildings of different style, (**b**) undamaged r.c. house (D0) close to the (**c**) monumental church (D3); (**d**) D1 damage in a church built on coarse-grained Pleistocene formation

The case histories of ground deformation and failure induced by the earthquake were relatively minor and mainly related to slope instability, collapse of some underground cavities (also referred to as 'sinkholes') and seismic compression of unsaturated soils. Most slope instabilities recorded after the major shocks can be classified into the following types:

- Failures of single rock blocks or small portions of the rock mass;
- Small slumps or slides in steep cut faces excavated in cataclastic limestones, slightly cemented
- Breccias/conglomerates, or debris;
- Fracturing and instability phenomena observed along the banks of Lake Sinizzo;
- Sinkholes and caves.

Flow slides involving debris, snow and ice on the southern flank of Mt. San Franco (NE of L'Aquila, epicentral distance ED = 18 km, I = VI MCS) were re-activated few days after the main shock. Dynamic actions possibly concurred with the prolonged intense rainfall occurred in the days following the main seismic event.

In the following sub-sections, only the main ground failures are described. Event descriptions and considerations reported in the following sub-sections have been originally collected during the surveys carried out by the GEER (Geoengineering Extreme Events Reconnaissance) Working Group.

### *1.5.1 Failures in Hard Rock Slopes*

Failures involved the limestone/sandstone bedrock and the cemented layers of the continental deposits (conglomerates, travertine, breccias).

In the bedrock formations, joints are often irregular, not regularly spaced and not much persistent. Under these structural conditions, large slides (requiring persistent weak surfaces) or large toppling failure (favoured by systematic jointing) cannot develop. Within recent continental units, sub-vertical discontinuities have generally wide spacing, which determines the failure of large isolated rock blocks, favoured by horizontal lithological variations. In all cases, only the shallower portions of the rock mass, where joints are opened and the rock mass is loosened, were involved in failures.

Failures of small blocks are widespread over a large area extending from the south-western flank of Gran Sasso ridge to the whole Aterno river valley, including the top of thc ridge bounding its eastern flank. The largest failures were those generating the rock fall at Lake Sinizzo (close to S. Demetrio) and the small rock avalanche at Fossa.

The north-eastern bank of lake Sinizzo (ED = 21 km, I = VI-VII MCS) is overlooked by a rock cliff formed by conglomerate layers with interbedded silty to sandy horizons (Fig. 1.55). The different strength and erodibility between the materials make the conglomerate layers to overhang. It is to be noted that the lake is on the prolongation of the seismogenic Paganica fault and its sides were affected by apparent ground ruptures that have not been explained yet.

The rock cliff failures at Sinizzo involved:

- An overhanging thick layer of conglomerates (Fig. 1.55b, right),
- A portion of the same layer subdivided into large blocks by irregular gapping fractures (Fig. 1.55b, left).

The village of Fossa (ED = 14 km, I = VII-VIII MCS) rises at the foot of a 300 m high limestone cliff, having the shape of a large amphitheatre (Fig. 1.56a). The cliff is the eastern side of a fault scarp formed by an irregularly fractured limestone. On the northern margin of the cliff, a rock slide generated a small rock avalanche

**Fig. 1.55** The rock cliff overlooking the NE side of Lake Sinizzo before (**a**) and after (**b**) the earthquake. Collapsed areas are circled in figure (**a**). Pre-failure picture, by courtesy of Dr. R. Giuliani, DPC

(Fig. 1.56b), that invaded the road below. Some out runner blocks threatened the outermost buildings of Fossa. A noticeable amplification at the cliff margin due to the particular morphology cannot be excluded.

Damages due to failures of single blocks were usually limited to the roads, but in some cases they hit buildings without causing casualties. A detailed report on failures of single blocks was drawn by ISPRA (2009a, b). The most impressive is the

a b

**Fig. 1.56** View of Fossa from mount Di Cerro: before (**a**) and after (**b**) the earthquake. The detachment area that generated the rock avalanche is circled. On top of the cliff (**b**), note the debris underneath the walls of Ocre castle produced by masonry failures

**Fig. 1.57** Block detached from a rock cliff above the Stiffe Caves (*left*). Impact marks (on the *right*) individuated by Scott Kieffer and Edward Button (GEER Working Group 2009)

rock fall at Stiffe Caves (ED = 20 km, I = VI-VII MCS), where a large block stopped its run against a building for tourist reception (Fig. 1.57). Details on the block run out can be found in the report by GEER Working Group (2009) and in the paper by Aydan et al. (2009).

**Fig. 1.58** Slides in (**a**) pervasively fractured limestone on the eastern flank of the Bazzano hill and (**b**) in cataclastic limestones uphill on Arischia

### *1.5.2 Failures in Soft Intensely Fractured Rocks and Coarse-Grained Materials*

Slides and slumps affected some abandoned quarry faces and cuts in pervasively jointed limestone rock masses, both in the Aterno Valley and northeast of L'Aquila along the high ridges of the Gran Sasso massif. A striking example is the eastern flank of the hill at Bazzano (Fig. 1.58a). Slides do not exceed few tens of $m^3$, but are widespread over the whole slope (ED = 10 km, I = VI MCS).

In the same area, small failures also affected cataclastic limestones. Cataclasites largely outcrop on the flanks of mountain slopes and are associated to major fault zones. The "rock" mass is highly non-homogeneous: intensely fractured, cohesionless or slightly cemented, alternate with partly structured portions. Figure 1.58b shows a slide uphill from Arischia, northeast of L'Aquila (ED = 10 km, I = VII-VIII MCS).

Small failures frequently occurred in the coarse grained slightly cemented continental deposits with outcrop extensively in the Aterno Valley. A relatively large rock face affected by slumps was found on a road cut excavated in breccias (Fig. 1.59) on the road from Barisciano to S. Stefano di Sessanio (ED = 20 km, I = VI MCS).

The failure involves an area wider than that shown in the picture, as demonstrated by the stepped ground surfaces with tension cracks behind the crown scarp. Inside villages rock cuts are often protected by dry stone masonry walls (mainly with lining function), frequently underwent failures which sometimes interacted with underground cavities.

**Fig. 1.59** Slump along the road from Barisciano to S. Stefano (Photo by D. Calcaterra and M. Ramondini, University of Naples)

**Fig. 1.60** Fractures along the NW perimeter delimiting slices and blocks lowered and rotated inward (**a**). Lowering of a large block detached from the western margin, evidenced by the submerged trees (**b**)

## *1.5.3 Failures at Lake Sinizzo*

After the earthquake, scarps and persistent and gaping fractures were observed along approximately 70–80% of Lake Sinizzo perimeter (Fig. 1.60). The lake, located to the east of San Demetrio ne' Vestini is roughly circular in plan view, with an average diameter of approximately 120 m and a maximum depth of about 10 m. It fills a natural karstic depression carved in a lacustrine deposit consisting of coarse grained materials (gravels) locally with fine plastic matrix (gravelly clay).

The limestone bedrock never outcrops in the vicinity of the lake and should be located few hundreds of meters below the lake bottom, according to the resistivity profiles reported by Bosi and Bertini (1970). During the earthquake, large slices of the shore lowered and rotated towards the lake (Fig. 1.60a) and slumps of large blocks into the lake occurred (Fig. 1.60b). The post-earthquake underwater morphology was soon detected by a bathymetric survey (Fig. 1.61) carried out by CNR-IAMC (GEER Working group 2009). Failures at the lake margin occurred also in the past, as demonstrated by the apparent scarps that can be observed along the whole perimeter.

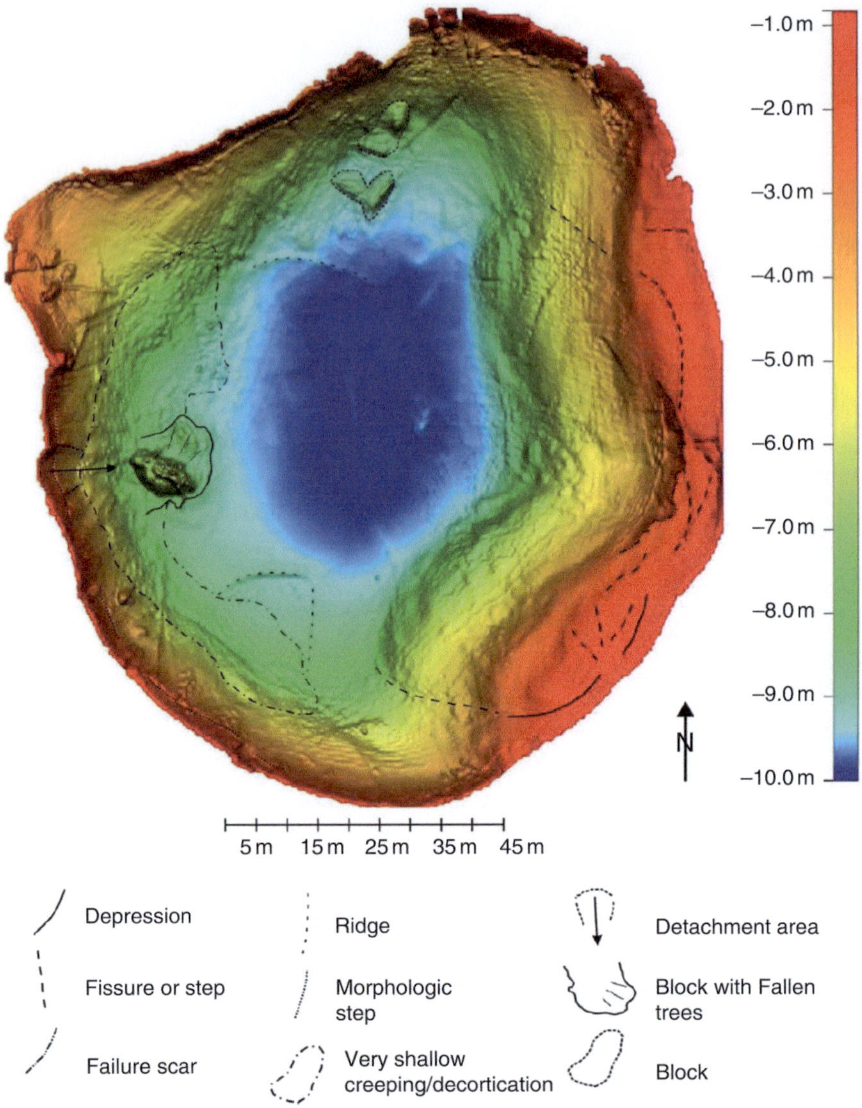


**Fig. 1.61** Bathymetry of Lake Sinizzo with major morphological elements (GEER Working group 2009)

At least two hypotheses, possibly concurring, can be advanced about the failure mechanism. They are to be analyzed through detailed geotechnical analyses.

The presence of a clayey matrix (possibly clayey levels) within the coarse grained lacustrine deposit and its saturation (seldom observed in the fine-grained materials outcropping along the slope Aterno valley) could have allowed the build-up of

**Fig. 1.62** Sinkholes in L'Aquila

noticeable excess pore pressures during the earthquake and triggered slumps along the submerged slopes of the lake.

Seismic loading and displacements at depth can have also induced instability of the karstic voids/conduits located below the lake, with subsequent sinking of the overlying sediments. In this respect, it is worth reminding that karstic structures are often involved in collapses even under lithostatic stresses.

In both cases, the vicinity of Lake Sinizzo to the prolongation of the main seismogenic structure related to the L'Aquila earthquake is to be considered.

### *1.5.4 Sinkholes and Caves*

The geology of L'Aquila involves limestone at the base, lacustrine clay and continental debris in the form of conglomerate and breccias and Holocene alluvial deposits from bottom to top. The L'Aquila breccia of Pleistocene age is known to contain karstic caves. Karstic caves are geologically well known and generally form along steep fault zones and fractures due to erosion as well as solution by groundwater.

The two large karstic caves shown in Fig. 1.60, which are both very close to the AQK strong motion station, were investigated. Karstic caves are a well known problem in L'Aquila. The remnant of karstic caves are easily noticed on a few rock slope cuts. Also, one may find a number of reports on the search of potential karstic caves by using various geophysical methods (i.e. Tallini et al. 2004a, b).

The L'Aquila earthquake caused a number of sinkholes in L'Aquila (Fig. 1.62 and Table 1.4). One of the sinkholes was well publicized worldwide. The width was about 10 m and its depth is not well known. A car was fallen into it. Another sinkhole had a slightly smaller size. Its length and width were estimated to be about 8 and 7 m respectively. The depth was about 10 m. The layers between the roof and road level were formed of breccia with calcareous cementation, breccia with clayey matrix and top soil, from bottom to top. A side trench was excavated to a depth of

**Table 1.4** Major parameters of sinkholes (values are approximate) (Aydan et al. 2009)

| L1 (m) | L2 (m) | h (m) | h* (m) | PGA-EW (g) | PGA-NS (g) | P-UD (g) | Location |
|---|---|---|---|---|---|---|---|
| 6 | 10 | 3 | 3 | 0.341 | 0.383 | 0.365 | L'Aquila-S1 |
| 7 | 6 | 3 | 3 | 0.341 | 0.383 | 0.365 | L'Aquila-S1 |
| 3 | 5 | 1.5 | 1 | – | – | – | Castelnuovo |

**Fig. 1.63** Outcropping of sand and water as evidence of the liquefaction phenomenon in Vittorito (L'Aquila) (courtesy of Mr. A. Civitareale)

3–4 m from the ground surface on the south side of the sinkhole. The reconnaissance team of the GEER also reported that a sinkhole occurred in Castelnuovo and it had a length and width of 5 and 3 m respectively, with a depth of 5 m. The roof thickness was about 1.5–2.5 m.

## 1.6 Liquefaction

The occurrence of liquefaction at some locations in the area of the village of Vittorito (L'Aquila prefecture) was suggested by a personal communication of Dr. Galadini, National Institute of Geophysics and Volcanology, Italy. A preliminary study for the area is reported hereafter. The obtained information should be considered of some relevance, due to the scarce occurrence of well documented case-histories of liquefaction in Italy in the recent years, when the main mechanisms ruling the phenomenon are somehow understood. The case-history is also important in the light of the new building code recently introduced in Italy (NTC 2008) that dedicates some specific sections to the engineering assessment of the liquefaction phenomenon.

The liquefaction phenomena in Vittorito were revealed trough a series of sand boils and sand volcanoes developed in free field, during and few hours after the main shock of L'Aquila earthquake.

Figure 1.63 shows a couple of photos made by a local farmer, eye-witnessing that some small sand volcanoes appeared in his artichokes field in the aftermath of

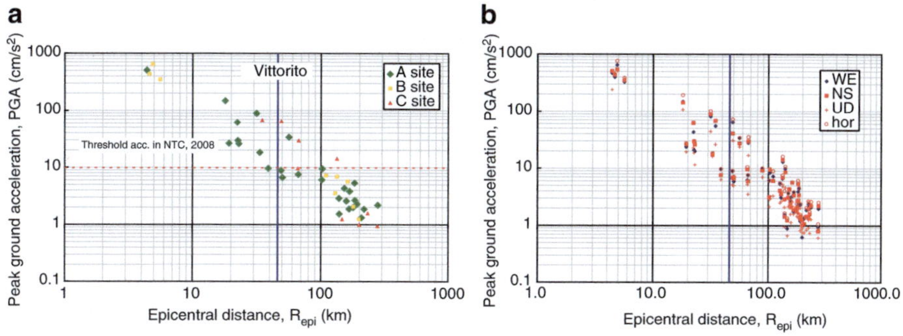


**Fig. 1.64** Attenuation of the peak ground acceleration with the epicentral distance for the April 6th 2009 Abruzzo earthquake: (**a**) data classified according to the subsoil conditions below the recording stations and (**b**) attenuation of all the components of the motion

the earthquake. Volcanoes were soon flatted fearing that they were an evidence of the outcropping of some radon, which was considered in the news as a strong evidence of a next coming earthquake.

Evidences of liquefaction related phenomena were also indicated by local newspaper and confirmed by experts during the post-earthquake field investigation, particularly because some water and sand was seen moving upwards during the execution of field tests and soil sampling.

The village of Vittorito is located approximately 40 km SE of the city of L'Aquila. The area where the phenomenon occurred is close the left side of the Aterno river, where the ground surface is relatively flat and no outstanding construction exists. Recent continental deposits outcrop in the area; the embankments of the Aterno River are constituted by holocene gravelly-sandy alluvial deposits that somehow might include some silty sand.

### *1.6.1 Ground Motion*

No single recording station of the Italian Strong Motion Network (RAN) is available in the proximity of the Vittorito area. However, the seismic motion characteristics produced by the main shock of the L'Aquila earthquake might be deduced by interpolating other available data as published in W.G. ITACA (2008).

Figure 1.64 reports the decay of peak ground acceleration versus the epicentral distance for the main shock occurred on April, 6.

The data points are distinguished according to the information available on the local geology for the RAN stations (Ameri et al. 2009). Some doubts still exist for the soil condition below the seismic stations; therefore some of the data reported could be influenced by site effects. The plot of Fig. 1.64b shows the three motion components plus the amplitude of the vector sum of EW and NS records.

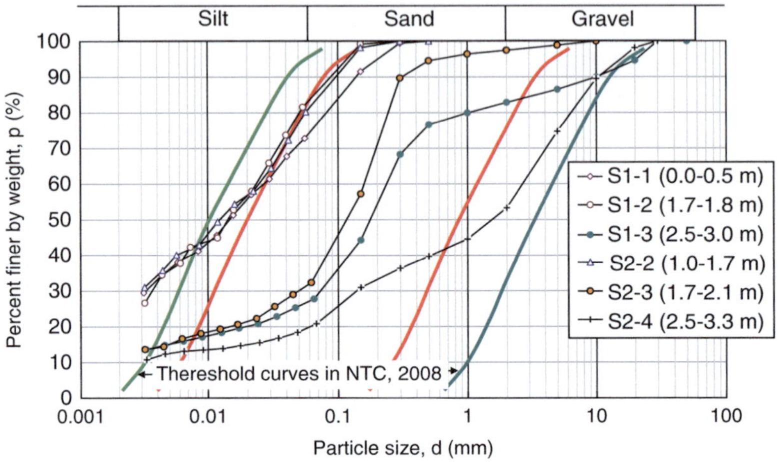


**Fig. 1.65** Grading curves of some materials in the potentially liquefied area and limit curves for a preliminary screening of the liquefaction potential for soils with high coefficient of uniformity (NTC 2008)

From the above attenuation laws, a mean horizontal peak ground acceleration as low as 0.065 g could be estimated on outcropping bedrock, and no significant differences are expected for different subsoil conditions and motion components.

The overall motion energy in the Vittorito area is attenuated: a value of the Arias intensity in the order of 0.01 m/s was estimated, while a significant duration (Trifunac and Brady 1975) of about 20 s can be expected at the potentially liquefied sites (Monaco et al. 2011).

### *1.6.2 Geotechnical Investigation*

A preliminary geotechnical investigation was carried out at three different sites in the area were the liquefaction phenomenon was observed. Two shallow boreholes were executed and three in situ seismic dilatometer tests SDMT were carried out.

Boreholes, having an average depth of about 5 m, showed that the subsoil profile is constituted by a first layer of vegetable soil having a thickness of about 1 m. This stratum lies above a sandy silt level having a thickness of about 2 m, below which some gravelly materials, with variable cementation degree, were found. Therefore, it appears that in the first meters of subsoil, the average particle size of the material increases with depth. The water table was detected at about 0.3 ÷ 0.5 m below the ground level.

Figure 1.65 shows the laboratory grading curves for the shallow soils. Two main materials are found: sandy silt and gravelly sand. In both cases, the uniformity coefficient is relatively high.

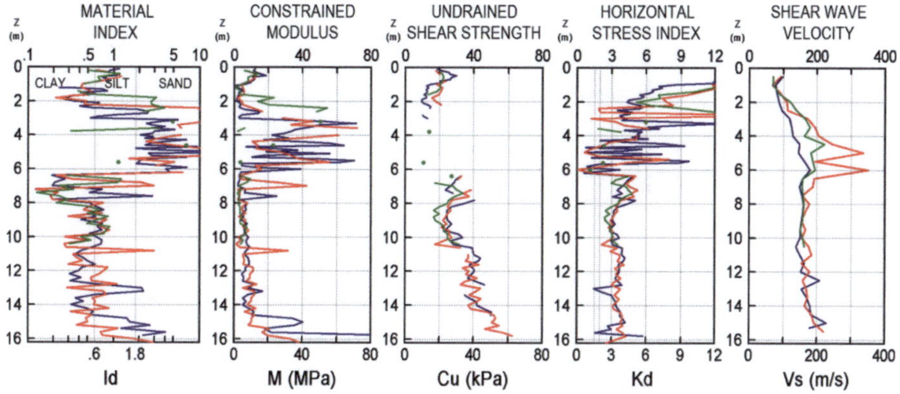


**Fig. 1.66** Results of three Seismic Dilatometer tests in the area of Vittorito

Figure 1.66 summarizes the results of the three seismic dilatometer tests SDMT performed in the area. It appears that soil profiles are almost similar in the three locations where tests were performed. At a first glance it appears also that, excluding the gravelly layer from approximately 3–6 m below the ground level, the mechanical properties of the soils are relatively poor. Therefore, the possibility that liquefaction phenomena happened might not be excluded a priori, despite the significant epicentral distance.

### *1.6.3 Liquefaction Analyses*

Following the approach introduced in Eurocode 8 (EN 1998-5 2004), NTC (2008) includes specific requirements to exclude the liquefaction phenomenon (see also Santucci de Magistris 2006). For some of the available data, the liquefaction case-history at Vittorito sites is borderline or against the specific criteria. First of all, the mean horizontal peak ground acceleration estimated for the area (0.065 g) is well below the threshold reported in the new Italian building code (0.1 g).

On the other hand, the magnitude of the main shock of L'Aquila earthquake, the shallow depth of the water table and the grading curves of the materials (see Fig. 1.65) suggested that some more specific analyses are needed. Monaco et al. (2011) discussed some details of these points.

Two simplified procedures were used to assess the occurrence of liquefaction. The first, suggested by Andrus and Stokoe (2000) and reviewed by Idriss and Boulanger (2004), uses the shear wave velocity to detect if the soil is liquefiable. The second refers to the horizontal stress index, $K_D$, measured in the dilatometer test (Monaco et al. 2005).

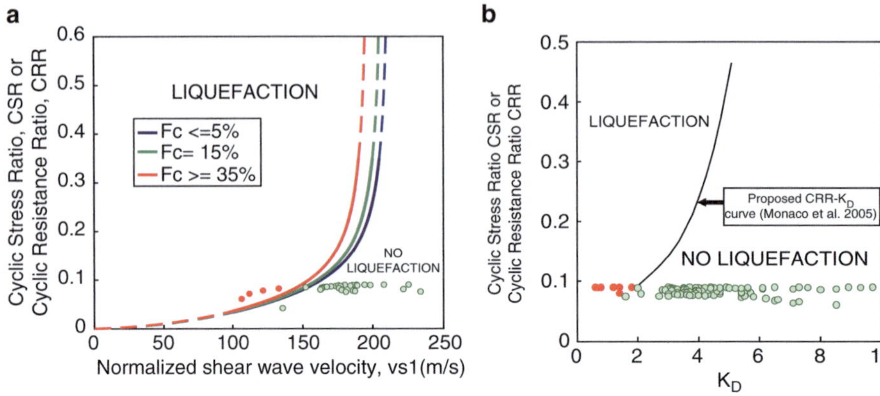


**Fig. 1.67** Example of verification charts from SDMT1: data from shear wave velocity profile (**a**); and from the horizontal stress index (**b**)

The SDMT routinely provides, among other measurements, pairs of profiles of $K_D$ and $V_S$, both correlated with the liquefaction resistance of sands. Hence, SDMT permits to obtain two parallel independent estimates of liquefaction resistance, expressed by the cyclic resistance ratio CRR: one from $K_D$ and the other from $V_S$, using CRR-$K_D$ and CRR-$V_S$ correlations, in the framework of the conventional simplified procedure developed by Seed and Idriss (1971).

For example, Fig. 1.67a shows the verification charts obtained plotting the data points obtained from the SDMT1 profile, in terms of cyclic stress ratio CSR versus the normalised shear wave velocity; the liquefaction limit curves by Idriss and Boulanger (2004) are drawn for comparison. The CSR-$K_D$ data points obtained from SDMT1 are also plotted in Fig. 1.67b, together with the liquefaction boundary curve. In both figures, red solid symbols indicate the conditions of likely occurrence of liquefaction.

In Fig. 1.68 the liquefaction potential index, IL (Iwasaki et al. 1982), for the three investigated sites is reported. Specifically, the plots in Fig. 1.67a are obtained starting from the shear wave velocity $V_S$, while data in Fig. 1.67b are derived from the horizontal stress index $K_D$.

It can be noticed that in both cases the IL values are relatively low, but compatible with the observed occurrence of the phenomenon of liquefaction. However, the identification of the sand layer which liquefied provided by the two methods is not consistent: while $V_S$ indicates possible liquefaction at very shallow depth (≈1–3 m), $K_D$ suggests that liquefaction may have occurred even in the deeper layer (≈4–6 m). This issue is currently under investigation.

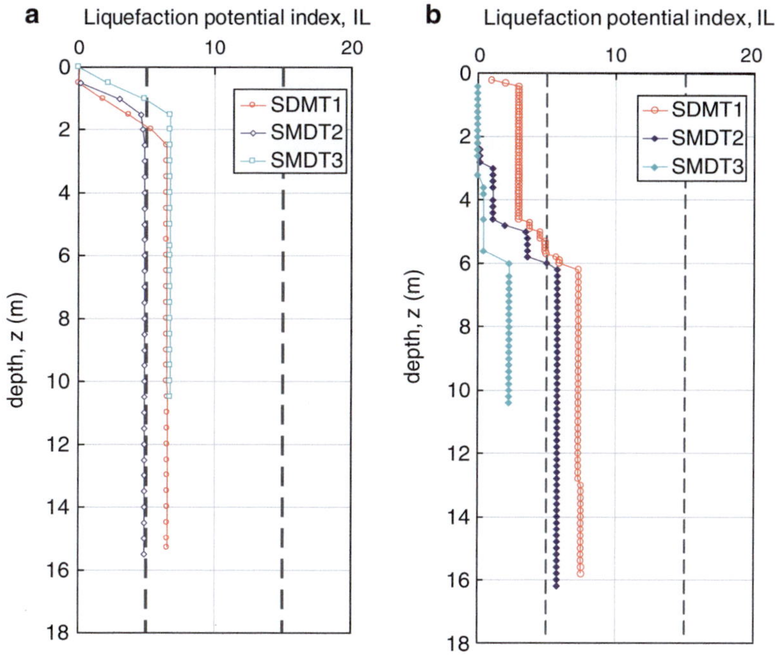


**Fig. 1.68** Profiles of liquefaction potential index for the three investigated sites in Vittorito: soil resistance from shear wave velocity (**a**) and from horizontal stress index (**b**)

## 1.7 Conclusions

The April 6, 2009 earthquake ($M_L = 5.8$ and $M_W = 6.3$) caused 308 victims, about 1,600 injured, 40,000 homeless and huge economic losses. The earthquake produced heavy damages in the city of L'Aquila, with a MCS Intensity I = VIII-IX, and even more severe damages in some nearby villages. A maximum MCS Intensity I = IX-X was experienced at Onna and Castelnuovo.

Several Italian researchers, on behalf of the Italian Geotechnical Society (Associazione Geotecnica Italiana, AGI), were involved in the field reconnaissance activities soon after the earthquake. In the following months, site investigation programs were planned at a number of sites selected for the location of new temporary houses for the homeless people (C.A.S.E. project); seismic microzonation studies promoted by the Department of Civil Protection are still in progress.

The L'Aquila seismic event is maybe the first Italian earthquake well recorded in the epicentral area. The April 6, 2009 main shock was generated by a normal fault with a maximum vertical dislocation of 25 cm and depth of the hypocentre of about 8.8 km. It was recorded by 56 digital strong motion stations, part of the Italian Strong Motion Network (Rete Accelerometrica Nazionale, RAN) owned and maintained by the Department of Civil Protection. Five of these strong motion stations were located within 10 km from the epicentre, on the hanging wall side of the fault. The recorded horizontal and vertical peak accelerations in the epicentral area were higher than 0.35 g, up to 0.65 g.

Considerable damage to structures was detected over a broad area of approximately 600 km$^2$, including the city of L'Aquila and several villages of the middle Aterno River valley. The distribution of damage within the affected area was irregular, creating speculation for rupture directivity effects as well as for local amplification phenomena. In the epicentral area, where the city of L'Aquila is located, an evaluation of site effects was obtained from the available strong motion recordings. In the middle Aterno valley, where such recordings were not available, a preliminary assessment of site effects was carried out based on the survey of the variable damage distribution. The evidence of site effects is illustrated in the paper by relating damage intensity with geological and morphological conditions in nearby villages at similar distance from the epicentre (village-to-village amplification), or within the same village (intra-village amplification), taking into account the different vulnerability of the buildings.

The geological setting of the area affected by the April 6, 2009 earthquake, located in a vast intra-Apennine basin elongated in NW-SE direction (parallel too many of the active normal faults), is very complex. The city of L'Aquila is settled on cemented breccias having thickness of some tens of meters, overlying lacustrine sediments resting on limestones. The Aterno valley is partly filled with lacustrine deposits, overlying the limestone bedrock and topped by alluvial deposits. The soils in the Aterno valley are characterized by a high variability of their geotechnical characteristics (grain size distribution, fabric/cementation, mechanical properties), both in horizontal and vertical direction. The heterogeneity of the soils in this area was already well known, before the earthquake, from the results of investigations carried out in the past by routine in situ and laboratory testing techniques.

A large amount of site investigations was carried out in the months following the earthquake at several sites, in the framework of the C.A.S.E. project and of the seismic microzonation project. The results obtained from in situ and laboratory tests carried out after the earthquake have permitted to improve significantly the static and dynamic characterization of the soils in the area of L'Aquila.

The after-earthquake investigations included measurements of the shear wave velocity $V_S$ by different in situ testing techniques, i.e. down-hole tests (DH), surface wave tests (MASW) and seismic dilatometer tests (SDMT). Comparisons at various test sites, illustrated in the paper, indicate that MASW, DH and SDMT results are in very good agreement, with surface wave tests allowing defining a deeper shear wave velocity profile.

Cyclic and dynamic laboratory tests were carried out on undisturbed samples from various sites, involving a network of research laboratories in Italy. The cyclic/dynamic shear tests – Resonant Column/Torsional Shear tests (RC-TS) and Double Sample Direct Simple Shear tests (DSDSS) – were aimed at characterizing the non-linear and dissipative behaviour (stiffness and damping) of medium- to fine-grained soils. The curves of the normalized shear modulus $G/G_0$ and damping ratio $D$ versus shear strain $\gamma$ obtained for the tested fine-grained materials, synthetically described in the paper, consistently reflect their different grain size distribution and plasticity. However, reference literature curves do not overlap the non-linear overall trend of the experimental curves, in some cases highlighting the peculiar nature of the soils of the Aterno valley.

Surveys carried out soon after the earthquake reported ground failures of various types in hard rocks, in soft intensely fractured rocks and in coarse-grained materials.

A case of liquefaction, triggered by the April 6, 2009 main shock, was detected in the area of Vittorito, approximately 45 km far from the epicentre. The interest in this case derives from the relatively limited amount of well documented case-histories of liquefaction in Italy and to its location far from the epicentre, beyond the threshold motion levels predicted by the Italian Technical Code and empirical relationships. The Vittorito liquefaction case, briefly described in the paper, was preliminarily analysed using the results of in-situ seismic dilatometer tests. Further research is in progress.

**Acknowledgments** The paper is the result of the joint efforts of many people and institutions cooperating together for the Italian Civil Protection Department and, most of all, for the people living in Abruzzo.

The working group was coordinated by the Italian Geotechnical Society, and the Authors wish to thank the President, Prof. Stefano Aversa, for the continuous and enthusiastic support.

The President of ReLuis Consortium, Prof. Gaetano Manfredi, is also acknowledged for the increasing interest shown in the survey and research activities.

Prof. Jonathan P. Stewart, as leader of the GEER Reconnaissance Team, is also thanked for the tremendous effort shown in the reconnaissance organization and reporting.

## References

AGI (1991) Geotechnical characterization of Fucino clay. In: Proceedings of the X ECSMFE, Firenze, vol 1, pp 27–40

Ameri G, Augliera P, Bindi D, D'Alema E, Ladina C, Lovati S, Luzi L, Marzorati S, Massa M, Pacor F, Puglia R (2009) Strong-motion parameters of the Mw=6.3 Abruzzo (Central Italy) earthquake. INGV internal report. (http://www.mi.ingv.it/docs/report_RAN_20090406.pdf)

Andrus DR, Stokoe KH II (2000) Liquefaction resistance of soils from shear-wave velocity ASCE. J Geotech Geoenviron 126(11):1015–1025

APAT (2006) Geological map of Italy at the scale 1:50,000 – Sheet No. 359 L'Aquila. S.EL.CA. Firenze

Aydan O, Kumsar H, Toprak S, Barla G (2009) Characteristics of 2009 L'Aquila earthquake with an emphasis on earthquake prediction and geotechnical damage. J Sch Marine Sci Tecnol, Tokai Univ 7(3):23–51

Bertini T, Bosi C, Galadini F (1989) La conca di Fossa S. Demetrio dei Vestini. In: CNR, Centro di Studio per la Geologia Tecnica, ENEA, P.A.S.: Elementi di tettonica pliocenico-quaternaria ed indizi di sismicita olocenica nell'Appennino laziale-abruzzese. Società Geologica Italiana, pp 26–58 (in Italian)

Bertini T, Farroni A, Totani G (1992a) Idrogeologia della conca aquilana. University of L'Aquila, Dipartimento Ingegneria Strutture Acque e Terreno, Publ. DISAT 92/6 (in Italian)

Bertini T, Totani G, Cugusi F, Farroni A (1992b) Caratterizzazione geologica e geotecnica dei sedimenti quaternari del settore occidentale della conca aquilana. University of L'Aquila, Dipartimento Ingegneria Strutture Acque e Terreno, Publ. DISAT 92/7 (in Italian)

Bosi C, Bertini T (1970) Geologia della Media Valle dell'Aterno. Memorie della Società Geologica Italiana, vol IX, pp 719–777 (in Italian)

Bouckovalas G, Kouretzis G (2001) Stiff soil amplification effects in the 7 Sept 1999 Athens (Greece) earthquake. Soil Dyn Earthq Eng 21:671–687

Bray JD, Stewart JP, (Coordinators and Principal Contributors), Baturay MB, Durgunoglu T, Onalp A, Sancio RB, Ural D (Principal Contributors) (2000) "Damage Patterns and Foundation Performance in Adapazari," Chapter 8 of the "Kocaeli, Turkey Earthquake of Aug 17, 1999 Reconnaissance Report", in Earthquake Spectra J Suppl A to vol 16, EERI, pp 163–189

Cavinato GP, Di Luzio E (2009) Rilievo geologico speditivo eseguito nel sito di Pianola. CNR-IGAG, Roma (in Italian)

Cavuoto G, Moscatelli M (2009) Rilevamento geologico preliminare del sito Bazzano. CNR-IGAG, Roma (in Italian)

Chioccarelli E, Iervolino I (2009) Direttività e azione sismica: discussione per l'evento dell'Aquila. In: Proceedings of the 13th Convegno "L'Ingegneria Sismica in Italia", Bologna. ANIDIS, Roma (in Italian)

D'Elia B, Lanzo G, Pagliaroli A (2003) Small strain stiffness and damping of soils in a direct simple shear device. Pacific Conference on Earthquake Engineering, Christchurch, New Zealand

De Luca G, Marcucci S, Milana G, Sanò T (2005) Evidence of low-frequency amplification in the city of L'Aquila, Central Italy, through a multidisciplinary approach including strong- and weak-motion data, ambient noise, and numerical modelling. Bull Seismol Soc Am 95(4):1469–1481

Di Capua G, Lanzo G, Luzi L, Pacor F, Paolucci R, Peppoloni S, Scasserra G, Puglia R (2009) Caratteristiche geologiche e classificazione di sito delle stazioni accelerometriche della RAN ubicate a L'Aquila. Report S4 Project (http://esse4.mi.ingv.it/), June 2009

Doroudian M, Vucetic M (1995) A direct simple shear device for measuring small-strain behaviour. Geotech Test J 18(1):69–85

DPC (2009) Down-hole tests, Internal technical report

EduPro Civil System, Inc., ProShake (1998) Ground response analysis program, Redmond, WA

EN 1998–5 (2004) Eurocode 8: Design of structures for earthquake resistance – part 5: foundations, retaining structures and geotechnical aspects. CEN European Committee for Standardization, Bruxelles, Belgium

Evangelista L (2009) A critical review of the MASW technique for site investigation in geotechnical engineering, PhD Thesis, University of Naples Federico II, Italy

Foti S (2005) Surface wave testing for geotechnical characterization. In: Lai CG, Wilmanski K (eds) Surface waves in geomechanics: direct and inverse modelling for soils and rocks, CISM Series, Number 481. Springer, Wien, pp 47–71

Foti S, Comina C, Boiero D, Socco LV (2009) Non uniqueness in surface wave inversion and consequences on seismic site response analyses. Soil Dyn Earthq Eng 29(6):982–993

Galli P, Camassi R (eds) (2009) Rapporto sugli effetti del terremoto aquilano del 6 aprile 2009. Rapporto congiunto DPC-INGV. http://www.mi.ingv.it/eq/090406/ (in Italian)

GEER Working Group (2009) Preliminary report on the seismological and geotechnical aspects of the Apr 6 2009 L'Aquila Earthquake in Central Italy. Report for web dissemination, Geotechnical Earthquake Engineering Reconnaissance (GEER) association, Report No. GEER-016, Version 2, (http://www.geerassociation.org/GEER_Post%20EQ%20Reports/Italy_2009/Cover_Italy2009_Rev.html)

Hepton P (1988) Shear wave velocity measurements during penetration testing. In: Proceedings of the penetration testing in the UK, ICE, Birmingham, pp 275–278

Idriss IM, Boulanger RW (2004) Semi-empirical procedures for evaluating liquefaction potential during earthquakes. In: Doolin D et al (eds) Proceedings of the 11th ICSDEE and 3rd ICEGE, vol 1, Berkeley, CA, pp 32–56

ISPRA (2009a) Tabella sintetica sopralluoghi 7-8-9-10 Aprile 2009
ISPRA (2009b) Documentazione Fotografica sopralluoghi (periodo 6–10 Aprile 2009)
Iwasaki T, Tokida K, Tatsuoka F, Watanabe S, Yasuda S, Sato H (1982) Microzonation for soil liquefaction potential using simplified methods. In: Proceedings of the 3rd international conference on microzonation, vol 3, Seattle, pp 1319–1330
Maraschini M, Ernst F, Boiero D, Foti S, Socco LV (2008) A new approach for multimodal inversion of Rayleigh and Scholte waves. In: Proceedings of the EAGE Rome, expanded abstract
Maraschini M, Comina C, Foti S (2009) Inversione multimodale delle onde di superficie per la caratterizzazione di siti della rete accelerometrica nazionale, IARG 2009, Rome (In Italian)
Marchetti S, Monaco P, Totani G, Marchetti D (2008) In Situ Tests by Seismic Dilatometer (SDMT). From research to practice in geotechnical engineering, ASCE Geotech. Spec. Publ. No. 180 Honouring John H. Schmertmann, pp 292–311
Martin GK, Mayne PW (1997) Seismic flat dilatometer tests in Connecticut valley varved clay. ASTM Geotech Testing J 20(3):357–361
Martin GK, Mayne PW (1998) Seismic flat dilatometer in Piedmont residual soils. In: Robertson PK, Mayne PW (eds) Proceedings of the 1st international conference on site characterization, Atlanta, vol 2. Balkema, Rotterdam, pp 837–843
Mayne PW, Schneider JA, Martin GK (1999) Small- and large-strain soil properties from seismic flat dilatometer tests. In: Proceedings of the 2nd international symposium on pre-failure deformation characteristics of geomaterials, Torino, vol 1, pp 419–427
McGillivray A, Mayne PW (2004) Seismic piezocone and seismic flat dilatometer tests at Treporti. In: Viana da Fonseca A, Mayne PW (eds) Proceedings of the 2nd international conference on site characterization, Porto, vol 2. Mill Press, Rotterdam, pp 1695–1700
Młynarek Z, Gogolik S, Marchetti D (2006) Suitability of the SDMT method to assess geotechnical parameters of post-flotation sediments. In: Failmezger RA, Anderson JB (eds) Proceeding 2nd international conference on the flat dilatometer, Washington, DC, pp 148–153
Monaco P, Marchetti S (2007) Evaluating liquefaction potential by seismic dilatometer (SDMT) accounting for aging/stress history. In: Proceedings of the 4th international conference on Earthquake Geotechnical Engineering ICEGE, Thessaloniki, Greece
Monaco P, Marchetti S, Totani G, Calabrese M (2005) Sand liquefiability assessment by Flat Dilatometer Test (DMT). In: Proceedings of the 16th ICSMGE, vol 4, Osaka, pp 2693–2697
Monaco P, Marchetti S, Totani G, Marchetti D (2009) Interrelationship between small strain modulus $G_0$ and operative modulus. In: Kokusho T, Tsukamoto Y, Yoshimine M (eds) Proceedings of the international conference on performance-based design in Earthquake Geotechnical Engineering (IS-Tokyo 2009), Tsukuba, Japan, June 15–17. Taylor & Francis Group, London, pp 1315–1323
Monaco P, Santucci de Magistris F, Grasso S, Marchetti S, Maugeri M, Totani G (2011) Analysis of the liquefaction phenomena in the village of Vittorito (L'Aquila). Bull Earthquake Eng (2011) 9:231–261
NTC (2008) Approvazione delle nuove norme tecniche per le costruzioni. Gazzetta Ufficiale della Repubblica Italiana, n. 29 del 4 febbraio 2008 – Suppl. Ordinario n. 30, http://www.cslp.it/cslp/index.php?option=com_docman&task=doc_download&gid=3269&Itemid=10 (in Italian)
Rovida A, Castelli V, Camassi R and Stucchi M (2009) Historical earthquakes in the area affected by the Apr 2009 seismic sequence. (http://www.mi.ingv.it/eq/090406/)
Santucci de Magistris F (2006) Liquefaction: a contribution to the Eurocode from the Italian Guideline "Geotechnical aspects of the design in seismic areas". ISSMGE ETC-12 workshop, NTUA Athens, Greece, (http://users.civil.ntua.gr/gbouck/gr/etc12/papers/paper8.pdf)
Seed HB, Idriss IM (1971) Simplified procedure for evaluating soil liquefaction potential. J Soil Mech Found Div ASCE 97(9):1249–1273
Socco LV, Strobbia C (2004) Surface wave methods for near-surface characterisation: a tutorial. Near Surface Geophys 2(4):165–185
Spadoni M, Sposato A (2009) Rilevamento geologico speditivo dell'area San Giacomo. CNR-IGAG, Roma, In Italian

Tallini M, Giamberardino A, Ranalli D, Scozzafava M (2004a) GPR survey for investigation in building foundations. In: Proceedings of the 10th international conference on ground penetrating radar, Delft, the Netherlands, pp 395–397

Tallini M, Ranalli D, Scozzafava M, Manocorda G (2004b) Testing a new low-frequency GPR antenna on karst environments of Central Italy. In: Proceedings of the 10th international conference on ground penetrating radar, Delft, the Netherlands, pp 133–135

Tallini M, De Caterini G, Di Eusebio F, Di Nisio C, Manetta M, Menis M, Zaffiro P (2009) Studio geologico finalizzato alla redazione della carta delle Microzone omogenee in prospettiva sismica di aree destinate all'urbanizzazione – Zona Roio Piano. University of L'Aquila (In Italian)

TC16 (2001) The Flat Dilatometer Test (DMT) in soil investigations – a report by the ISSMGE committee TC16. Reprint in "Flat Dilatometer Testing". In: Failmezger RA, Anderson JB (eds) Proceeding 2nd international conference on the flat dilatometer, Washington, DC, Apr 2–5, 2006, pp 7–48

Totani G, Monaco P, Marchetti S, Marchetti D (2009) Vs measurements by Seismic dilatometer (SDMT) in non penetrable soils. In: Proceedings of the 17th international conference on soil mechanics and geotechnical engineering, Alexandria, Egypt, vol 2, 5–9 Oct 2009, pp 977–980

Trifunac MD, Brady AG (1975) A study on the duration of strong earthquake ground motion. Bull Seismol Soc Am 65:581–626

Vucetic M, Dobry R (1991) Effects of the soil plasticity on cyclic response. J Geotech Eng ASCE 117(1):89–107

Working Group ITACA (2008) Data base of the Italian strong motion data. http://itaca.mi.ingv.it

Zywicki DJ (1999) Advanced signal processing methods applied to engineering analysis of seismic surface waves, PhD thesis, Georgia Institute of Technology

# Chapter 2
# Geotechnical Aspects of 2008 Wenchuan Earthquake, China

**Ikuo Towhata and Yuan-Jun Jiang**

**Abstract** The 2008 Wenchuan earthquake of China produced significant damage in the local community. Since the affected area was so vast, it became very difficult to apply previous experiences to the restoration of the community. The present paper concerns findings during technical visits to the damaged area and lessons for the future of the area. From the geotechnical viewpoints, this earthquake is characterized by many slope failures with a large range of size. Many landslide dams were formed as a consequence of big slope failures to be a threat to the downstream area. Smaller slope failures, that were much more in number, are important as well, because many local roads were destroyed by them and post-earthquake rescue and restoration activities were affected. It was interesting that different types of slope stabilization measures functioned to different extents, avoiding total collapse. Some retaining walls translated due to significant seismic earth pressure. Mountain slopes that were thus disturbed by strong seismic shaking will cause further problems in the coming decades. As have been experienced after big quakes at different places in the world, those slopes will produce further failures and debris flows, and affect safety of local communities and transportations. Finally, geotechnical problems other than slopes are discussed.

## 2.1 Introduction

The 2008 Wenchuan earthquake of China hit Sichuan Province on May 12th at 14:28 by local time. The seismic magnitude is reportedly 8.0 by Richter scale. The total number of victims exceeded 80,000; most of them were killed by collapse of

---

I. Towhata (✉) • Y.-J. Jiang
Department of Civil Engineering, University of Tokyo, Tokyo, Japan
e-mail: towhata@geot.t.u-tokyo.ac.jp; yuanjun.jiang@geot.t.u-tokyo.ac.jp

M.A. Sakr and A. Ansal (eds.), *Special Topics in Earthquake Geotechnical Engineering*, Geotechnical, Geological and Earthquake Engineering 16,
DOI 10.1007/978-94-007-2060-2_2, 

houses and buildings and slope failures. The authors visited the damaged areas several times as an activity of the Japanese Geotechnical Society and the University of Tokyo. Those visits took place in late June and middle October, 2008, and middle April, 2009. The present text is going to report findings during those visits and discussions for the future of the damaged area.

## 2.2 Geological and Topographical Conditions of Damaged Area

China may be considered to be relatively inactive in earthquake sense. There are, however, areas in which earthquakes produced significant damage in the past. For example, Xi-an was shaken in AD 1487 and historical buildings were severely affected. The 1920 Haiyuan earthquake in Northwest Ningxia Autonomous Province claimed more than 200,000 victims and also caused many failures in loess slopes. The 1976 Tangshan earthquake occurred to the east of Beijing and destroyed many buildings. Liquefaction and failure of embankment were many in number as well.

The seismically affected part of Sichuan Province is situated along the boundary between the low and flat land of Sichuan Basin and the mountain ranges to the west (Fig. 2.1). This particular area is tectonically affected by the collision of Indian subcontinent against Eurasian continent; hence high mountains have been generated which are underlain by big faults. Thus, it may be stated that seismic risk in this area is high, although few earthquakes have been known during the historical era.

The local geology in the mountainous area is composed of such sedimentary rocks as mudstone and sandstone together with occasional limestone and granite. The granite formation occurs along the Longmenshan fault system because of the following reason. Since the strong collision between Indian and Eurasian continents, the tectonic movement along the Longmenshan fault system is controlled by the thrust of Tibet plateau and the stubborn resistance of Yantze plate. In the subduction process of Yantze plate, the earth temperature and pressure changes caused the crust to melt and magma to elevate and extrude through crust; finally granite formed in the region.

Since the average precipitation per year is in the range of less than 700 mm in Wenchuan to 990 mm in Qingchuan and 1,300 mm in Beichuan, the rate of erosion is high. Hence, mountain slopes are generally steep, making slope failure easy to occur whether due to earthquakes or due to rainfalls.

The earthquakes in 2008 were caused by a right lateral type of strike-slip movement in a combination of three faults which are namely Longmenshan fault system (Lou et al. 2009) (F2 to F4 in Fig. 2.1). The total length of the entire fault was as long as 240 km. The fault rupture started from the southwest part near Yingxiu and propagated mainly towards the northeast direction. Until the time of writing this paper, much information is not yet released to the public about the acceleration records during this earthquake.

**Fig. 2.1** Earthquake-damaged area and historical earthquakes (Drawn after Wenchuan earthquake disaster maps. Institute of Geology, China Earthquake Administration 2009)

The Wenchuan earthquake is characterized by its enormous number of victims. While many of them were killed by collapse of buildings, the number of those who were killed by slope failures is not small. Yin et al. (2009) stated that the total number of geohazards was 15,000, which claimed 20,000 casualties.

In Sichuan Province there are records of historical earthquakes. For instance the 1933 Diexi earthquake in Maoxian County with magnitude 7.5, where intensity at the epicentre was ten in Chinese Scale, caused 6,865 deaths and 15,324 house damage. The huge landslides occurred in this earthquake and blocked three places along Minjiang River. Until now there are still two famous dammed lakes which are named Dahaizi and Xiaohaizi, where are famous tourism places in Sichuan Province. On August 16th, 1976 the other big earthquake occurred between the northwest of Songpan county and Pingwu county with magnitude 7.2. Just 6 days later another two big earthquakes sequentially occurred on August 22nd and 23rd. Their magnitudes were 6.7 and 7.2 respectively. Fortunately before earthquakes the local Earthquake Administration gave long-term and middle term earthquake forecasts and the large amount of casualties were avoided.

## 2.3 Slope Failures of Large Size

This section concerns slope failures of a large size. For their locations, refer to Fig. 2.2. Figure 2.2 indicate that most slope failures occurred along the strike of the fault. While Huang et al. (2008) reported a gigantic Daguangbao slope failure that was of 742 million $m^3$, this site is in a remote mountain area and the authors have not been able to reach it. First, Fig. 2.3 indicates the slope failure at Xiejiadian, where debris was produced at higher elevations in the mountain and flowed downstream as a debris flow. One village was buried under the debris deposit and 100 people were killed. The involved materials are mostly sandstone and mudstone. The size of the debris deposit was approximately 300 m in length, 300 m in width, and 50 m in maximum thickness, and the entire volume was assessed to be 4 million $m^3$. The gradient angle between the end of the deposit and the top of the source was 26°.

The debris from the Xiejiadian slope stopped at the bottom of the valley and formed a hill of stones. Figure 2.4 shows that one layer of debris deposit measured approximately 4 m in thickness, while the size of the maximum stone therein was about 2 m. See small surface failures as well to the left in this figure.

The second site is Donghekou in Qingchuan County that is located near the northeast end of the rupture zone (Fig. 2.5). The total volume of the failed mass was 18 million $m^3$ (Wang et al. 2009). The angle between the top of the source area and the bottom of the debris deposit was 20°. The bottom part of the slide, however, is of much smaller angle at its surface, suggesting more fluid nature of the material (Fig. 2.5). In the same figure, the mountain in front is made of sedimentary rocks, including slate, while the one in the back is composed of limestone. The debris deposit here stopped the stream of Hongshihe and formed a lake. Near this site, moreover, the stream of Qingzhujiang was stopped by a smaller slope failure and a natural reservoir was formed by an 80 m-thick deposit of debris (Fig. 2.6). This dam was drained later by opening a water channel by blasting. The total number of major natural dams was 34 and their possibility of collapse formed big threats to the safety of communities in the downstream areas.

**Fig. 2.2** Sites of slope failures (Drawn after Wenchuan earthquake disaster maps. Institute of Geology, China Earthquake Administration 2009)

## 2.4 Slope Failures of Smaller Size

There occurred many smaller slope failures (Fig. 2.7 for example); their total number is uncountable. They may not attract research concerns because bigger ones appear more interesting. However, those small ones affected the public in a severe manner and need to be studied further.

Figure 2.8 shows the continuous occurrence of failure in the slope surface along the Minjiang Valley. See the steep slope in this figure. Although the individual failed mass was thin, this event blocked the transportation in a local important road connecting Yingxiu and Wenchuan; Fig. 2.9. The geology in this valley is made of granite. Figure 2.9 shows that the length of a failed slope is significant and that this

Fig. 2.3 Slope failure at Xiejiadian site in Pengzhou city

Fig. 2.4 Debris deposit at Xiejiadian site at the entrance to Yinchanggou

slope is from now on vulnerable to further failures such as collapse and rock falls. Hence, the road traffic is subject to slope failure hazards under both fine and rainy weathers. Figure 2.10 illustrates a collapsed shape of a mountain. The entire slope was affected by shear failure of the surface materials. With many examples of a similar nature, the significant damage in mountain bodies is attributed to concentration

**Fig. 2.5** Donghekou slope failure in Qingchuan county

**Fig. 2.6** Site of Shibangou natural dam

**Fig. 2.7** Surface failures near Yingxiu

**Fig. 2.8** Continuous failures in Minjiang valley slope in Yingxiu

of earthquake energy that is put in at the bottom of a mountain and is concentrated towards the top, generating significant intensity of shaking. Finally, in Figs. 2.7–2.10, attention should be paid to the fact that slope failure started from the top of slopes, suggesting intensified shaking near top of slopes.

**Fig. 2.9** Road destroyed by surface failure near Yingxiu

**Fig. 2.10** Surface failure from top of mountain on the Zipingpu dam reservoir

**Fig. 2.11** Different extent of slope damage due to different geological settings

**Fig. 2.12** Failure in a slope with geological stratification parallel to surface.

Figure 2.11 was taken near Beichuan County. It is important that the slope on the left-hand side was subjected to significant slope failure most probably because the geological stratification was parallel to the slope surface, making the sliding easier (Fig. 2.12). On the contrary on the opposite side, there was little slope instability because the geological stratification was normal to the surface. It is also possible that the steep slope angle therein received less amount of rain water and the extent of weathering was less.

## 2.5 Effects of Slope Stabilization During Earthquake

The authors visited sites of slope stability reinforcement along the highway between Dujiangyan and Yingxiu. Although many roads were blocked by stone falls and slope failure during the earthquake, traffics got recovered once failed materials were removed from the road. Thus, slope stabilization measures functioned in an appropriate way. Figure 2.13 illustrates a distorted shape of shot crete coverage with metal mesh that was installed to prevent deterioration of relatively good rock slope. Although the shot crete surface deformed substantially as shown in the figure, slope failure was successfully prevented and road traffic was possible after the quake. It therefore seems that this stabilization measure was successful and that the induced deformation is allowable from the viewpoints of seismic performance and cost vs. benefit.

Figure 2.14 exhibits a case of a slope that was originally less stable than the one in Fig. 2.13 and was reinforced by RC frames equipped with ground anchors. This measure functioned in an appropriate way as well because the slope did not fail. Similarly, Fig. 2.15 indicate stabilization by bored piles with panels. Although the

**Fig. 2.13** Post-quake appearance of rock slope reinforced by shot crete and metal fence

**Fig. 2.14** Post-quake appearance of slope surface reinforced by RC frame

**Fig. 2.15** Post-quake appearance of slope that was stabilized by bored piles and panels

induced earth pressure made one of the piles tilt probably during the quake, the overall stability of the slope was maintained and the road traffic was not affected.

Figure 2.16 illustrates minor distortion of a road embankment that was reinforced by metal strips. Upon the earthquake, surface panels fell down and some

**Fig. 2.16** Post-quake appearance of reinforced embankment

metal strips came out of the soil. Hence, some people claimed that reinforcement was not successful here. This negative idea is not right because, as shown in Fig. 2.16, the main body of the embankment was not seriously affected and the road traffic was maintained after the quake. From the viewpoints of seismic performance and cost vs. benefit, the distortion at this site is within an allowable limit.

## 2.6 Discussion on Long-Term Instability of Seismically Disturbed Mountain Slopes

Past experiences suggest that the 2008 Wenchaun earthquake has triggered a long-term instability of seismically disturbed mountain slopes, which will produce rock falls and debris flows. The first example of the past experiences is found in Muzaffarabad City of Kashmir in Pakistan where a big earthquake in 2005 disturbed mountain slopes substantially. In addition to many slope failures during the earthquake, new slope failures and debris flows started due to rains and erosion. Figure 2.17 shows one of the problems.

A similar problem is going on in the central mountain of Taiwan where the 1999 Chichi earthquake significantly disturbed slope materials. Figure 2.18 shows a consequence in 2008 where a gigantic debris flow was induced by rainfall and thereby a bridge was destroyed.

**Fig. 2.17** Slope erosion near Muzaffarabad of Pakistan

**Fig. 2.18** Debris flow in seismically disturbed valley in Taiwan in 2008

A more dramatic example of long-term instability is found in the Abe River of Japan. Figure 2.19 illustrates the most upstream area of this river where the gigantic Ohya Landslide was caused probably by the 1706 Ho-Ei earthquakes in the subduction zone of the Pacific Ocean. Since this earthquake, the seismically-disturbed mountain slope has been generating debris flows frequently upon many heavy rainfalls. As a consequence, the river channel of the Abe River is today filled with debris deposits (Fig. 2.20), that have been raising the elevation of the river bed continuously.

**Fig. 2.19** Ohya landslide of Abe river in Japan

**Fig. 2.20** Abe river channel filled with frequent debris flow

The long-term instability has already become reality in China. The town of Beichuan has been attacked by rainfall-induced slope failures several times after the quake (Fig. 2.21) and the remaining buildings have been destroyed. Thus, the reconstruction of the town became more and more difficult and finally the entire town was

**Fig. 2.21** Post-earthquake expansion of Wangjiayan slide in Kushan town of Beichuan county

**Fig. 2.22** Endangered Yingxiu-Wenchuan road along Min Jiang river (Photograph by Zhao Yu)

abandoned. Moreover, the road that connects Yingxiu and Wenchuan are vulnerable to occasional slope failures upon rainfalls (Fig. 2.22), and therefore is often closed during bad weathers. It is thought therefore that many efforts will be needed from now on to prevent or reduce the amount of falling stones and soil mass that may

**Fig. 2.23** Rainfall-induced debris flow in Yinchanggou

severely affect the safety of local communities. Finally, Fig. 2.23 illustrates debris flow that attacked Yinchanggou resort which is close to Xiejiadian in July during heavy rainfall 2 months after the quake.

Such efforts are not so easy, however, because the size of the affected area is so vast and the number of unstable slopes is big. It is therefore impossible to carry out elaborate and perfect stabilization measures everywhere. Design of safety measures should be based on cost-benefit or performance principles in which a minor risk in future is allowed.

## 2.7 Other Geotechnical Problems

This section addresses geotechnical problems other than slope failures. First, Fig. 2.24 show a failure shape of Xiaoyudong Bridge. The major part of the bridge fell down completely because, in accordance with local regulation, no seismic design was practiced. Figure 2.25 shows the movement of bridge abutment at Xiaoyudong Bridge. It is evident that the forward movement of the abutment caused shear failure of the bridge structure. Figure 2.26 indicates a fault movement near the road embankment connected to this bridge. Since the movement of this fault was left-lateral, opposite from that of the main fault, this fault is not a part of the main fault. This movement consisted of 1 m of vertical and 55 cm of left-lateral displacements. This ground movement affected the road embankment and a stone

**Fig. 2.24** Collapse of Xiaoyudong bridge (Photograph by Huang Dong)

**Fig. 2.25** Seismic displacement of abutment of Xiaoyudong bridge

wall (Fig. 2.27) that rested upon this fault. Figure 2.28 indicates the effects of main fault in Yingxiu that affected a retaining wall upon the Minjiang River.

Figure 2.29 exhibits minor displacement of a bridge in Yingxiu. This bridge rotated in the horizontal direction because of the seismic inertia force. This movement

**Fig. 2.26** Fault movement near Xiaoyudong bridge

**Fig. 2.27** Fault–induced damage in road embankment and stone wall

was fortunately stopped by small and simple lateral constraints that were installed on the bridge piers.

Figure 2.30 exhibits the downstream side of the Zipingpu Dam. This rockfill dam measures 150 m in height and was constructed by the surface concrete facing.

Fig. 2.28 Fault-induced damage in retaining wall in Yingxiu

Fig. 2.29 Successful result of lateral constraint to prevent falling of bridge

Although the concrete facing in the upstream slope developed minor damage upon the quake, it is said that this damage was limited to the part above the reservoir water level. Hence, the water level, which was once lowered for inspection immediately after the quake, was raised again within a few months after the quake.

**Fig. 2.30** Successful behaviour of Zipingpu dam near Dujiangyan. (Photograph from www.chinawater.com.cn)

Thus, the idea of surface pavement, which converts the water pressure in the reservoir to effective stress in the dam body, worked successfully.

## 2.8 Conclusions

This paper described findings and learnt lessons that were obtained from field investigations conducted in Sichuan Province of China after the gigantic Wenchuan earthquake. Since many slope failures were triggered by this quake, this paper is rather focused on the slope problems. The major conclusions drawn from the study are described in what follows.

1. This devastating earthquake was caused by faults of which no previous action had been known during the historical time.
2. Some huge slope failures exhibited high fluid-like nature and debris travelled over a long distance.
3. Smaller failures at slope surface are as important as bigger ones in that they hindered transportations after the quake.
4. Mountain slopes thus disturbed by the seismic shaking will be unstable for a long time in future, producing a large amount of debris upon heavy rainfalls.
5. Many bridges were prevented from falling down by small constraining mechanisms.
6. A big dam with surface pavement functioned in a reasonable manner and avoided collapse.

**Acknowledgement** One part of the present study was carried out as a joint investigation of five societies which are namely the Japanese Geotechnical Society, the Japan Society of Civil Engineers, Architectural Institute of Japan, Japan Association of Earthquake Engineering, and Seismological Society of Japan in collaboration with Southwest Jiaotong University in Chengdu of China. The remaining part of study was conducted with Chengdu Institute of Mountain Hazards and Environment as well. The authors deeply appreciate the supports and helps provided from these joint activities.

## Appendix Second Author's Experience

Although more than 1 year has passed after the quake, the shocking memory given by the Wenchuan Earthquake still gives me frequent stimulus to my motivation to engage in disaster prevention study. In May, 2008, I was doing my master course in Southwest Jiaotong University, Chengdu, Sichuan Prov. China. At 14:28, on May 12th, 2008, I was still in bed for taking a nap which is a common habit in Sichuan Province. The sudden and strong jump of my bed woke me up. After a few seconds I still blamed this was a shaking caused by the experiment inside nearby National Power and Traction Lab. With uproar and yell coming out from the inside and outside of my dormitory, my intuition told me this was not an easy shaking. Without caring how much clothes still covering my body, I jumped onto the floor. At this moment the ponderous but loud noise created by the left and right shaking of the building hustled running pace of mine and my roommates. We crashed out of the door, and ran and jumped at a desperate speed from the fifth floor to the first floor. When we surged out the gate of our dorm, there were still more than two seconds shaking. Shamefully to say that for me fear and excitement coexisted in this one and half minutes. I never thought about the catastrophic destruction caused by this big "shaking", because this was my first time to experience a real and devastating earthquake.

Since Chengdu City was around 100 km away from Yingxiu County, and also the epicentre was out of Sichuan basin, the destruction in Chengdu was not severe and nearly no building collapsed. However for the sake of students' safety, all the universities arranged students camp outside in open areas or stadiums. After a few days I got a chance as a volunteer to join the rescues work in Deyang City, and mainly involved in logistic works, including transporting and distributing disaster relief supplies. In this short 2 days I visited Shifang County, where ruins and bodies of victims were excavated from the collapsed buildings. It was the first time for me to realize what a catastrophic earthquake it was. At that time I had no tear for them. What I did was to focus my full strength on the voluntary rescue work.

## References

Huang RQ, Pei XJ, Li TB (2008) Basic characteristics and formation mechanism of the largest scale landslide at Dagungbao occurred during the Wenchuan earthquake. J Eng Geol 16(6):730–741 (in Chinese)

Lou H, Wang CY, Lu ZY, Yao ZX, Dai SG, You HC (2009) Deep tectonic setting of the 2008 Wenchuan Ms8.0 earthquake in south-western China —Joint analysis of teleseismic P-wave receiver functions and Bouguer gravity anomalies. Sci China Ser D 52(2):166–179

Wang FW, Cheng QG, Highland L, Miyajima M, Wang HB, Yan CG (2009) Preliminary investigation of some large landslides triggered by the 2008 Wenchuan earthquake, Sichuan Province, China. Landslides 6(1):47–54

Wenchuan earthquake disaster maps (2009) Institute of Geology, China Earthquake Administration (in Chinese)

Yin YP, Wang FW, Sun P (2009) Landslide hazards triggered by the 2008 Wenchuan earthquake, Sichuan, China. Landslides 6(2):139–152

# Chapter 3
# Effects of Fines and Aging on Liquefaction Strength and Cone Resistance of Sand Investigated in Triaxial Apparatus

**Takaji Kokusho, Fumiki Ito, and Yota Nagao**

**Abstract** Innovative miniature cone penetration and subsequent liquefaction tests were carried out in a modified triaxial apparatus on sand specimens containing fines. It has been found that one unique curve relating cone resistance $q_t$ and liquefaction strength $R_L$ can be established, despite the differences in relative density and fines content, the trend of which differs from the current liquefaction evaluation practice. In order to examine an aging effect on the relationship, sands containing fines added with a small amount of cement are tested to emulate a long geological period in a short time. The addition of cement to the fines tends to increase the liquefaction strength $R_L$ much more than the penetration resistance $q_t$, resulting in obvious upward shift of the $q_t \sim R_L$ curve from the curve obtained for specimens without cement. Thus, it is revealed that the cementation effect by aging which dominantly occurs in fines can explain why higher fines content leads to higher liquefaction strength for the same cone resistance.

## 3.1 Introduction

Liquefaction strength is evaluated using penetration resistance of standard penetration tests (SPT) or cone penetration tests (CPT) in engineering practice. If sand contains a measurable amount of fines, liquefaction strength is normally raised in

T. Kokusho (✉)
Faculty of Science & Engineering, Chuo University, Kasuga 1-13-27, Bunkyo-ku, Tokyo, 112-8551, Japan
e-mail: kokusho@civil.chuo-u.ac.jp

F. Ito • Y. Nagao
Chuo University, Tokyo, Japan

M.A. Sakr and A. Ansal (eds.), *Special Topics in Earthquake Geotechnical Engineering*, Geotechnical, Geological and Earthquake Engineering 16,
DOI 10.1007/978-94-007-2060-2_3, © Springer Science+Business Media B.V. 2012

accordance to fines content $F_c$ in most of liquefaction evaluation methods, such as the road bridge design code using SPT $N$-values (Japan Road Association 1996). In the case of CPT, Suzuki et al. (1995) carried out in situ penetration tests and soil samplings by in situ freezing technique from the same soil deposits and compared the tip resistance $q_t$-value and liquefaction strength $R_L$ from lab tests. The comparison showed that higher fines content tends to increase $R_L$ for the same $q_t$-value. In contrast to their finding, however, quite a few laboratory tests have shown that $R_L$ clearly decreases with increasing $F_c$ of low plasticity fines having the same relative density $D_r$ (e.g. Kokusho 2007). Thus, a wide gap remains between the current practice for liquefaction potential evaluation and laboratory soil tests for liquefaction strength in sands containing fines despite its importance in engineering design.

In establishing empirical correlations between penetration resistance and liquefaction strength, calibration chamber tests were sometimes carried out to develop relationships between the penetration resistance and relative density at first and then, liquefaction tests were separately conducted for the same sand reconstituted in the same way as in the chamber tests. On the other hand, in situ penetration tests were combined with laboratory liquefaction tests on intact samples recovered from the same ground. In both experimental methods, the soils tested in the penetration test and in the liquefaction test are not exactly the same and may change due to heterogeneity of in situ soils or other reasons.

In order to establish more direct correlations between cone resistance and liquefaction strength ($q_t - R_L$ curve) considering the effect of fines, a systematic experimental study was undertaken by the present research group, in which a miniature cone penetration test and subsequent cyclic loading test were carried out on the same triaxial test specimens (Kokusho et al. 2005). An innovative simple mechanism originally introduced in a normal cyclic triaxial apparatus by Kokusho et al. (2003) was used enabling a miniature cone to penetrate the sand specimen at a constant speed. Results from the two serial tests on the same specimen were compared to develop direct $q_t - R_L$ correlations for sands containing various amounts of fines.

In the first part of this paper, major results of the previous experimental research (Kokusho et al. 2005) are explained here again, in which a unique correlation is recognized between cone resistance $q_t$ and liquefaction strength $R_L$ despite differences in fines content and relative density. In the latter part, series of accelerated tests using the same apparatus for soils added by small amount of cement are carried out in order to discuss the aging effect on the $q_t$ - $R_L$ correlations.

## 3.2 Test Apparatus and Test Procedures

In a triaxial apparatus used in this research, the specimen size is 100 mm diameter and 200 mm height. In liquefaction tests, the soil specimen can be loaded cyclically by a pneumatic actuator from above as a stress-controlled test. In order to carry out a cone penetration test in the same specimen prior to cyclic loading, a metal pedestal below the soil specimen was modified as shown in Fig. 3.1, so that a miniature cone

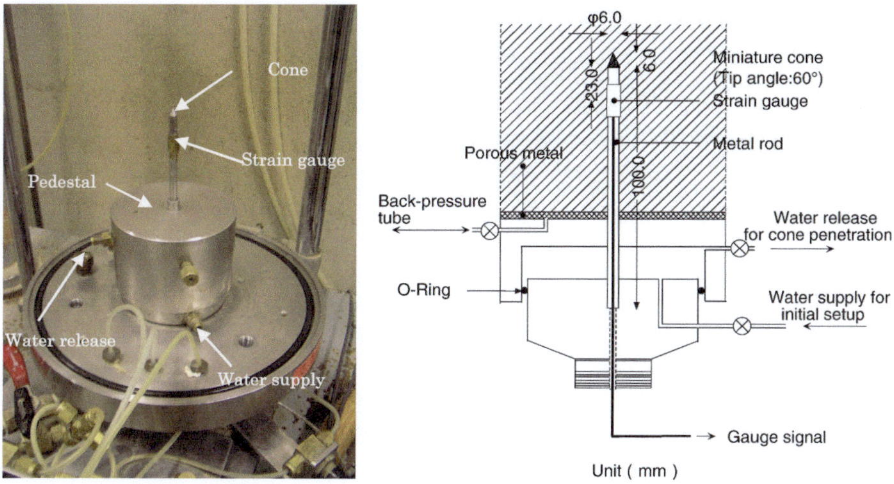


**Fig. 3.1** Photograph (*left*) and cross-section (*right*) of the modified pedestal in the lower part of the triaxial apparatus

can penetrate in a simple mechanism into the specimen from below. For that goal, the pedestal consists of two parts, a fixed circular base to which the cone rod is fixed and a movable metal cap, through the centre opening of which the cone rod penetrates in the upward direction into the overlying specimen. The annulus between the two parts is sealed by O-rings, enabling the cap to slide up and down by pressure supplied into a water reservoir in between the two parts. During the test, the pedestal cap is initially set up at the highest level by supplying water inside, and specimen is constructed on it. In this stage, the cone already projects into soil specimen by 47 mm. By opening a valve at the start of the cone penetration, the water in the reservoir is drained by the chamber pressure, resulting in settlement of a total body of the specimen together with the disc and piston at the top and realizing the relative upward penetration of the cone by 25 mm. The penetration rate is almost constant, though it is much slower than prototype CPT tests.

The miniature cone is 6 mm diameter, 6 mm height and 60° tip angle, about 1/35 smaller in the cross-sectional area than the 10 cm$^2$ prototype cone normally used in engineering practice. The strain gauges are glued at the top (23 mm lower than the base of the cone) of outside face, meaning that the measured vertical load includes not only the tip resistance but also some skin friction at the top portion of the rod. Though the separation of the tip load is actually needed, the total load is considered as the cone resistance (denoted here by $q_t$) in the present study because the contribution of the skin friction seems to be considerably smaller than the tip load.

In the test sequence, the penetration test was first carried out after consolidating the specimen. Then, after releasing pore pressure and reconsolidating it under the same confining pressure, the same specimen was cyclically loaded in undrained condition. The cone penetration was basically carried out under the undrained condition in the

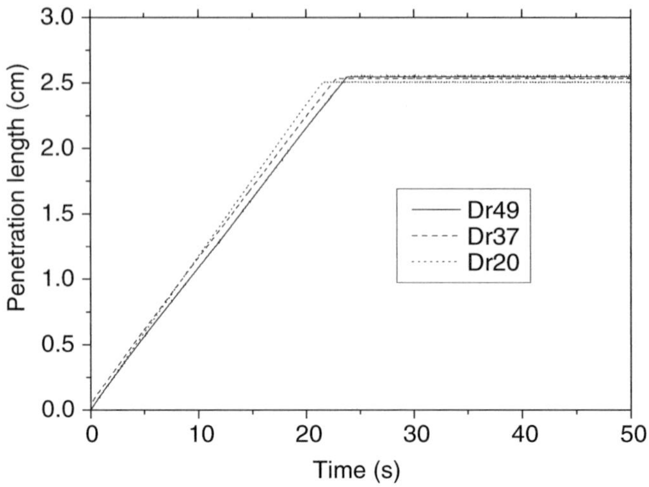


**Fig. 3.2** Penetration length versus time relationships for the miniature CPT in triaxial specimens under undrained condition

triaxial specimen, though a few drained penetration tests was also conducted to compare with as will be explained later. In undrained conditions, the pore pressure of the specimen was also measured by the electric piezometer in the same way as in normal undrained triaxial tests.

In Fig. 3.2, relationships between penetration lengths versus elapsed time obtained during preliminary tests by Toyoura sands of different densities are shown. The penetration rate is about 1 mm/s, much slower than prototype CPT, and is almost constant despite the difference in relative density of the sand specimen $D_r$ = 20% to 49%.

The cyclic axial load of a sinusoidal wave was applied with the frequency of 0.1 Hz. The cell pressure and the pore-water pressure were measured with electric piezometers with the maximum capacity of 490 kPa and the axial deformation is measured with LVDT of 50 mm maximum capacity outside the pressure chamber.

It may well be suspected that, in such a test sequence, the liquefaction strength is possibly influenced by the preceding cone penetration and subsequent reconsolidation. In order to examine the influence of the test sequences, preliminary comparative tests were conducted in which results by normal liquefaction tests in the same triaxial apparatus without the cone rod were compared with those in this test sequence. Specimens of Toyoura sand with the relative density of around 40–50% were prepared by wet tamping and tested. Figure 3.3 shows the comparison of the results with or without the cone, indicating no clear difference between the two corresponding best-fit curves (the solid curve with cone and the dashed curve without cone) correlating stress ratios $R_L$ and number of loading cycles $N_c$ reaching 5% double amplitude strain.

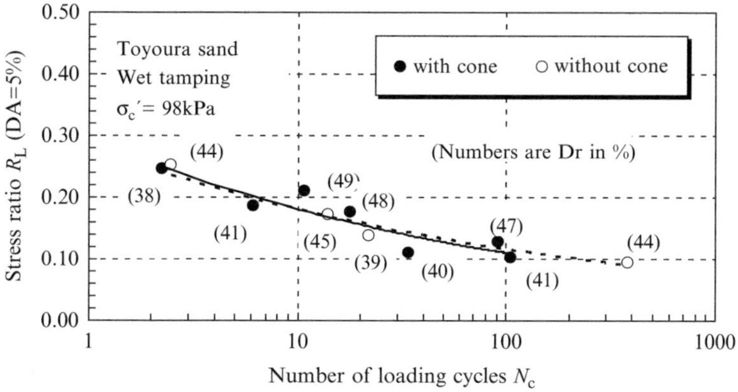


**Fig. 3.3** Comparison of liquefaction strength ($R_L$ versus $N_c$ curve) between specimens with or without cone rod

In this test series, a river sand with the mean grain size $D_{50}=0.169$ mm and the uniformity coefficient $C_u=1.44$ consisting of sub-round particles of hard quality was used. Fines mixed with sand was silty and clayey soils with low plasticity index of Ip ≃ 6 sieved from decomposed granite (called Masa soil) in reclaimed ground of the Kobe city, Japan. Soil specimens of 100 mm diameter and 200 mm height were prepared by wet tamping method to meet prescribed relative densities as close as possible. The specimen was fully saturated by using $CO_2$-gas and de-aired water so that the Skempton's B-value larger than 0.95 was attained, and isotropically consolidated to an effective stress of 98 kPa with the back-pressure of 196 kPa. In the test, relative density $D_r$ and fines content $F_c$ of the specimens were parametrically changed to investigate their effects on penetration resistance $q_t$ and undrained cyclic strength $R_L$.

## 3.3 Penetration Tests

Miniature cone penetration tests were conducted on the clean river sand having loose or medium density under undrained conditions. Figure 3.4 a and b show variations in cone resistance $q_t$ and excess pore-pressure $u$, respectively, plotted versus the penetration length. As observed, higher $q_t$-values and lower positive pore-pressures are measured for the denser sands. The cone resistance tends to monotonically increase and asymptotically converges to an ultimate value, but in some cases, starts to decrease after the peak values. In any case, the maximum value during 25 mm penetration was taken as a representative $q_c$-value to be correlated to liquefaction strength. The pore-pressure tends to increase in the positive direction, almost linearly with the penetration length in all cases. The pore pressure increases up to 5% to 60% of the initial effective confining stress for Dr ≈ 20% to 50%, respectively.

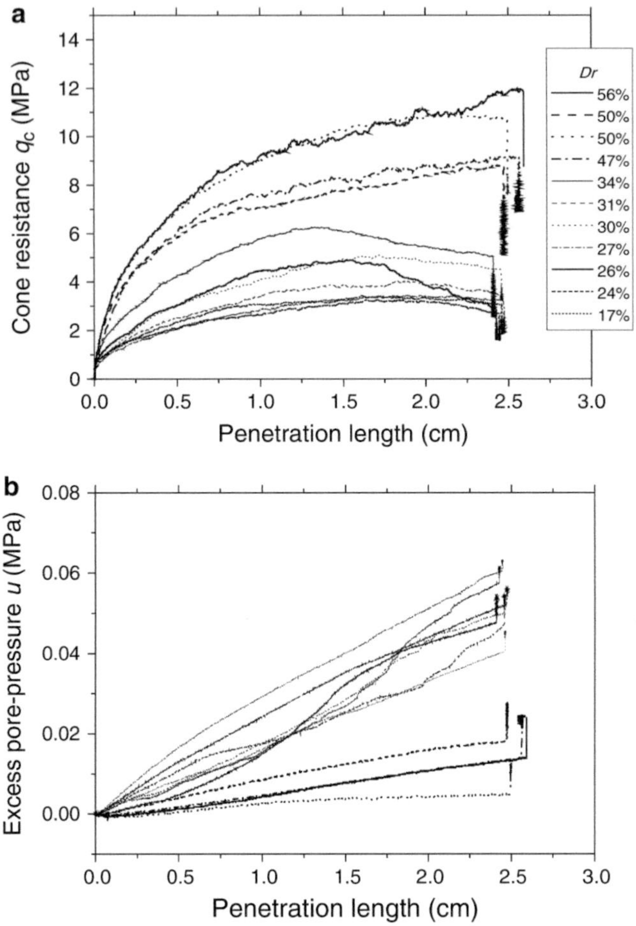


**Fig. 3.4** Cone resistance (**a**) or excess pore-pressure, (**b**) versus penetration length for river sand specimens with different densities

In Figs. 3.5a, b, variations in cone resistance $q_t$ and excess pore-pressure $u$ are shown as a function of the penetration length for specimens of $D_r \approx 50\%$ with different fines content of $F_c = 0$–$30\%$. The increase in fines content tends to develop larger pore-pressure increase and reduce penetration resistance during penetration. The effect of fines content is pronounced even at a small value of $F_c = 5\%$ particularly on the pore-pressure build-up and tends to asymptotically converges to an ultimate curve as it approaches to $F_c = 30\%$.

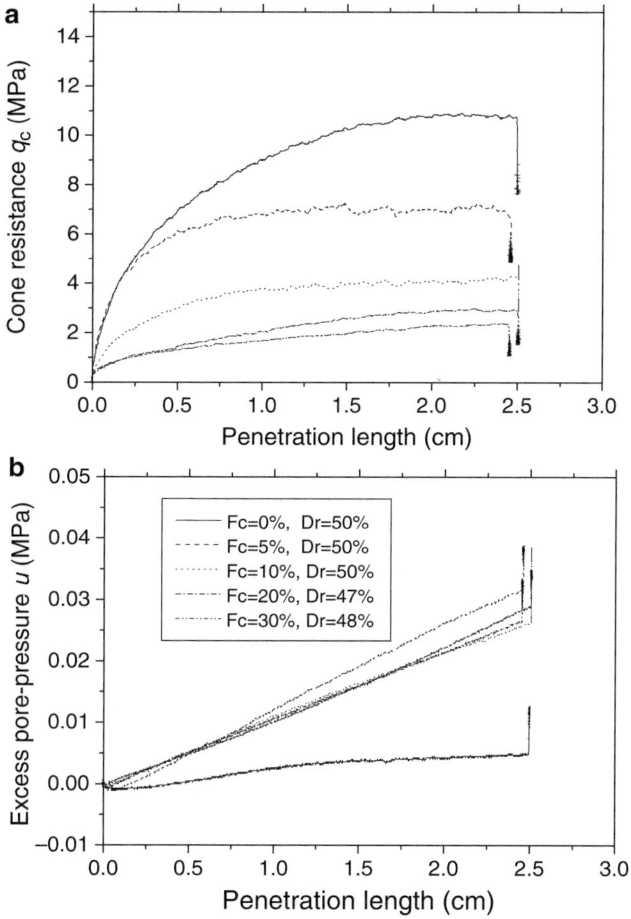


**Fig. 3.5** Cone resistance (**a**) or excess pore-pressure, (**b**) versus penetration length for river sand specimens of Dr≈50% with different fines content

## 3.4 Penetration Resistance Versus Liquefaction Strength

Figures 3.6a–e show relationships on log-log charts between the stress ratios $R_L$ and the number of loading cycles $N_c$ for 5% double amplitude axial strain obtained by the series of undrained cyclic loading tests on sand specimens having $D_r \approx 50\%$ with fines content, $F_c = 0\%$, 5%, 10%, 20% and 30%, respectively. The close circles on the charts are the results from undrained cyclic loading tests conducted on the same test specimens after the cone penetration tests, while the open circles are those

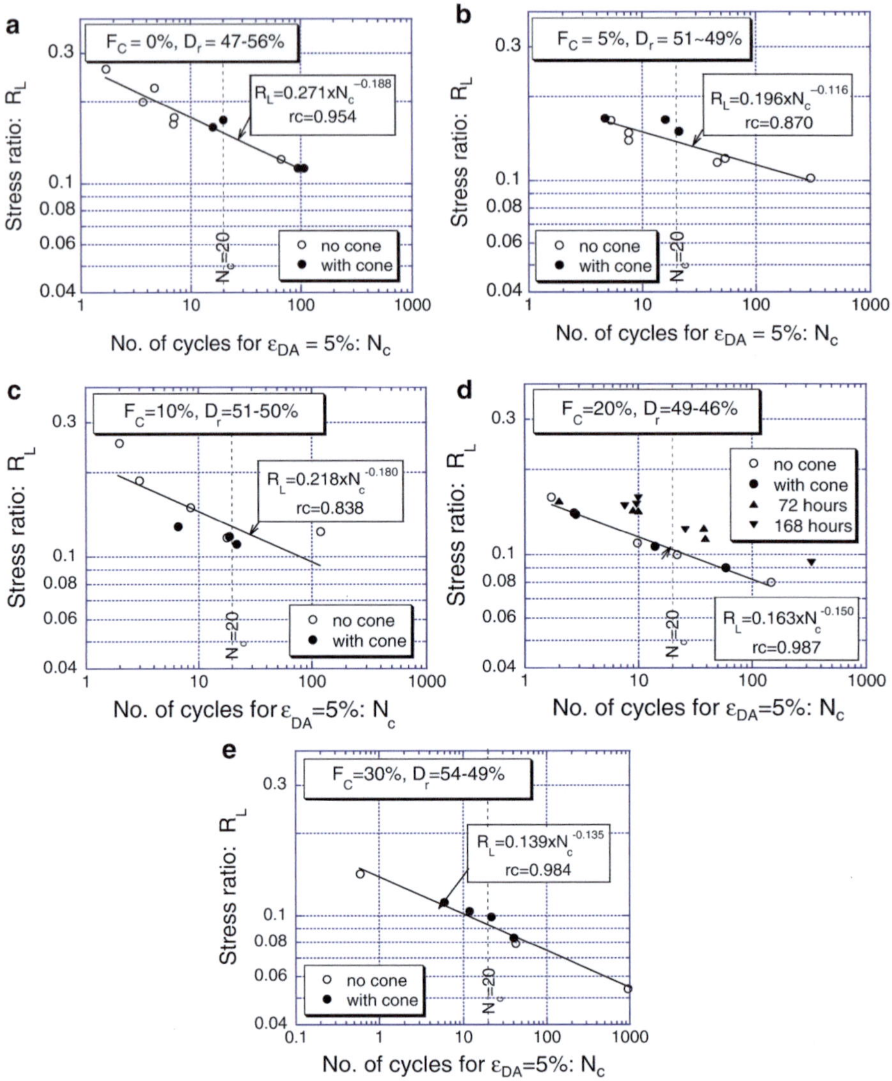


**Fig. 3.6** Liquefaction strengths for sands of $Dr \approx 50\%$ with fines content, *Fc*, (**a**) 0%, (**b**) 5%, (**c**) 10%, (**d**) 20% and (**e**) 30%. In (**d**), the effect of sustained consolidation is considered

without preceding cone tests, demonstrating again that the history of prior cone penetration is unlikely to significantly affect the liquefaction strength subsequently measured, although the relative density actually increased only by 2% on average between the initial consolidation and the reconsolidation. The plotted data points can be approximated by the straight lines on the log-log diagrams with relatively

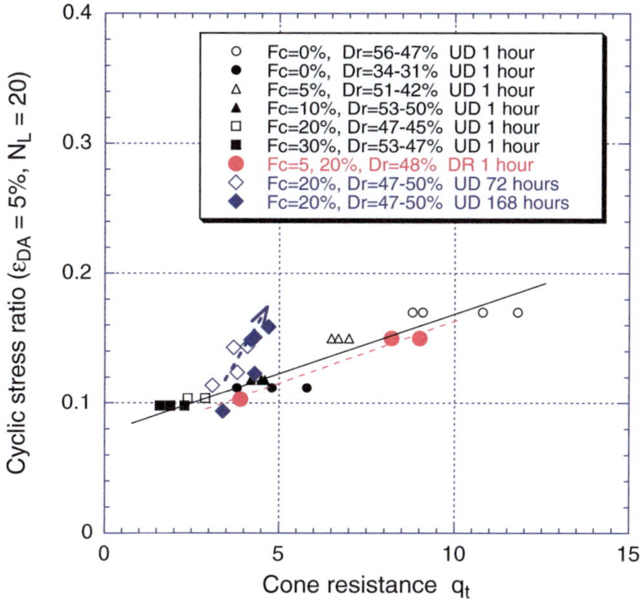


**Fig. 3.7** Direct relationship between stress ratios for liquefaction (DA 5%, Nc=20) and cone resistances for sands of $Dr \approx 30$–50% with fines content, $Fc = 0$–30%

high regression coefficients *rc* as indicated in the charts. By comparing the stress ratios $R_L$ for double amplitude strain, $\varepsilon_{DA,}$ equal to 5%, the $R_L$ -value evidently decreases with increasing fines content.

In order to correlate penetration resistances to corresponding liquefaction strengths for the soil specimens with different $F_c$, the stress ratio $R_L$ for $\varepsilon_{DA=}5\%$ in $N_c = 20$ are estimated from the line regressed through the close circles. In Fig. 3.7, the liquefaction strengths $R_L$ thus obtained are plotted on the vertical axis versus cone resistances $q_t$ on the horizontal axis. The data points concentrate in a narrow area which may be represented by a single straight line drawn in the chart (the regression coefficient=0.95). As maybe observed, a unique $q_t$ -$R_L$ relationship exists despite the large differences in relative density and fines content. This finding seems quite contradictory to the present state of practice in which liquefaction strength corresponding to a given CPT resistance is increased by a certain amount according to increasing fines content.

In Fig. 3.8 the same test results are plotted again to compare with field investigation data obtained by Suzuki et al. (1995) combining prototype cone tests in situ and undrained cyclic triaxial tests on intact samples recovered from the same soil deposits by in situ freezing technique. In the horizontal axis, $q_{t1}$ is taken to normalize in situ cone tip resistance $q_t$ as $q_{t1} = q_t / \sigma_v'^{0.5}$, where $\sigma'_v =$ the vertical overburden stress. Needless to say, considerable differences in terms of cone size, penetration rate, drainage condition, consolidation effects, etc. exist between the miniature cone

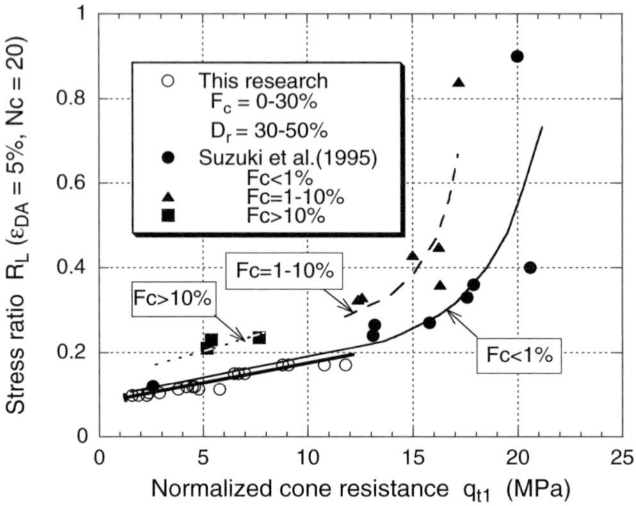


**Fig. 3.8** $R_L \sim q_t$ relationship obtained in this research compared with that based on in situ CPT and lab tests of undisturbed samples by Suzuki et al. (1995)

tests conducted here and prototype tests. Nevertheless, the correlation by the present research is in surprisingly good agreement with in situ data for $F_c < 1\%$ in the interval of $q_{t1} \approx 0$–12. It should be noted however that the correlation looks quite insensitive to fines content in the present study whereas the in situ data indicate its clear influence. In order to examine the contradiction between the two experimental studies more closely, the effects of drainage during cone penetration and long-time consolidation or aging were further investigated.

## 3.5 Effect of Drainage on Cone Penetration Test

The drainage condition during cone penetration tests depends on several influencing factors, such as the penetration rate, the cone size, the fines content, etc. In this test series, completely undrained condition was chosen for the sake of simplicity of test conditions and data interpretation. The drainage condition may depend on the soil permeability which largely changes with fines content and also on the penetration speed. The reality may be in between the undrained and drained conditions. Consequently, it is worth carrying out the other extreme, drained penetration tests.

Figures 3.9a, b compare typical results of undrained and drained tests in terms of cone resistance $q_c$ and excess pore-pressure $u$, respectively, plotted as a function of the penetration length for specimens of $D_r \approx 50\%$ and $F_c = 5\%$. While undrained $q_c$-value tends to asymptotically converge to an ultimate value, drained one keeps increasing even at the maximum penetration length of 25 mm. The difference is

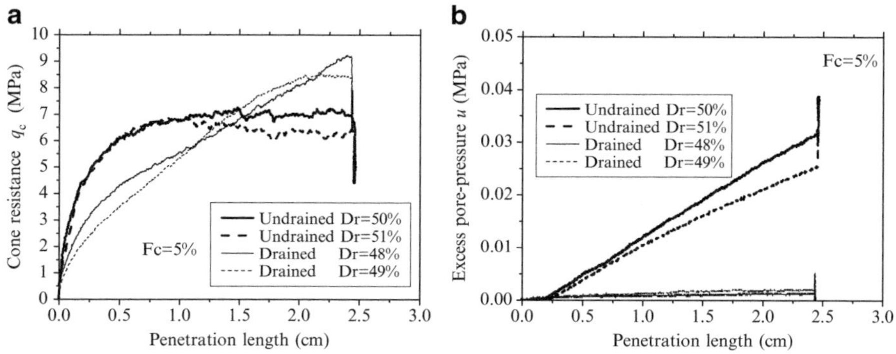


**Fig. 3.9** Cone resistance (**a**) or excess pore-pressure, (**b**) versus penetration length for river sand specimens of Dr≈50% and Fc=5% under drained and undrained conditions

certainly remarkable in the pore pressure $u$. The drained $q_t$-values at the maximum penetration length were taken and directly correlated to corresponding stress ratio, $R_L$ for $\varepsilon_{DA}=5\%$, $N_c=20$. In Fig. 3.7, data points of the drained condition are plotted with large solid circles and compared with the undrained data of small symbols. Although the number of data is not enough to make a final judgment, the two extreme drainage condition seems to shift the $q_t - R_L$ fitting line downward approximately by 1 MPa, which is not large compared to the data scatter involved in the test series. Considering this fact, it may be said that the drainage condition during the cone penetration test is not responsible for the insensitiveness to fines content in the $q_t - R_L$ relationship for sands containing fines in this research.

## 3.6 Effect of Sustained Consolidation

In the test series, the consolidation sustained about 1 h before the cone test. In order to investigate the effect of longer consolidation time, specimens of $F_c=20\%$ were chosen, because higher fines content is likely to strengthen the long time consolidation effect in comparison with clean sands. Specimens with the relative density around $Dr=50\%$ were consolidated for three different durations; 1 h, 72 h (3 days) or 168 h (7 days).

Figures 3.10a, b compare the typical results of cone resistance $q_c$ and excess pore-pressure $u$, respectively, plotted versus the penetration length. It is obvious that even several days of sustained consolidation makes measurable difference in the cone resistance. Cyclic triaxial test results for the long term consolidation effect shown in Fig. 3.6d with solid triangles, though somewhat scattered, clearly indicate that the sustained consolidation tends to raise the $R_L \sim N_c$ fitting line, too. In order to correlate penetration resistances to corresponding liquefaction strengths for the individual test specimens, the stress ratio $R_L$ of 5% DA axial strain for $N_c=20$ were

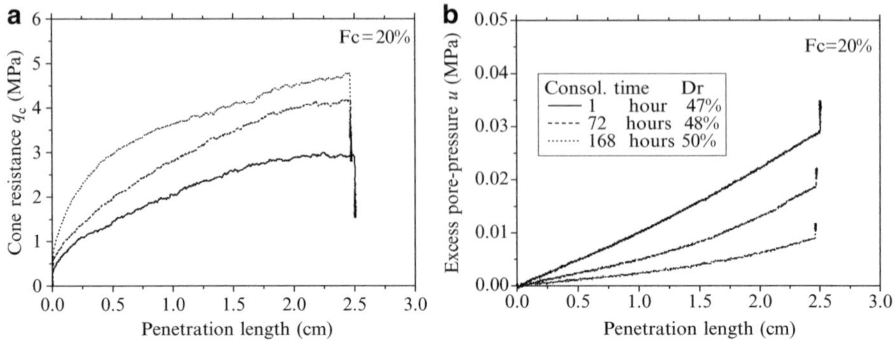


**Fig. 3.10** Cone resistance (**a**) or excess pore-pressure, (**b**) versus penetration length for river sand specimens of Dr≈50% with different consolidation time

estimated by drawing straight lines through the solid triangles in parallel to the line statistically determined as explained before. In Fig. 3.7, data points thus obtained for the sustained consolidation are plotted for $F_c$=20% with larger diamond symbols to compare with all the other data for 1 h consolidation. It may well be judged that, despite some data dispersion, the sustained consolidation tends to shift the $R_L$ versus $q_c$ correlation, leading higher liquefaction strength for the same $q_c$-value. Consequently, although the correlation for very young test specimens looks to be uniquely determined irrespective to soil density and fines content; it is likely that long time consolidation in situ makes the difference. However, much more systematic research is certainly needed to draw more clear conclusions considering geological periods of longer than thousands years.

## 3.7 Effect of Cementation by Accelerated Test

One may agree that the long-time geological effect is mainly attributable to cementation or chemical bonding between soil particles. In order to emulate it in short laboratory tests, a small amount of cement were added to the fines (Masa soil) mixed with sand.

In this series of test, the miniature cone was replaced by a new type with strain gages glued inside the rod tube 23 mm lower than the foot of the cone. This means the cone became slender without any outside projection in contrast to the original one with the strain gauges glued at the rod outer face and coated with some protection material. The tested sand was also changed from what used in the previous test series (explained above) to a new marine sand with $D_{50}$=0.21 mm and $C_u$=1.9. A prescribed quantity of Portland cement was completely mixed with the Masa soil to make the fines with different degree of chemical activity. The cement content $C_c$, the weight ratio of cement to soil (including fines), varied from 0% to 1%, and the fines

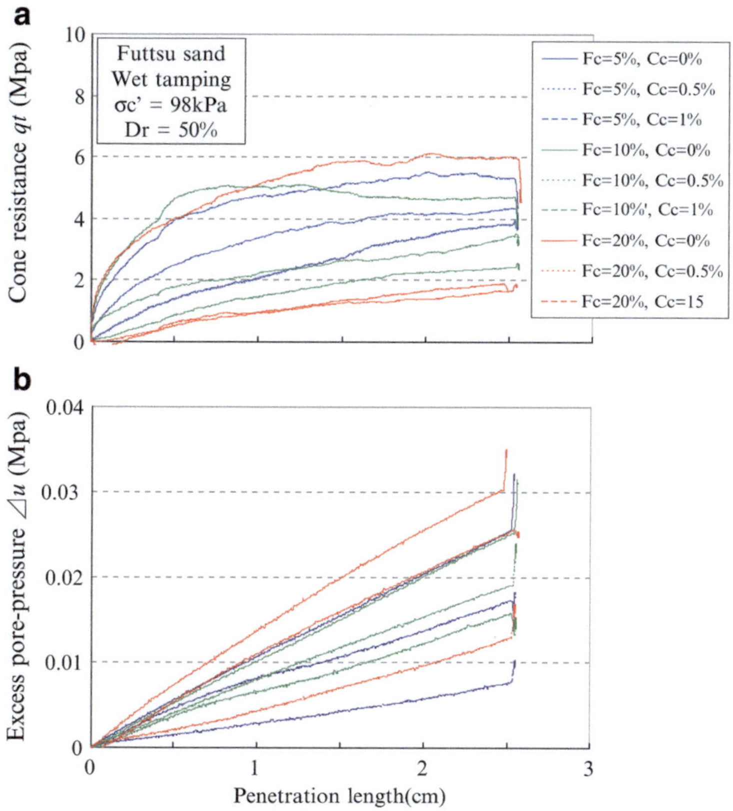


**Fig. 3.11** Cone resistance (**a**) or excess pore-pressure, (**b**) versus penetration length for river sand specimens of Dr≈50% with different fines content *Fc* and cement content *Cc*

content $F_c$ including the cement changed from 0% to 20%. This means that the ratio $C_c/F_c$, a parameter representing chemical activity of fines, varied from 0% to 20%. Then the sand specimen with a given value of $F_c$ was prepared by wet tamping method to make a given relative density $D_r$, which was changed in three steps, about 30, 50 and 70%. The specimen was then completely saturated with de-aired water, and consolidated under the isotropic effective stress of 98 kPa with the back-pressure of 196 kPa. If cement is added to the fines, the consolidation time before the cone test was controlled exactly 24 h from the wetting during sample preparation, while for tests without cement it was about 1–2 h.

Figure 3.11 shows cone resistance $q_t$ and excess pore-pressure $u$, respectively, plotted versus the penetration length for this test series. It is clearly observed that $q_t$ increases with increasing $C_c$ under the same fines content $F_c$. In Fig. 3.12 the stress ratios $R_L$ are plotted versus number of loading cycles $N_c$ for all specimens for the test series. Obviously, the increase in fines content $F_c$ tends to reduce liquefaction

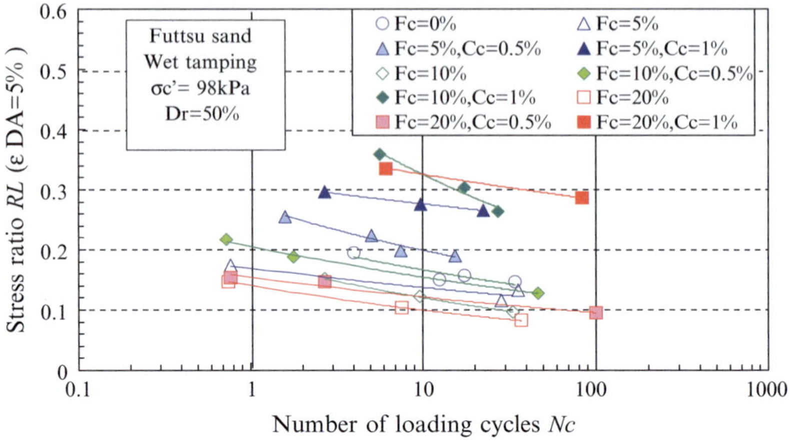


**Fig. 3.12** stress ratios $R_L$ versus number of loading cycles $N_c$

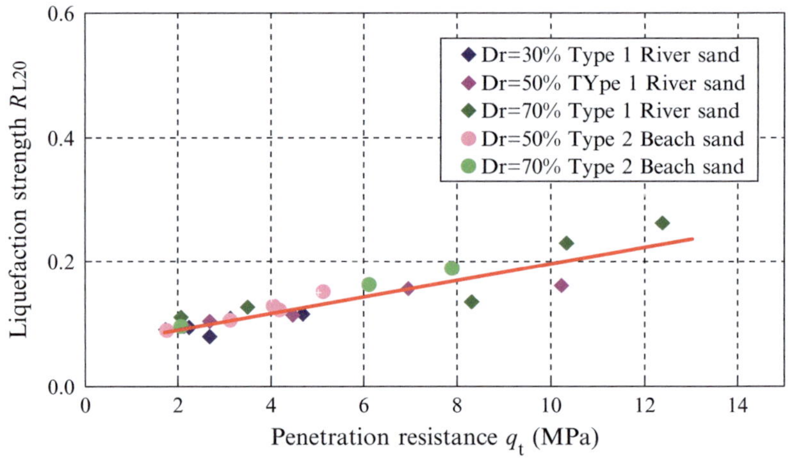


**Fig. 3.13** $R_L$ versus *qt* for a new cone compared with the old cone

strength $R_L$ in specimens without cement as already demonstrated by previous researches. Also indicated in the figure is that the stress ratio $R_L$ increases with increasing cement content for all fines content $F_c$ and the increment is particular large for $F_c$ =20%.

Figure 3.13 shows the relationship between $R_L$ and $q_t$ obtained by the reference test without cement carried out at the top of this new test series. The cone resistance $q_c$ obtained by the new-type cone is obviously lower than the old-type cone for the same stress ratio $R_L$ as compared in the figure. However, the $R_{L\sim}\ q_t$ relationships obtained from the two test series are not so much separated from each other that they can be correlated by a single regression line as shown in Fig. 3.13.

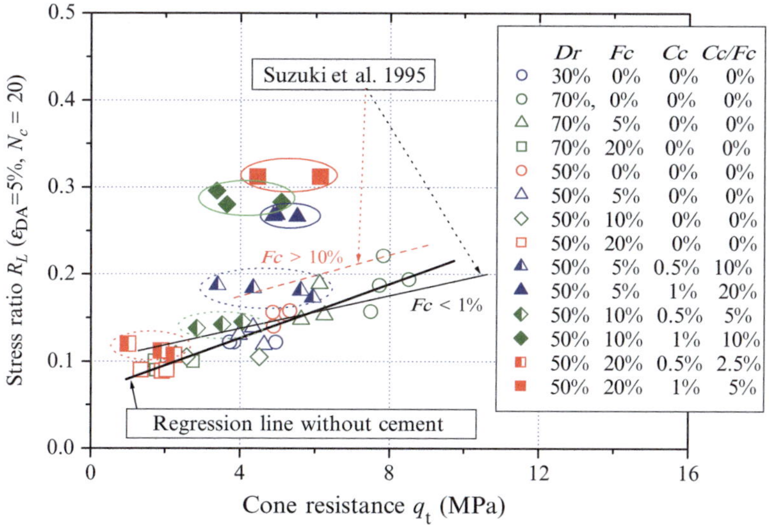


**Fig. 3.14** Liquefaction strength $R_L$ plotted versus $q_t$ for all test results in the new test series

In Fig. 3.14, all the test results in the new test series with different values of $D_r$, $F_c$ and $C_c$ are plotted on the $q_t$ ~ $R_L$ diagram with different symbols. There are 2–4 data points of the same symbol, indicating the 2–4 test results under the same $D_r$, $F_c$ and $C_c$ but with different cyclic stress ratios in the liquefaction test. The open symbols corresponding to specimens without cement can be represented by the solid regression line (a reference line) shown in the figure despite the differences in $D_r$ and $F_c$, as already mentioned. The half-close and full-close symbols are all for specimens containing cement. Among them, the triangles, for instance, represent the case $F_c$=5%, and they move up in the diagram from the open symbol ($C_c$=0) to the half-close one ($C_c$=0.5%) further to the full-close one ($C_c$=1.0%) as $C_c/F_c$ changes from 0% to 20% for the same value of $F_c$=5%. In the similar manner, the diamonds and the squares move up with increasing $C_c$ or $C_c/F_c$ for the same fines content of $F_c$=10% and $F_c$=20%, respectively. Specimens with higher $C_c/F_c$ may represent longer geological age because higher chemical activity is exerted in the same soil. This indicates that the aging effect tends to push up the data points in the $q_t$ ~ $R_L$ diagram from the reference line and gives higher liquefaction strength under the same cone resistance.

On the other hand, if data points with the same $C_c/F_c$-value of 10% are concerned, they move up as indicated in the chart from the reference line to the half-close triangle ($F_c$=5.0%) further to the full-close diamond ($F_c$=10%) with increasing fines content. The same trend can also be seen for $C_c/F_c$-value of 5%. This indicates that, under the same $C_c/F_c$-value (the same geological age), higher fines content results in higher liquefaction strength for the same cone resistance.

Based on the data shown in Fig. 3.14, the increment rates of the cyclic stress ratio $R_L$ and the cone resistance $q_t$ are plotted versus the $C_c/F_c$-value with the fines content $F_c$ as a parameter in Fig. 3.15. It clearly indicates that the $R_L$-value increases more

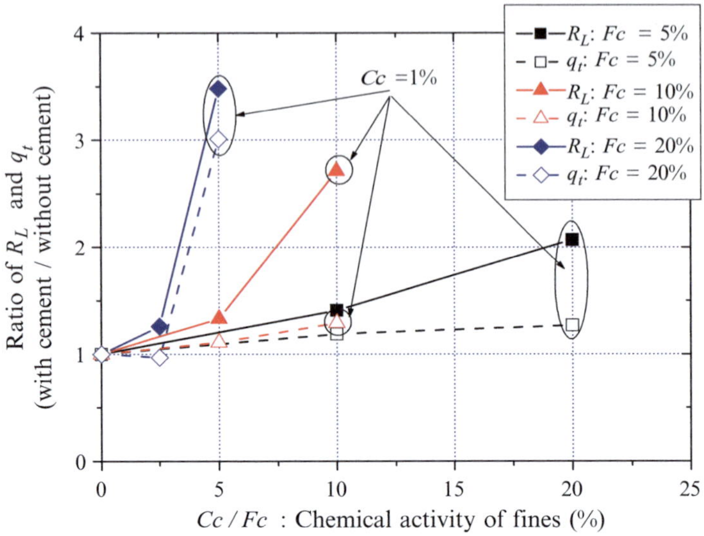


**Fig. 3.15** Rate of increase of $R_L$ and *qt* versus *Cc/Fc*

than the $q_t$-value with increasing chemical activity $C_c/F_c$. This is probably because the liquefaction strength tends to be very sensitive to chemical bonding between sand particles, whereas the cone penetration tends to break it easily without much increase in the penetration resistance.

Also noted in Fig. 3.15 is that not only the $C_c/F_c$-value but also the fines content $F_c$ tends to considerably increase the rate of liquefaction strength in particular even under the same cement content as indicated with the arrows for $C_c$ = 1.0% in the figure. It is probably because that higher fines content means larger soil surface area on which greater chemical reaction can occur even under the same cement content. For the cone resistance, the effect of $F_c$ are considerably large only for the case of $C_c/F_c$ = 5% and $F_c$ = 20% in which some unaddressed effect may be responsible for the drastic increase not only in $R_L$ but also in $q_t$. In all other cases, the effect of $F_c$ on cone resistance is evidently smaller than on liquefaction strength. This seems to be the major reason why the $R_L \sim q_t$ curves tend to shift upward with increasing fines content in the field, which is not observed for reconstitute soils without cement.

## 3.8 Conclusions

A series of experimental study by miniature cone penetration tests and subsequent cyclic loading tests are carried out in the same reconstituted specimen with parametrically changing relative density $D_r$ and fines content $F_c$ yielded the following major findings;

- Cone penetration tests performed prior to cyclic undrained tests have little effect on liquefaction strength of specimens, demonstrating that direct and reliable comparison between penetration resistance and liquefaction strength is possible in the innovative triaxial test.

- With increasing fines content $F_c$, both cone resistance $q_t$ and liquefaction strength $R_L$ decrease, resulting in a unique single curve on the $q_t \sim R_L$ chart despite the difference in $F_c$ and $D_r$. This indicates that, for reconstituted soils, liquefaction strength for a given cone resistance is constant despite the difference in fines content, which is contrary to the current liquefaction evaluation practice.
- If specimens with certain fines content are consolidated for longer time up to a week, both cone resistance and liquefaction strength increase by measurable amounts. If the results are plotted on the $q_t \sim R_L$ chart, they seem to be located slightly above the unique curve obtained by the short period consolidation tests.

In order to understand the long term geological effect more clearly, accelerated tests have been conducted by adding a small quantity of cement (cement content $C_c$=0–1%) to fine soils, yielding the followings;

- Specimens with higher value of $C_c/F_c$ (emulating longer geological age) tend to give higher liquefaction strength for the same cone resistance, indicating that the aging effect tends to raise the $R_L \sim q_t$ line from that for soils of short time consolidation.
- For the same $C_c/F_c$-value (emulating the same geological age), higher fines content results in higher liquefaction strength for the same cone resistance, which is consistent with the trend previously found in the field investigation by Suzuki et al. (1995).
- The liquefaction strength $R_L$ increases more than the $q_t$-value due to chemical bonding between sand particles and this effect is more pronounced for higher fines content. This is why the $q_t \sim R_L$ curves tend to shift upward due to the increase of fines content in the field, which is not observed for reconstitute soils by short-term consolidation in the laboratory.

## References

Japan Road Association (1996) Liquefaction potential evaluation in sandy ground Road Bridge Design Code: Japan Road Association Chapter 7.5 (in Japanese), pp 91–94

Kokusho T, Murahata K, Hushikida T, Ito N (2003) Introduction of miniature cone in triaxial apparatus and correlation with liquefaction strength. In: Proceedings of the annual convention of JSCE, III-96, Toyota, pp 191–192 (in Japanese)

Kokusho T, Hara, T,, Murahata K (2005) Liquefaction strength of fines-containing sands compared with cone-penetration resistance in triaxial specimens. In: Proceedings of the 2nd Japan-US workshop on geomechanics, ASCE Geo-Institute Publication No. 156, pp 356–373

Kokusho T (2007) Liquefaction strengths of poorly-graded and well-graded granular soils investigated by lab tests. In: Proceedings of the 4th international conference on earthquake geotechnical engineering, Thessaloniki. Springer, pp 159–184

Suzuki Y, Tokimatsu K, Taya Y, Kubota Y (1995) Correlation between CPT data and dynamic properties of in situ frozen samples. In: Proceedings of the 3rd international conference on recent advances in geotechnical earthquake engineering and soil dynamics, vol 1, St. Louis, pp 249–252

# Chapter 4
# Full Scale Testing and Simulation of Seismic Bridge Abutment-Backfill Interaction

**Ahmed Elgamal and Patrick Wilson**

**Abstract** At high levels of seismic excitation, passive earth pressure at the abutments may provide resistance to excessive longitudinal bridge deck displacement. Full scale tests and Finite Element (FE) simulations are performed in order to investigate this resistance. The static passive earth pressure force-displacement relationship is recorded in two tests, by pushing a model wall into dense sand with silt ($c$-$\varphi$) backfill. Based on an analysis of the recorded data, a calibrated FE model is developed and used to generate passive force-displacement relationships for a range of typical backfill soil properties and abutment heights. In an additional testing phase, the wall-backfill system is subjected to shake table excitations in order to document the corresponding dynamic earth pressure forces and mechanisms. At high g-levels of excitation, the instantaneous passive resistance is shown to also depend on the inertial backfill forces caused by ground shaking. Based on the testing and simulation results, simplified abutment models are presented, which include the experimentally observed backfill inertial effects along with the static passive force-displacement resistance. Numerical FE simulations are finally used to demonstrate the influence of these abutment models within the overall bridge system configuration, and highlight the salient dynamic response characteristics as a function of the various modeling parameters.

A. Elgamal (✉)
University of California, San Diego, La Jolla, CA 92093-0085, USA
e-mail: elgamal@ucsd.edu

P. Wilson
Earth Mechanics, Inc, Fountain Valley, CA 92708, USA
e-mail: p.wilson@earthmech.com

M.A. Sakr and A. Ansal (eds.), *Special Topics in Earthquake Geotechnical Engineering*, Geotechnical, Geological and Earthquake Engineering 16,
DOI 10.1007/978-94-007-2060-2_4, 

## 4.1 Introduction

In seismic design (Caltrans 2004; AASHTO 2007; Shamsabadi et al. 2007), an abutment system relies on the soil backfill to provide resistance to longitudinal bridge deck displacement (Fig. 4.1). During strong shaking, if the deck impacts the abutment (Fig. 4.1), a sacrificial portion (the back wall) is designed to easily break-off and translate into the backfill. Resistance to further displacement of the deck and back wall is then provided by passive earth pressure within the densely compacted backfill soil (Shamsabadi et al. 2007).

The above abutment resistance mechanism decreases the demand placed on other seismic components such as the bridge columns and foundation (AASHTO 2007). Current available models for the resulting backfill passive force-displacement relationship include bilinear representations of limited data from earlier static tests (Maroney 1995; Caltrans 2004; AASHTO 2007). As a result, more accurate representation of the abutment backfill resistance may lead to a safer and economic design of the overall bridge system.

To aid in this undertaking, experiments performed on a wall-backfill system, full scale in height (1.7 m or 5.5 ft), are described. Static passive earth pressure force-displacement tests were performed first. Finite element (FE) analysis was then employed to further investigate the static backfill passive force-displacement resistance considering a range of representative backfill soils and depths. Next, the soil container-back wall-backfill system was subjected to shake table excitation in order to record the effect of shaking on the mobilized passive resistance. Based on the results from these full scale tests and FE simulations, simplified bridge models were then developed to simultaneously include the static passive force-displacement resistance along with the experimentally observed inertial effects. Reflecting the salient findings, a set of conclusions and recommendations is finally presented.

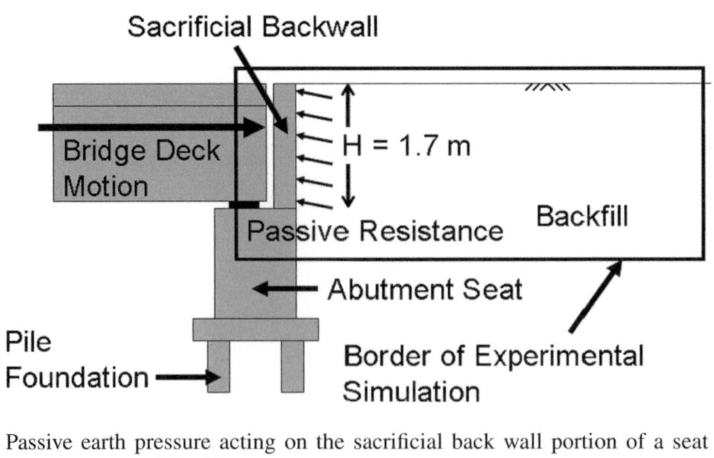


**Fig. 4.1** Passive earth pressure acting on the sacrificial back wall portion of a seat abutment (After Wilson and Elgamal 2010b)

## 4.2 Full Scale Tests

The experiments were performed inside a large soil container (Fig. 4.2) attached to the outdoor shake table at the Englekirk Structural Engineering Center of the University of California in San Diego (UCSD). By employing this test configuration, static passive earth pressure load-displacement experiments were performed first (Test 1 and Test 2), and dynamic excitations were also imparted (Dynamic Excitation Tests).

### *4.2.1 Test Setup*

Primary components of the experimental configuration (Wilson and Elgamal 2010a) included a large soil container (Fig. 4.2a), a vertical test wall section (Fig. 4.2b), a loading mechanism (Fig. 4.2e) and a compacted sand backfill (Fig. 4.2d). The inside dimensions of the soil container were 2.9, 6.7 and 2.4 m in width, length and height, respectively. A restraining system in the form of two stiff towers was connected to the container sides (Fig. 4.2a) to constrain relative translation of its individual laminate frames.

Well-graded sand with 7% silt content was compacted (Fig. 4.2c) in compliance with Caltrans (2006) standard specifications for structure backfill. The compacted

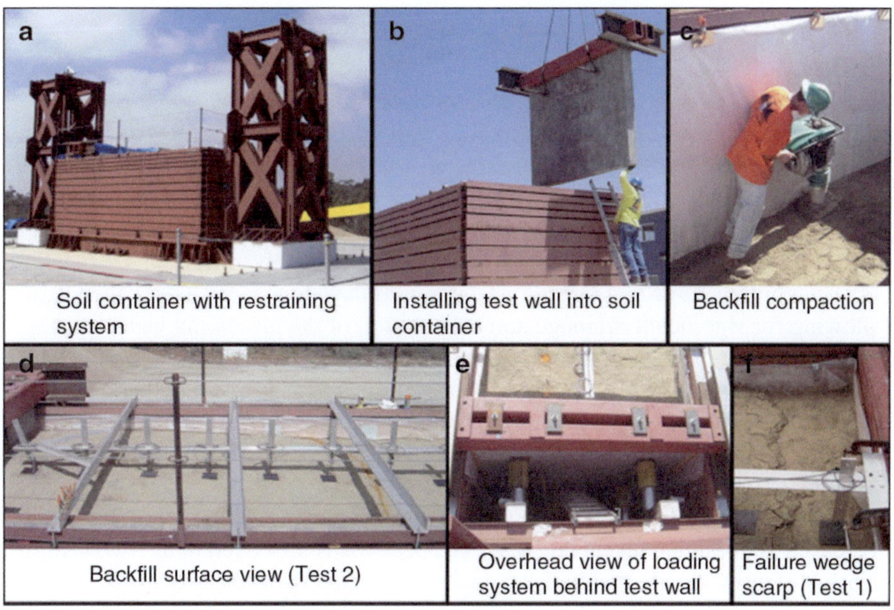


**Fig. 4.2** Test configuration photographs (Wilson and Elgamal 2010a)

backfill total unit weight was approximately 20.6 kN per cubic meter. Backfill dimensions were 2.9, 5.6 and 2.15 m in width, length and height, respectively.

Direct shear and triaxial tests (Wilson 2009, Wilson and Elgamal 2010a) were performed on samples remolded as closely as possible to the experimental backfill placement conditions. The direct shear test results suggest peak and residual (*r*) friction angles of $\varphi$=48, and $\varphi_r$=35°, with cohesion intercepts $c$=14 kPa, and $c_r$=8 kPa, respectively. Compared to the direct shear tests, the triaxial results suggest a lower peak $\varphi$=44°, slightly higher residual $\varphi_r$=36°, the same $c$=14 kPa, and slightly lower $c_r$=6 kPa.

The reinforced concrete test wall was suspended from a stiff beam resting on rollers (Fig. 4.2b), and engaged the upper 1.7 m (5.5 ft) of backfill (Fig. 4.1). Hydraulic jacks reacted through load cells onto concrete-filled steel posts (Fig. 4.2e) to push the wall into the backfill while measuring the applied load during the passive pressure tests. The load cells (Fig. 4.2e) also measured the total force (including the test wall inertia and the earth pressure) during the dynamic excitation tests. Six transducers attached to the wall at three different depths, measured the pressure between the test wall and the backfill. Wall and backfill displacements were measured by linear potentiometers (Fig. 4.2d).

Additional instrumentation (not shown) included breakable foam cores to identify the passive failure wedge shape in Test 1. Accelerometers were also placed throughout the soil layer and on the test wall and soil container during the dynamic excitation tests (Wilson 2009).

### 4.2.2 Passive Pressure Load-Displacement Tests

Earlier tests to record the bridge abutment passive force-displacement relationship have been performed by pushing a model wall into a static backfill (e.g., Maroney 1995; Bozorgzadeh 2007; Lemnitzer et al. 2009). Similarly, the hydraulic jacks (Fig. 4.2e) were used to push the wall into the backfill in Test 1 and Test 2. After Test 1, the backfill was fully removed and re-compacted before conducting Test 2. As discussed below, the two test results were noticeably different, as the time between constructions and testing was about 20 days longer for Test 1 than Test 2 (allowing for significant drainage and evaporation of the placement water content).

When the lateral jack load was applied in Tests 1 and 2, the test configuration dictated that the wall was permitted to move upwards as it was displaced laterally into the backfill. In this configuration, low wall-soil friction was mobilized ($\delta_{mob}$ of 2–3°), and the resulting failure wedge shape was triangular (Wilson and Elgamal 2010a).

In Test 1 and Test 2, the passive earth resistance increased up to a peak and then decreased to a residual level (Fig. 4.3). The peak loads measured in the two tests are clearly different, but the residual resistance is virtually the same (Fig. 4.3). With the mobilized backfill width of 2.87 m and height $H$ of 1.7 m (Fig. 4.1), the peak load measured in Test 1 was 1,105 kN (251 kips) at a displacement of 46 mm (1.8 in), or

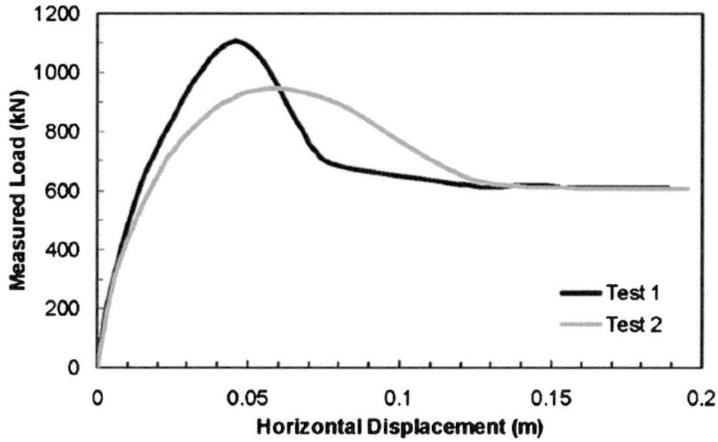


**Fig. 4.3** Total measured load versus wall horizontal displacement backbone relationships for Test 1 and Test 2 (Wilson and Elgamal 2010a)

about 2.7% of the supported backfill height. Peak load measured in Test 2 was 936 kN (213 kips) at a 51 mm (2 in.) displacement, or about 3.0% of the supported height. After the peak, the resistance decreased until a residual level of about 608 kN (55% and 65% of the peak from Test 1 to Test 2, respectively) was reached in both tests, at a lateral displacement of about 8% of the supported backfill height.

The backfill soil state was closest to the lab direct shear and triaxial test conditions in Test 2, and more dried-out in Test 1 (Wilson and Elgamal 2010a). For Test 2, the Log Spiral (Terzaghi et al. 1996) and Coulomb peak estimates (essentially equal due to the low $\delta_{mob}$ of 2–3°) from triaxial test parameters were about 10% lower than the measured result, while the direct shear test value predictions were about 10% higher. Compared with Test 2, the drier backfill condition of Test 1 led to about 20% greater maximum resistance and more rapid shear strength degradation beyond the peak (Fig. 4.3).

#### 4.2.2.1 FE Simulation of Tests 1 and 2

Tests 1 and 2 were simulated next (Wilson and Elgamal 2010b), using a 2-D plane strain model (Plaxis 2004). In the model, stiff plate elements (overall, 1.7 m tall) supported the soil laterally (Fig. 4.4), with the same weight as the experimental wall and supporting beam (Fig. 4.2b). The backfill behavior was simulated by using the Hardening Soil (HS) model (Plaxis 2004) with 15-node triangular backfill elements (Fig. 4.4). The HS model uses the Mohr-Coulomb failure rule, and a nonlinear hyperbolic stress-strain relationship. All analyses were restricted to the pre-peak loading range, with numerically stable solutions.

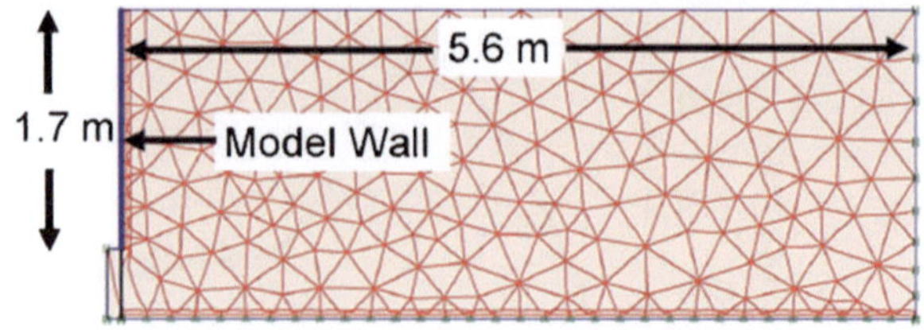


**Fig. 4.4** FE model mesh for simulation of the full scale passive earth pressure tests (Wilson and Elgamal 2010b)

For simulation of Test 2 (Soil T2, closest to the lab test condition), backfill shear strength parameters ($\varphi$ and $c$) were determined based on the direct shear and triaxial tests (Table 4.1). The reference stiffness parameter $E_{50}^{ref}$ was selected at a reference stress $p_{ref}$ = 100 kPa from the triaxial test stress-strain data (Wilson and Elgamal 2010a) according to a power law with $m$ = 0.5 (Plaxis 2004). Soil total unit weight was specified as $\gamma$ = 20.6 kN/m$^3$, according to the field condition. A failure ratio value of $R_f$ = 0.75 was also adopted, which is within the range recommended by Duncan and Mokwa (2001). User manual recommendations and internal adjustments made by Plaxis (2004) determined the remaining HS model parameters.

For simulation of Test 1 (Soil T1, drier condition), backfill shear strength parameters (Table 4.1) were adjusted based on analysis of the observed passive failure wedge (Wilson and Elgamal 2010a). A larger $E_{50}^{ref}$ accounted for the experimentally observed higher stiffness compared with Test 2 (Fig. 4.3).

Numerical FE simulations of the soil container passive earth pressure experiments were made by prescribing a horizontal displacement boundary condition along the wall plate (left side of Fig. 4.4), while allowing free vertical displacement. Using this configuration, the wall moved upwards with the backfill in accordance with the experiments. As shown in Fig. 4.5, the FE models provide a satisfactory representation of the experimental load-displacement behavior.

**Table 4.1** FE model soil parameters where Soil D-S is dense (clean) sand, MD-SM is medium-dense silty sand, and MD-SC is medium-dense clayey sand (Wilson and Elgamal 2010b)

| FE model | Soil | | | | | |
|---|---|---|---|---|---|---|
| parameter | T1[a] | T2[a] | D-S[b] | MD-SM[b] | MD-SC[b] | Units |
| $\phi$ | 52 | 46 | 38 | 33 | 23 | degrees |
| $c$ | 13 | 14 | 0 | 24 | 95 | kPa |
| $\psi$ | 22 | 16 | 8 | 3 | 0 | degrees |
| $p^{ref}$ | 100 | 100 | 100 | 100 | 100 | kPa |
| $m$ | 0.5 | 0.5 | 0.5 | 0.5 | 0.5 | ~ |
| $E_{50}^{ref}$ | 50,000 | 40,000 | 35,000 | 30,000 | 30,000 | kN/m$^2$ |
| $E_{oed}^{ref}$ | 50,000 | 40,000 | 35,000 | 30,000 | 30,000 | kN/m$^2$ |
| $K_0^{nc}$ | 0.4 | 0.4 | 0.4 | 0.45 | 0.6 | ~ |
| $\gamma$ | 20.6 | 20.6 | 20 | 19 | 19 | kN/m$^3$ |
| $R_f$ | 0.75 | 0.75 | 0.75 | 0.75 | 0.75 | ~ |

[a]Current experimental study
[b]Earth Mechanics Inc. (2005)

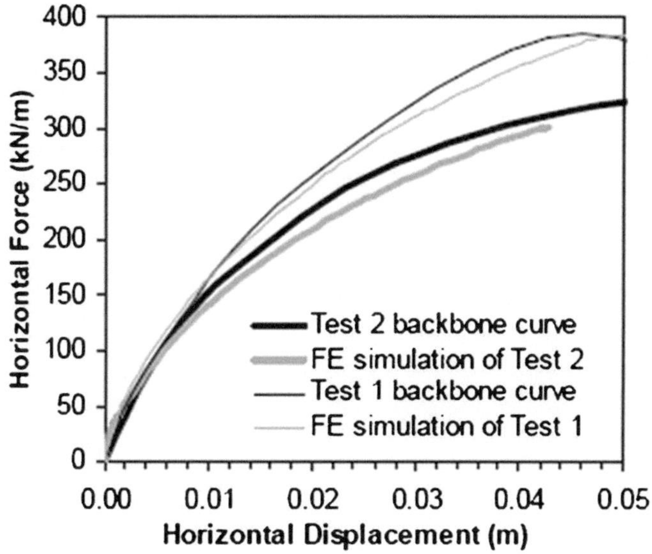


**Fig. 4.5** FE model results compared with the full scale tests (Wilson and Elgamal 2010b)

#### 4.2.2.2 FE Simulation for a Range of Backfill Soils and Abutment-Wall Heights

On the basis of the satisfactory match with the full scale test behavior (Fig. 4.5), additional FE simulations were performed, considering abutment wall heights ranging from 1–3 m. Additional soils (Table 4.1) were investigated to cover a range of likely backfill properties (Earth Mechanics 2005).

In order to consider taller wall configurations, and to minimize potential interference with the model boundaries, these additional simulations were performed on extended and deepened backfill domains (Wilson and Elgamal 2010b). Considering the potentially large friction force between the end of the bridge deck and the backwall which may prevent vertical wall movement (Lemnitzer et al. 2009), a wall-backfill friction angle $\delta = 0.35\varphi$ interface $R_{inter}$ (Plaxis 2004) was provided along the plate-soil boundary for the FE simulations. A horizontal displacement was again ascribed to the plate that represented the wall, but the vertical displacement was assigned as zero.

Simulations were performed considering four soils (Soils T2, D-S, MD-SM, and MD-SC). Model parameters for these soils (Table 4.1) were determined based on laboratory test results (Earth Mechanics 2005, Wilson and Elgamal 2010a). Soil T2 represents the placement condition backfill from Test 2 of the current experimental study. Soils D-S, MD-SM and MD-SC represent three categories of sandy soils found in California in an extensive investigation of actual bridge abutment backfills (Earth Mechanics 2005). Soil D-S is dense (clean) sand, MD-SM is medium-dense silty sand, and MD-SC is medium-dense clayey sand.

Figure 4.6 shows the simulated passive force-displacement response, per meter of wall length, for the four soils mentioned above considering wall heights H ranging from 1–3 m. These curves are limited to the numerically stable solution range, and may not fully define the peak passive resistance.

From Fig. 4.6, there is clearly a wide range in backfill strength and stiffness, depending on both the soil type and the wall height. According to the FE model simulation with clean sand backfill (Soil D-S), the passive resistance with H=3 m reached nearly ten times that of the H=1 m case (Fig. 4.6b). The stiffness also increased rapidly as the wall became taller for Soil D-S (Fig. 4.6b). For instance, at a horizontal wall displacement of 0.01 m, about 4.4 times the passive resistance was mobilized for H=3 m, compared with H=1 m (Fig. 4.6b). In contrast, for the high $c$ and lower $\varphi$ Soil MD-SC (Fig. 4.6d), the passive resistance with H=3 m reached only about four times that of the H=1 m case, and the stiffness increase for taller walls was also less pronounced.

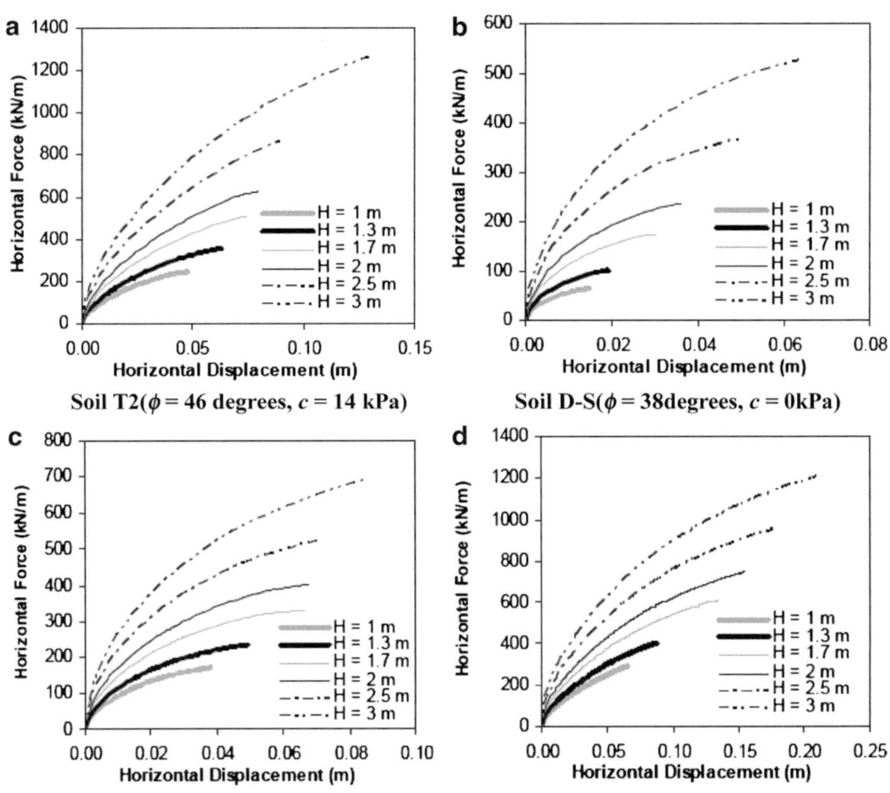


**Fig. 4.6** Passive earth pressure backbone curves for four soil types over a range of wall heights per meter of wall length (Wilson and Elgamal 2010b)

### *4.2.3 Dynamic Excitation Tests*

During a strong earthquake, inertial forces due to ground acceleration may act on the abutment backfill (Fig. 4.7). Such inertial forces may cause the instantaneous passive resistance to be different than that of the corresponding static case. In order to illustrate this effect, a series of shake table testing events (Wilson and Elgamal 2009) was executed after first mobilizing a significant level of passive resistance behind the test wall (Fig. 4.8). As mentioned earlier, these dynamic excitation tests were conducted using the same configuration from Tests 1 and 2, but with a removed and re-compacted backfill and an array of accelerometers added (Wilson and Elgamal 2009).

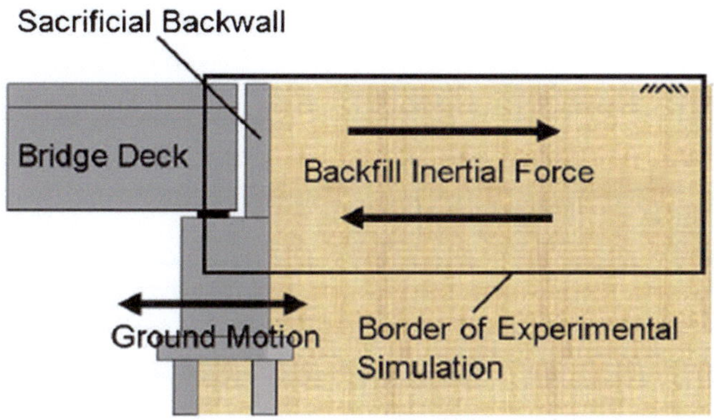


**Fig. 4.7** Schematic bridge elevation view showing backfill inertial force (Wilson and Elgamal 2009)

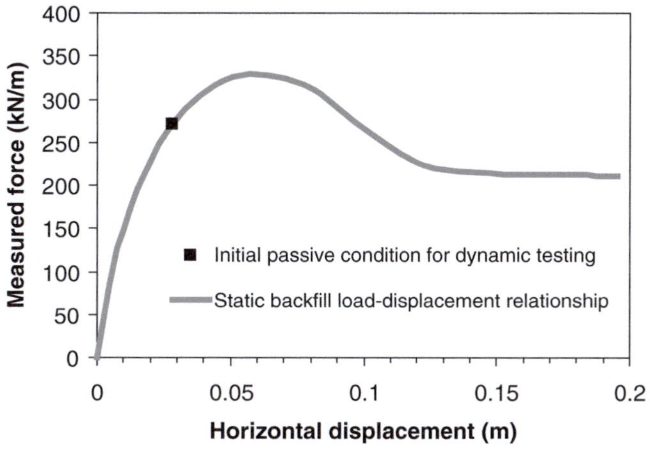


**Fig. 4.8** Passive force-displacement curve showing approximate initial testing condition for dynamic excitation tests

As part of a collaborative seismic bridge-ground investigation (Saiidi et al. 2008), scaled versions of a modified Century City Station record of the 1994 Northridge earthquake were developed for input excitation (Fig. 4.9). Before these input excitations were applied, the hydraulic jacks (Fig. 4.2e) were used to push the wall into the backfill in order to mobilize a significant portion (about 270 kN/m) of the expected peak passive resistance (Fig. 4.8). Hydraulic oil flow to the jacks was then locked, and the wall remained in this configuration throughout the series of shaking events.

Figure 4.9 shows representative results from two of the conducted shaking events (with peak input accelerations of about 2/3 and 1 g) in the form of recorded input (base) acceleration and total load cell force (sum of the load cell readings normalized by the length of the wall) time histories. Situated on the hypothetical bridge-deck side of the abutment wall (Figure 4.2e) the total load cell force represents the combined resultant passive thrust and test wall inertia during the shaking events.

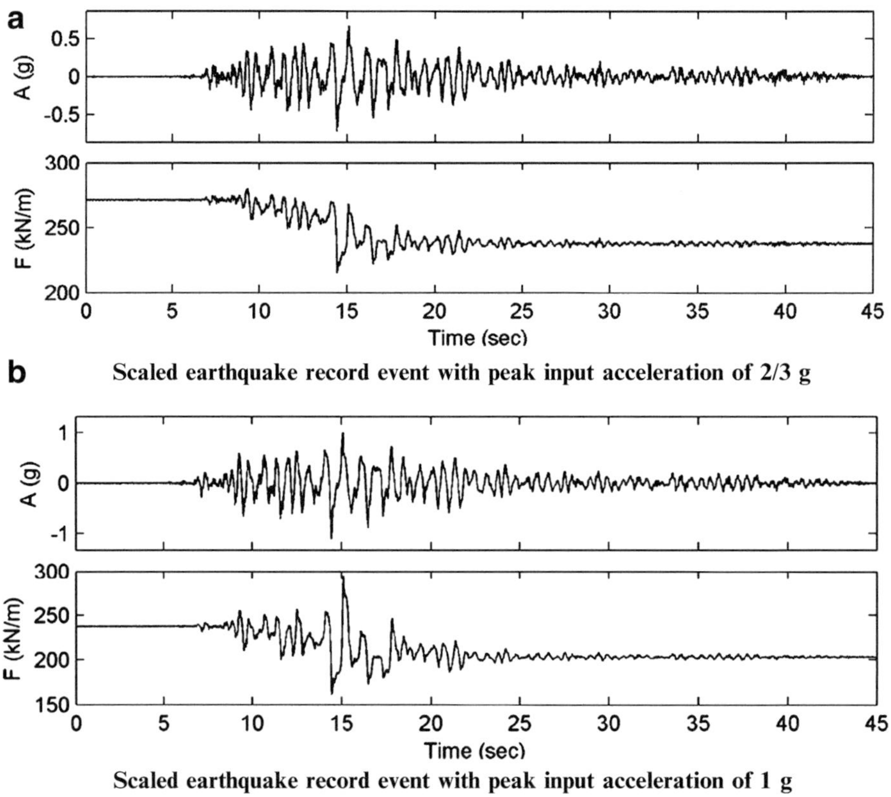


**Fig. 4.9** Time histories showing the input (base) acceleration (A) and the resultant passive force (F) measured by the load cells (Fig. 4.2e) per meter of wall length. (**a**) Scaled earthquake record event with peak input acceleration of 2/3 g. (**b**) Scaled earthquake record event with peak input acceleration of 1 g

Based on the response shown in Fig. 4.9, at the instants of large acceleration, available passive resistance to bridge deck displacement may substantially depend on the magnitude and direction of the excitation. For instance, compared to the static level of passive force (at time equal to zero), a possible increase of about 30% was observed at the 1 g instant of peak acceleration (Fig. 4.9b). Substantial instantaneous reduction in pressure may also occur at such high g-levels (Fig. 4.9b). Equally noteworthy, is that a portion of the mobilized static passive stress within the backfill was permanently relieved during to the strong shaking phase (Fig. 4.9), on account of the lateral stress being much larger than the vertical.

## 4.3 Abutment-Bridge Interaction Modeling

In this section, a 2-D bridge model is described, which was developed using the Pacific Earthquake Engineering Research (PEER) Center FE analysis code, Open System for Earthquake Engineering Simulations, or OpenSees (Mazzoni et al. 2006). A spring material which can be used to represent the static backfill passive force-displacement resistance is presented first. Next, an additional simple mass-spring system is added to the bridge model to account for the effect of the backfill inertia. Finally, results from earthquake record simulations are presented and discussed.

### *4.3.1 Static Passive Force-Displacement Resistance*

Within the overall collaborative framework of the seismic bridge-ground investigation mentioned above (Saiidi et al. 2008), a new spring material has recently been implemented and is available for use in OpenSees. For that purpose, a "HyperbolicGapMaterial" (Fig. 4.10) was developed with the help of Matthew Dryden of the University of California at Berkeley, as a part of his PhD study under the supervision of Professor Gregory Fenves (Wilson and Elgamal 2008; Dryden 2009).

The model (Fig. 4.10) follows a hyperbolic backbone curve to represent the passive force-displacement relationship (Cole and Rollins 2006, Shamsabadi et al. 2007) for virgin loading, based on the following equation (Duncan and Mokwa 2001):

$$F(y) = \frac{y}{\dfrac{1}{K_{\max}} + R_f \dfrac{y}{F_{ult}}} \tag{4.1}$$

Where $F$ is the resisting force, $y$ is the horizontal displacement, $K_{max}$ is the initial (linear) stiffness, $F_{ult}$ is the maximum passive resistance, and $R_f$ is a failure ratio. Hyperbolic model parameters were fitted to match the simulated response for the four soil types and range of wall heights shown above in Fig. 4.6 (Wilson 2009; Wilson and Elgamal 2010b).

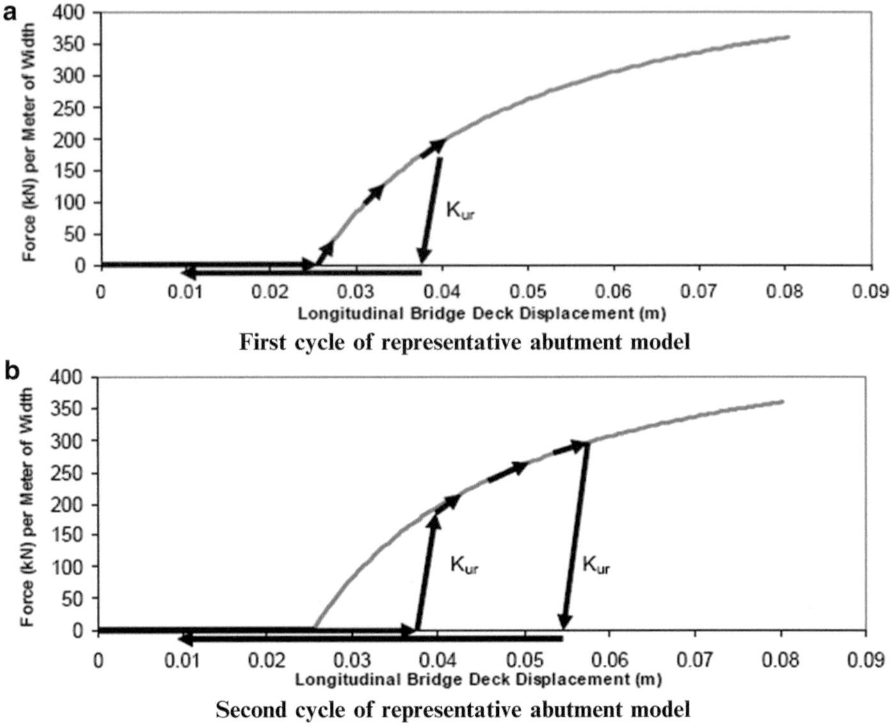


**Fig. 4.10** Schematic of representative abutment passive resistance spring model behavior (Wilson and Elgamal 2008). (**a**) First cycle of representative abutment model. (**b**) Second cycle of representative abutment model

The Hyperbolic Gap Material (Fig. 4.10) includes an adjustable expansion gap (between the end of the bridge deck and the abutment wall). An unloading and reloading stiffness $K_{ur}$ is included for subsequent response cycles (Fig. 4.10b). The model assumes that if the bridge deck pushes the abutment back wall into the backfill and then retreats, the wall essentially remains at its furthest penetration. On subsequent loading cycles, it is assumed that the abutment loses its resisting capacity up to the point of prior unloading (Fig. 4.10b). For the implemented model, a high $K_{ur}$ value, such as the initial stiffness of the employed hyperbolic model curve, may be adopted. Moderate changes in $K_{ur}$ were not found to produce noticeable changes in the overall bridge model response (Wilson 2009).

Figure 4.11 schematically illustrates the implementation of the Hyperbolic Gap Material as a spring at the abutments in a simple 2-D (longitudinal direction) two span elastic bridge model (using OpenSees). Bridge dimensions and associated model parameters were set to approximately match the "Route 14 bridge (R14)" highway overpass in California, as reported by Aviram et al. (2008). The bridge deck ends are supported on rollers allowing free lateral movement until the expansion gap is closed (Fig. 4.11). The bridge columns were modeled as fixed at the base and at the connection to the bridge deck (deforming in double-curvature). With this configuration, the first mode in the longitudinal direction had a first natural period of about 0.7 s.

The abutment force-displacement response from a bridge simulation using a representative strong motion record is shown in Fig. 4.12 in order to demonstrate the overall response. The model successfully captures the effects of the expansion gap and the nonlinear passive force-displacement resistance, with a simplified linear unloading and reloading stiffness to allow for simulation of cyclic loading.

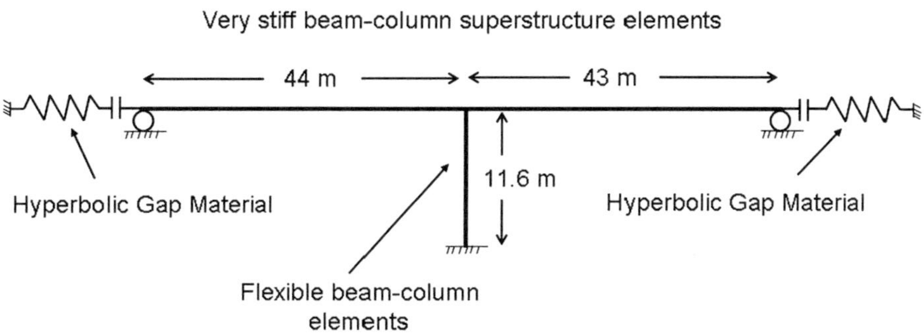


**Fig. 4.11** 2D bridge model implementation of the abutment passive resistance spring model behavior (Wilson and Elgamal 2010c)

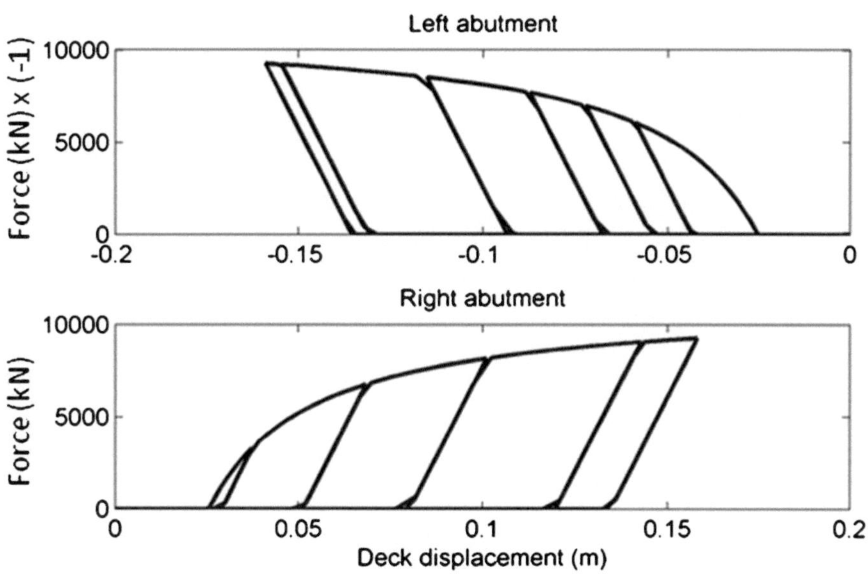


**Fig. 4.12** Representative abutment spring model response during strong earthquake simulation (Wilson and Elgamal 2010c)

A scaled version (with peak base acceleration of about 0.35 g) of the modified 1994 Northridge earthquake record motion (Fig. 4.13) used in the dynamic earth pressure experiments was also employed in order to demonstrate the model response. Five cases were included in the demonstration as follows (Table 4.2):

1. Neglecting the abutment passive earth pressure resistance (with rollers in the longitudinal direction at the bridge deck ends); and including the abutment static passive resistance (for a 1.7 m back wall height) by using the Hyperbolic Gap Material according to:
2. The tested dense sand with silt soil passive force-displacement backbone curve (Soil T2, Fig. 4.6a);
3. The dense clean sand backbone curve (Soil D-S, Fig. 4.6b);
4. The medium-dense silty sand backbone curve (Soil MD-SM, Fig. 4.6c); and
5. The medium-dense clayey sand backbone curve (Soil MD-SC, Fig. 4.6d).

The maximum column deformation (and the related ductility demand) is often a critical consideration in seismic bridge design. As such, the displacement at the connection between the top of the column and the bridge deck, relative to the base, is used as a basis for comparison of the model response. Figure 4.13 shows the input acceleration and displacement at the column-bridge deck connection for Case 1 (no abutment contribution). From Fig. 4.13, with a peak input acceleration of about 0.35 g; the columns displaced as much as 14 cm, relative to the base. The response shown in Fig. 4.13 are used as a basis for comparing the outcome when considering the typical passive backfill soil resistance contributions (Fig. 4.6).

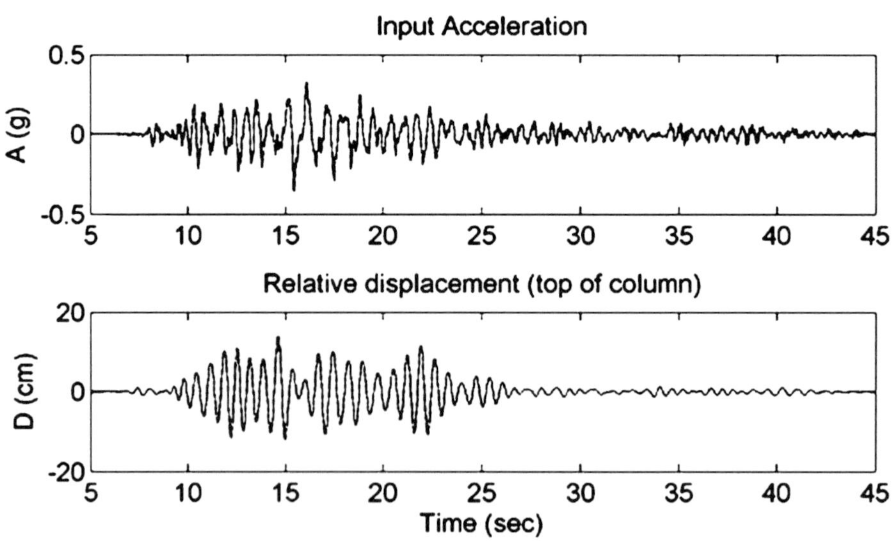


**Fig. 4.13** FE simulation results without abutment spring (Case 1)

**Table 4.2** Comparison of bridge model response considering four different backfill soils (Wilson 2009)

| Case | | Spring model parameters | | | Max column disp | |
|---|---|---|---|---|---|---|
| abutment backfill soil type | | $K_{max}$(kN/m/m) | $F_{ut}$(kN/m) | $R_f$ | (cm) | (% of Case 1) |
| 1 | None | 0 | 0 | ~ | 14.0 | 100 |
| 2 | T2 (dense well graded silty sand) | 24,000 | 550 | 0.8 | 8.2 | 59 |
| 3 | D-S (dense clean sand) | 24,000 | 190 | 0.8 | 11.3 | 81 |
| 4 | MD-SM (medium-dense silty sand) | 21,000 | 325 | 0.75 | 10.5 | 75 |
| 5 | MD-SC (medium-dense clayey sand) | 15,000 | 650 | 0.75 | 8.3 | 59 |

The peak column displacement from each case is listed in Table 4.2. From Table 4.2, including the abutment contribution considering the backfill soil used in the full scale tests (Soil T2) resulted in a 40% reduction in the column displacement demand. Including a weaker backfill (Soil D-S) reduced the demand by only 20%. The range of backfill soil force-displacement resistance models (Cases 2 through 5) resulted in variations of the maximum column deflection of up to 3 cm, when subject to the input acceleration shown in Fig. 4.13.

### *4.3.2 Including the Backfill and Back Wall Inertia*

In order to include an approximation of the observed backfill inertial effects from the dynamic tests, a simple mass-spring system is added to the bridge model (Fig. 4.14). The participating lumped backfill mass (Fig. 4.14) is determined based on the size of the passive failure wedge (Wilson 2009) and the back wall mass. The required additional spring stiffness (Fig. 4.14) is estimated based on the shear stiffness of the backfill. The force exerted by the mass on the bridge deck was also validated against the recorded changes in passive thrust from the dynamic tests (Fig. 4.9).

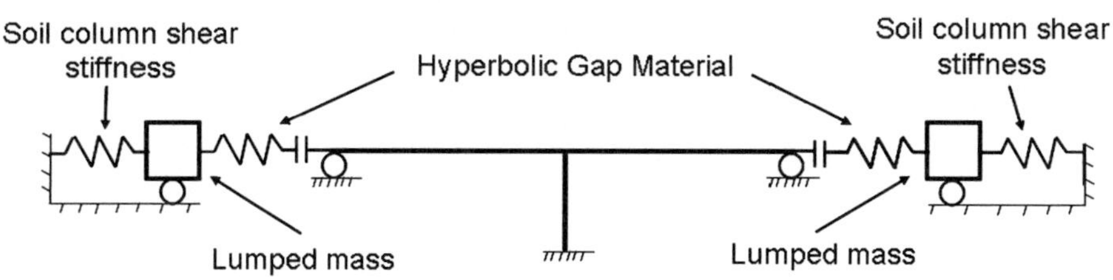


**Fig. 4.14** 2D bridge model implementation of the backfill inertia effect using a lumped mass (Wilson and Elgamal 2010c)

The abutment force–displacement response from a bridge simulation using the same representative strong motion as that of Fig. 4.12 (without the lumped mass) are shown in Fig. 4.15 (with the lumped mass). This model adds the effect of the backfill inertia which causes the passive force-displacement relationship to deviate (Fig. 4.15) from the static resistance behavior shown in Fig. 4.12.

The simulated bridge response is also compared in terms of the column deformation for two earthquakes in Figs. 4.16 and 4.17 using three different abutment models: (i) no resistance provided by the abutments (using roller supports without any springs), (ii) considering only the static backfill passive force-displacement (Spring), and (iii) including both the static force-displacement resistance and the backfill inertia effects (Spring-mass).

From Figs. 4.16 and 4.17, by including the effect of the static abutment resistance (as opposed to only using rollers), the predicted column displacement demand was reduced substantially. However, it may be seen in Fig. 4.16 that the additional spring-mass abutment-inertia model was of marginal consequence. On the other hand, during the Rinaldi input record (Northridge 1994) of Fig. 4.17, influence of the backfill inertia was more pronounced. In Fig. 4.17, the sudden, strong acceleration jolt at about 2.5 s resulted in a larger displacement (about 20% more) when the backfill inertial effect was included. At such instances, the backfill inertial force reduced the available abutment resistance (detrimental effect). In some other cycles (e.g. around 5 s), the backfill inertia increased the abutment resistance, and reduced

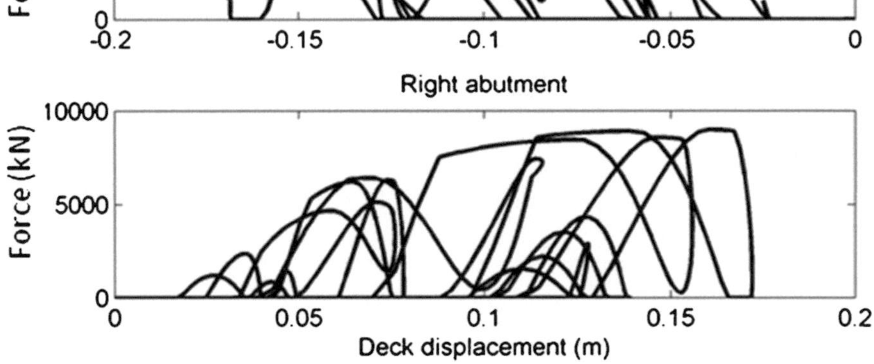


**Fig. 4.15** Force-displacement response at the bridge deck-abutment interface using the lumped mass model (Wilson and Elgamal 2010c)

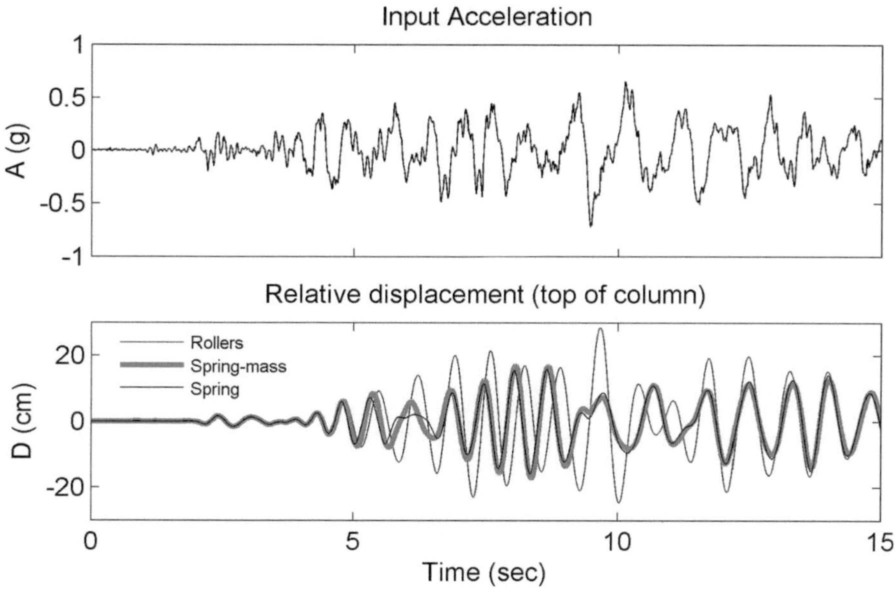


**Fig. 4.16** Comparison of bridge displacement response (Century City Station earthquake record) considering different abutment modeling possibilities (Wilson and Elgamal 2010c)

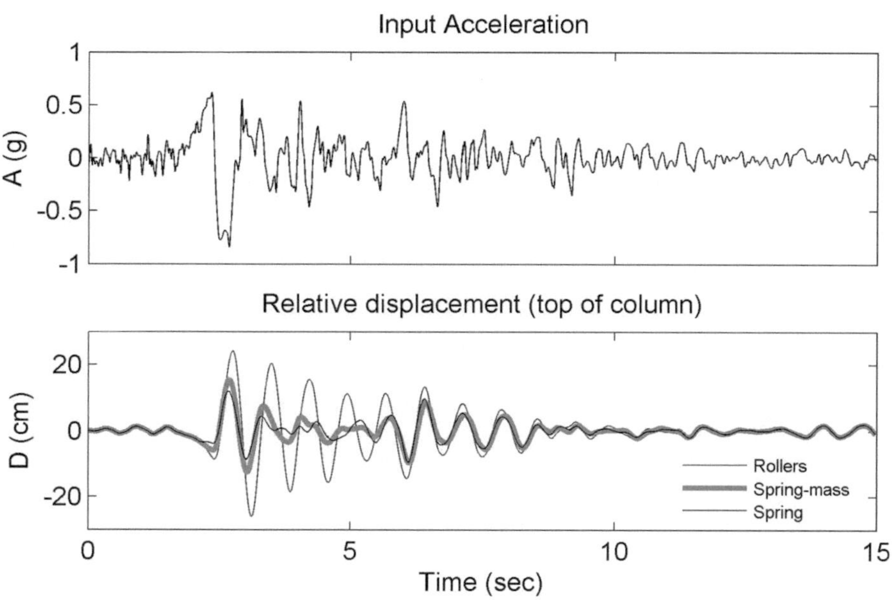


**Fig. 4.17** Comparison of bridge displacement response (Rinaldi earthquake record) considering different abutment modeling possibilities (Wilson and Elgamal 2010c)

the corresponding bridge displacement (beneficial effect). As such, additional studies are warranted to quantify the influence of this response mechanism on overall bridge response.

## 4.4 Conclusions

Full scale tests were performed in order to investigate the abutment backfill resistance to earthquake induced bridge deck displacement. Results from the conducted passive earth pressure load-displacement tests were used to calibrate FE models. The calibrated FE models were then used to develop backbone passive force-displacement curves for a wide range of backfill soil types and abutment wall heights. Dynamic excitation tests were performed next, with passive pressure mobilized in the backfill. Results from these tests illustrated the potential for inertial forces in the backfill to influence the available passive resistance.

Based on the results and insights gained from the full scale tests and simulations, modeling techniques for the observed key aspects of response were developed and demonstrated. A spring model was implemented in the FE program OpenSees which can represent the passive earth pressure load-displacement resistance to bridge deck displacement. An additional lumped-mass dynamic system was also implemented to simulate the effects of inertial forces in the backfill.

These abutment resistance models were also employed in overall bridge response simulations using OpenSees. Including a contribution which represents the abutment resistance was shown to significantly reduce the predicted column displacement demand. In addition, the type of backfill soil was shown to have a noticeable impact on the model response.

The added dynamic mass to simulate the effects of inertial forces in the backfill caused instants of both increased and decreased abutment resistance during shaking. Depending on the bridge and input ground motion configurations, the overall influence of backfill inertia may be small. In such cases, simply including the static backfill passive force-displacement resistance would be sufficient. However, additional research is needed in order to more comprehensively address the effects of backfill shaking on the abutment passive force-displacement resistance, and its potential influence on the bridge deck displacement.

**Acknowledgements** Support for this research was provided by the National Science Foundation (NSF) NEES-R grant number 0420347 under the overall direction of Professor M. Saiid Saiidi of the University of Nevada at Reno (UNR). Financial support for the second author in the form of the UCSD Robert and Natalie Englekirk Fellowship is also gratefully acknowledged. Dr. Chris Latham, Dr. Azadeh Bozorgzadeh, and Dr. Anoosh Shamsabadi provided much help and insight during the planning stages. The success of the experimental phase would not have been possible without the help of Alex Sherman, Lonnie Rodriguez, Mike Dyson and the entire Englekirk Structural Engineering Center staff. Dr. Arul Arulmoli of Earth Mechanics, Inc., Joe Vettel of Geocon Inc., and James Ward and Allan Santos of Leighton and Associates are all gratefully acknowledged for services donated regarding backfill material testing.

## References

AASHTO (2007) Proposed guide specifications for LRFD seismic bridge design. American Association of State Highway and Transportation Officials, Washington, DC

Aviram A, Mackie K, Stojadinovic B (2008) Effect of abutment modeling on the seismic response of bridge structures. Earthq Eng Vibration 7(4):395–402

Bozorgzadeh A (2007) Effect of structure backfill on stiffness and capacity of bridge abutments. PhD thesis, Department of structural engineering, University of California, San Diego, CA

Caltrans (2004) Seismic design criteria, version 1.3. California Department of Transportation, Sacramento, CA

Caltrans (2006) Standard specifications. California department of transportation, Sacramento, CA

Cole R, Rollins K (2006) Passive earth pressure mobilization during cyclic loading. J Geotech Geoenviron Eng 132(9):1154–1164

Dryden GM (2009) The integration of experimental and simulation data in the study of reinforced concrete bridge systems including soil-foundation-structure-interaction. PhD thesis, Department of Civil and Environmental Engineering, University of California, Berkeley, CA

Duncan M, Mokwa R (2001) Passive earth pressures: theories and tests. J Geotech Geoenviron Eng 127(3):248–257

Earth Mechanics, Inc (2005) Field investigation report for abutment backfill characterization. UCSD Report Number. SSRP-05/02, Department of Structural Engineering, University of California, San Diego, La Jolla, CA

Lemnitzer A, Ahlberg E, Nigbor R, Shamsabadi A, Wallace J, Stewart J (2009) Lateral performance of full-scale bridge abutment back wall with granular backfill. J Geotech Geoenviron Eng 135(4):506–514

Maroney BH (1995) Large-scale abutment tests to determine stiffness and ultimate shear strength under seismic loading. PhD thesis, Department of Civil Engineering, University of California, Davis, CA

Mazzoni S, McKenna F, Fenves G (2006) Open system for earthquake engineering simulation user manual. Pacific Earthquake Engineering Research Center, University of California, Berkeley

Plaxis (2004) Plaxis V8. In: Brinkgreve R, Broere W (eds) PLAXIS, AN DELFT, Delft, the Netherlands

Saiidi M, Nelson R, Zadeh M, Buckle I (2008) Seismic performance of a large scale four-span bridge model subjected to shake table testing. In: Proceedings of the national concrete bridge conference, St. Louis, MO

Shamsabadi A, Rollins K, Kapuskar M (2007) Nonlinear soil-abutment-bridge structure interaction for seismic performance-based design. J Geotech Geoenviron Eng, ASCE 133(6):707–720

Terzaghi K, Peck R, Mesri G (1996) Soil mechanics in engineering practice, 3rd edn. Wiley, New York

Wilson P (2009) Large scale passive force-displacement and dynamic earth pressure experiments and simulations. PhD thesis, Department of Structural Engineering, University of California, San Diego, CA

Wilson P, Elgamal A (2008) Full scale bridge abutment passive earth pressure tests and calibrated models. In: Proceedings of the 14th world conference on earthquake engineering, Beijing, China

Wilson P, Elgamal A (2009) Full-scale shake table investigation of bridge abutment lateral earth pressure. Bull NZ Soc Earthq Eng, NZSEE 42(1):39–46

Wilson P, Elgamal A (2010a) Large scale passive earth pressure load displacement tests and numerical simulation. J Geotech Geoenviron Eng, ASCE 136(12):1634–1643

Wilson P, Elgamal A (2010b) Passive earth pressure load-displacement models for a range of backfill soils and depths. In: Proceedings of the 5th international conference on recent advances in geotechnical earthquake engineering and soil dynamics, San Diego, CA

Wilson P, Elgamal A (2010c) Seismic bridge simulations with calibrated abutment models. In: Proceedings of the 9th US National and 10th Canadian conference on earthquake engineering, Toronto, Canada

# Chapter 5
# Macro Element for Pile Head Cyclic Lateral Loading

**Michael Pender, Liam Wotherspoon, Norazzlina M. Sa'don, and Rolando Orense**

**Abstract** Interaction between laterally loaded piles and the surrounding soil is a complex phenomenon, particularly when nonlinear soil behaviour is involved; so complex that usually design calculations rely on computer software based on discrete spring formulations using empirically derived nonlinear p-y relationships. This chapter explores a macro element, Davies and Budhu (1986), as an alternative which uses relatively simple formulae that are available for evaluating the lateral stiffness of long elastic piles embedded in elastic soil and an extension to handle nonlinear soil-pile interaction. The predictions of these equations are confirmed using the three dimensional finite element software OpenSeesPL, Lu et al. (2010), as well as data from field lateral load testing on driven piles in a stiff residual soil at a North Auckland site. Furthermore, in this chapter an extension of the macro element to cyclic loading is presented and this is shown to model well the field data and also the predictions of OpenSeesPL. The pile head macro element method is not completely general as it applies only to a homogeneous soil profile, but, since we deal with long piles, the soil homogeneity needs to extend only over the pile shaft active length. Measured lateral load response of the piles at the Auckland site indicates that it is necessary to distinguish the "operational" modulus of the soil from the small strain modulus; the field data indicates a value of about one third to one quarter of the small strain value.

M. Pender (✉) • L. Wotherspoon • R. Orense
Department of Civil and Environmental Engineering, University of Auckland, Auckland, New Zealand
e-mail: m.pender@auckland.ac.nz

N.M. Sa'don
Department of Civil Engineering, University Malaysia Sarawak, Sarawak, Malaysia

M.A. Sakr and A. Ansal (eds.), *Special Topics in Earthquake Geotechnical Engineering*, Geotechnical, Geological and Earthquake Engineering 16,
DOI 10.1007/978-94-007-2060-2_5, 

## 5.1 Introduction

This chapter deals with the cyclic lateral loading of long piles embedded in stiff cohesive soil. The intention is to extend a relationship for predicting the nonlinear static lateral load deformation response of piles, Davies and Budhu (1986), to handle cyclic lateral loading. A two-fold verification of the method is presented. First, nonlinear lateral load data obtained during field testing of driven piles in an Auckland residual soil profile is used. Second, the three-dimensional nonlinear finite element software OpenSeesPL, Lu et al. (2010), is used to confirm that the Budhu and Davies equations predict monotonic static lateral loading satisfactorily and that the extension to cyclic lateral loading also produces reasonable results.

The term macro element is used in the chapter title. A macro element is a computational entity which provides a simplified method for calculating the response of a single point in a system that would usually require a finite element analysis. As such macro elements provide useful tools for preliminary design and also a means for checking the output of sophisticated numerical modelling.

Earthquake demands on piles come from two sources. Earthquake waves travel up through the soil profile and interact with the pile shaft – this produces kinematic loading. The structure supported by the pile is excited by the kinematic loading arriving at the pile head and then applies additional loads to the pile; a kind of feedback known as inertial loading. The material presented in this chapter, when it is developed further to handle dynamic rather than cyclic pile head response, will be relevant to evaluating the inertial phase of the earthquake response of pile foundations.

Prior to the field testing of piles in Auckland residual soil a detailed site investigation was carried out which included measurement of the shear wave velocity of the soil using several methods. The velocities obtained all indicated a small strain shear modulus Gmax for the soil between 30 and 40 MPa. However, it will be seen that the measured load-deformation behaviour of the piles indicates that the "operational" shear modulus at "working level" pile loads is considerably less than that obtained from $G_{max}$. It is suggested that the provisions in Table 4.1 of Part V of Eurocode 8 (CEN 2003) offer useful ways of converting from small strain soil modulus to an "operational" modulus.

## 5.2 Lateral Load and Moment Response of a Long Vertical Elastic Pile Embedded in an Elastic Soil

A pile is long when the deflections induced by actions applied at the pile head approach zero before the bottom of the pile is reached. The length specifying this is known as the active length; when the pile is longer than the active length the pile head stiffnesses are independent of pile shaft length. Several researchers have proposed convenient expressions for the active length as well as pile head flexibility coefficients, or the components of the pile head stiffness matrix, for long piles: Kuhlemeyer (1979), Randolph (1981), Gazetas (1984, 1991), and Davies and Budhu (1986). These expressions are for static loading, although Gazetas (1984) has shown that pile head dynamic lateral and rotational stiffnesses for piles in a constant modulus soil profile are not particularly sensitive to loading frequency (although for dynamic situations damping needs to be included which does, of course, increase the pile resistance for velocity dependent damping).

Davies and Budhu present design equations for three different soil profiles: over consolidated clay, soft clay, and sand. For the latter two soil profiles Young's modulus is assumed to increase linearly with depth from zero at the ground surface, for the first the Young's modulus is constant with depth. Since this chapter is concerned with piles in stiff cohesive soil profiles the Davies and Budhu equations for a soil profile having a constant Young's modulus with depth, or at least constant for depths up to the active length, will be presented.

For an elastic pile embedded in an elastic soil the pile shaft displacement $u_{gl}$ and rotation $\theta_{gl}$ at the ground-line (ground surface) are given by:

$$u_{gl} = f_{uH} H_{gl} + f_{uM} M_{gl} \tag{5.1}$$

$$\theta_{gl} = f_{\theta H} H_{gl} + f_{\theta M} M_{gl} \tag{5.2}$$

Where: $H_{gl}$ is the horizontal load applied to the pile shaft at the ground-line, $M_{gl}$ is the moment applied to the pile shaft at the ground-line, and $f_{uH}$, $f_{uM}$, $f_{\theta H}$, $f_{\theta M}$ are flexibility coefficients.

Equations 5.1 and 5.2 give the pile shaft deflections at the ground-line. Frequently the loading is applied to a projection of the pile shaft above ground level, and it is here that the displacements are required. These are obtained by adding to the ground-line deflections the cantilever contributions from the pile shaft projection.

The flexibility coefficients for a long pile are expressed in terms of the pile shaft diameter and a modulus ratio:

$$K = \frac{E_p}{E_s} \quad \text{for constant soil modulus with depth} \tag{5.3}$$

where: $E_p$ is the Young's modulus for the pile material, and $E_s$ is the Young's modulus of the soil.

The equations were derived for piles of solid circular section. For sections other than this an equivalent value of $E_p$ is obtained by equating $E_pI_p$ for the actual pile section with that for a solid pile having a diameter equal to the width of the actual pile in the loading direction.

For static loading the active length of the pile shaft is:

$$L_a = 0.50DK^{0.36} \tag{5.4}$$

where: D is the pile shaft diameter.

If the pile shaft length is greater than that given by the above equation, then the pile is "long" and the following equations by Davies and Budhu can be used for the ground-line flexibility coefficients:

$$\begin{aligned} f_{uH} &= \frac{1.3K^{-0.18}}{E_s D} \\ f_{uM} = f_{\theta H} &= \frac{2.2K^{-0.45}}{E_s D^2} \\ f_{\theta M} &= \frac{9.2K^{-0.73}}{E_s D^3} \end{aligned} \tag{5.5}$$

The location of the maximum moment in the pile section is given by:

$$L_{M_{max}} = 0.40L_a \tag{5.6}$$

The magnitude of the moment is obtained from:

$$\begin{aligned} M_{max} &= I_{MH} DH_{gl} \\ I_{MH} &= aK^{b} \\ a &= 0.12 + 0.24f + 0.1f^2 \\ b &= \exp(-13 - 0.34f) \end{aligned} \tag{5.7}$$

with: $f = \dfrac{M_{gl}}{DH_{gl}}$

If $I_{MH}$ is greater than 6 the maximum moment is equal to the ground-line moment, that is $I_{MH} = f$.

The above equations were used to calculate the unrestrained ground-line free-head lateral and rotational stiffnesses for pile shaft diameters ranging between 0.2 and 2.0 m. Additional calculations were done using the Randolph, Gazetas and Kuhlemeyer equations. The results are plotted in Fig. 5.1a for the lateral stiffness and Fig. 5.1b for the rotational stiffness. Not surprisingly, it is clear that all four approaches give very similar values. Values of the largest lateral stiffness of the four, Kuhlemeyer, are 7% greater than the smallest, Davies and Budhu; for the rotational stiffnesses the largest is 4% greater than the smallest.

## 5.3 Three Dimensional Modelling of Pile-Soil Interaction in Nonlinear Soil: Openseespl

OpenSeesPL (Lu et al. (2010) and https://neesforge.nees.org/projects/OpenSeesPL) has been developed within the OpenSees (Mazzoni et al. (2009) and http://opensees.berkeley.edu) framework specifically for calculations involving pile-soil interaction.

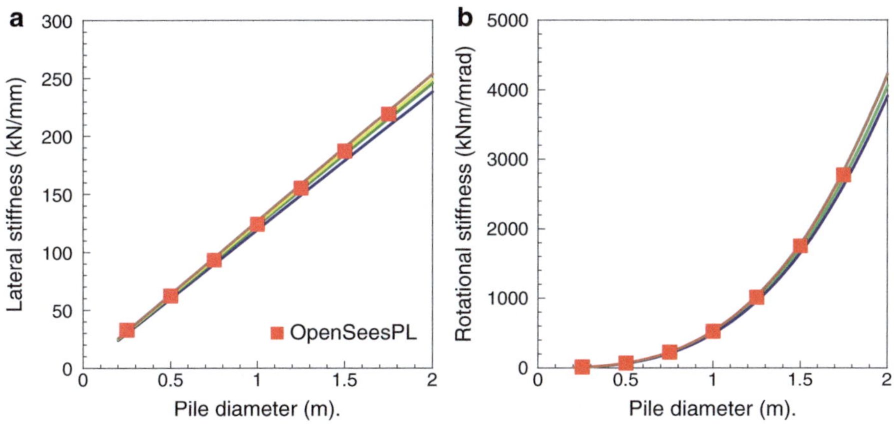


**Fig. 5.1** Unrestrained ground-line free-head lateral and rotational pile head stiffnesses given by Budhu and Davies, Gazetas, Randolf and Kuhlemeyer, and OpenSeesPL plotted against pile shaft diameter; (**a**) lateral stiffness, (**b**) rotational stiffness

The pile is vertical but the ground surface does not have to be horizontal, the meshing within the "soil box" is semi-automated so that the element size increases with distance from the pile shaft. Actions can be applied to the pile head, both static and dynamic, earthquake motions can be applied to the base of the finite element mesh, and a range of boundary conditions is available. Eight node and 20 node brick elements are available for the soil discretisation. The software also has the facility to account for nonlinear soil and nonlinear pile behaviour. OpenSeesPL differs from OpenSees in that a graphical user interface is provided for input of model details and post-processing is provided for plotting the output data.

The first application of OpenSeesPL in the work under discussion was to model an elastic pile in an elastic medium. The purpose of this modelling was to find out how fine the 3D finite element mesh needs to be to give good modelling of the elastic stiffness results. The unrestrained pile head lateral and rotational stiffnesses were evaluated for a range of pile diameters and from the results plotted in Fig. 5.1a, b it is apparent that OpenSeesPL duplicates these results obtained from the four methods most effectively. A view of the finite element mesh used to achieve this close agreement is shown in Fig. 5.2. The dimensions of the mesh box are: length 100 m and width 50 m; the pile is 20 m long, the mesh has 32 elements in the radial direction (closely spaced near to the pile shaft), in the vertical direction there are 26 elements, and in the circumferential direction 4 elements. As can be seen from Fig. 5.2, symmetry in the direction of the pile head shear force means that only half the pile-soil mesh needs to be modelled. Poisson's ratio for the soil was set to 0.49. Finite element calculations with Poisson's ratio set at 0.30 for the soil have pile head displacements about 6% greater than those for a value of 0.49 which confirms the conclusions of Davies and Budhu, Gazetas, Randolph, and Kuhlemeyer that Poisson's ratio for the medium in which the pile is embedded has only a minor effect on pile head stiffness.

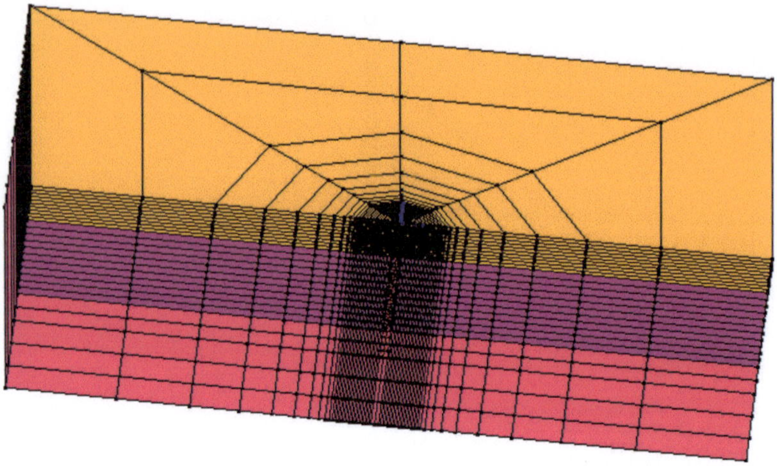

**Fig. 5.2** Finite element mesh used for the OpenSeesPL results plotted in Fig. 5.1

At this stage we have reached the conclusion that the four sets of simple equations for predicting the lateral and moment load response of a long vertical elastic pile in an elastic medium are mutually consistent. Furthermore we found that the OpenSeesPL finite element software is capable of calculating pile head stiffness values which are very close to those calculated with the other methods. Figure 5.1 presents the data on which this conclusion is based.

## 5.4 Davies and Budhu Extension to Nonlinear Soil-Pile Interaction

Davies and Budhu (1986), using a boundary element method, estimated the effect of local failure at the pile-soil interface, based on the undrained shear strength of the clay. They calculated a modification factor to be applied to elastic predictions of pile behaviour. For a free head pile they give the following equations for the ground-line pile shaft displacements and maximum pile shaft moment:

$$\begin{aligned} u_y &= I_{uy} u_E \\ \theta_y &= I_{\theta y} \theta_E \\ M_{My} &= I_{My} M_{ME} \end{aligned} \tag{5.8}$$

Where: $I_{uy}$, $I_{\theta y}$, and $I_{My}$ are yield influence factors, $u_E$ is the elastic pile head displacement from Eqs. 5.1 and 5.5, $\theta_E$ is the elastic pile head rotation from Eqs. 5.2 and 5.5, $M_{ME}$ is the maximum elastic pile shaft moment from Eq. 5.7.

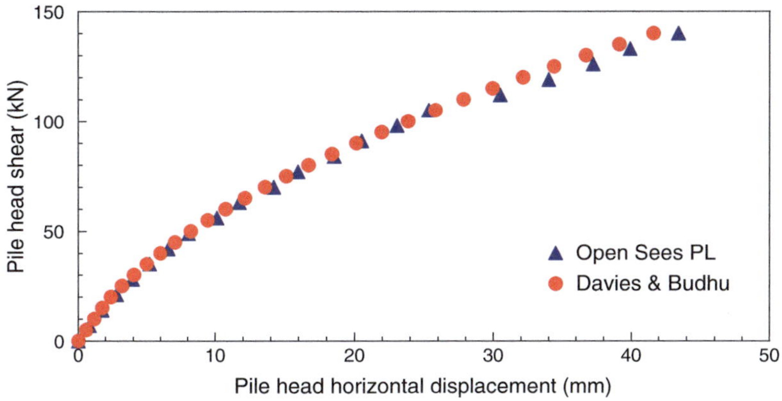


**Fig. 5.3** Comparison of the nonlinear lateral load – lateral displacement curves obtained with OpenSeesPL and the Davies and Budhu equations

The yield influence factors are given by:

$$\begin{aligned} I_{uy} &= 1+\frac{h-2.9k^{0.2}}{10.5k^{0.45}} \\ I_{\theta y} &= 1+\frac{h-2.9k^{0.2}}{12.5k^{0.33}} \\ I_{My} &= 1+\frac{h-2.9k^{0.2}}{20.0k^{0.29}} \end{aligned} \tag{5.9}$$

Where: $h = H_{gl}/s_u D^2$ ($s_u$ being the undrained shear strength of the soil) and $k = K/1{,}000$.

The above term for h includes the applied horizontal load but not the moment, however the pile head moment is not neglected as it is used in calculating the elastic displacements to which $I_{uy}$ and $I_{\theta y}$ are applied. For small pile loads the yield influence factors calculated with the above equations may be less than unity, in which case a value of unity is adopted.

Equations 5.8 and 5.9 have been used to calculate a nonlinear lateral load deformation curve for a long steel tube pile and OpenSeesPL for the same situation; the two curves are compared in Fig. 5.3. It is very clear from these that the deflections obtained from the Davies and Budhu approach are comparable to those obtained from OpenSeesPL. This is an independent verification of the equations given by Davies and Budhu, although other verification has been done by Davies and Budhu (1986) and Pranjoto (2000).

The pile flexibility coefficients in Eq. 5.5 are specified in terms of the Young's modulus of the soil. OpenSeesPL has input in terms of the shear wave velocity of

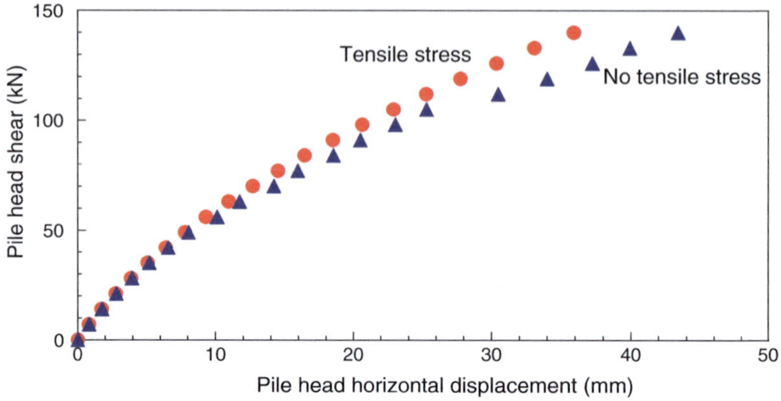


**Fig. 5.4** Nonlinear pile head lateral load – lateral displacement curve calculated with OpenSeesPL showing the effect of allowing or preventing the development of tensile stresses in the soil surrounding the pile shaft

the soil, and the site investigation work, to be discussed below, determines the shear wave velocity. Consequently, herein the soil property is expressed in terms of the shear modulus.

Despite the close comparison in Fig. 5.3 the two approaches perform very different calculations. As mentioned above, the Davies and Budhu equations are based on boundary element calculations. Interface stresses, both normal and shear, between the pile shaft and the surrounding soil are evaluated. When these stresses reach limiting values, defined by well-established limiting equilibrium solutions, the contact stresses in these regions are not further increased although the soil beyond the interface remains elastic. Results from these boundary element calculations were synthesised to obtain Eq. 5.9. (Eqs. 5.5–5.7 were obtained in a similar manner using displacement obtained from calculations with elastic piles embedded in elastic soil.) On the other hand, OpenSeesPL considers the possibility of nonlinear behaviour right through the whole soil domain. The OpenSees constitutive model, used to represent the undrained deformation of the clay, has a series of nested cylindrical yield loci with progressive decrease in soil stiffness as each yield locus is engaged.

In Fig. 5.3 the Davies and Budhu and OpenSeesPL results have been calculated by assuming that the full pile-soil interface contributes to the stiffness right from the ground surface. However, two questions arise. Does a gap form near the ground surface between the rear of the pile shaft and the surrounding soil? Are tensile stresses generated in the soil that lead to the formation of cracks in the soil behind the pile near the ground surface? The second of these questions can be addressed with OpenSeesPL as there is a no-tension facility available. The result of this calculation is shown in Fig. 5.4, from which it is clear that the elimination of tensile stresses in the soil behind the pile has a significant effect on the pile shaft ground-line displacements.

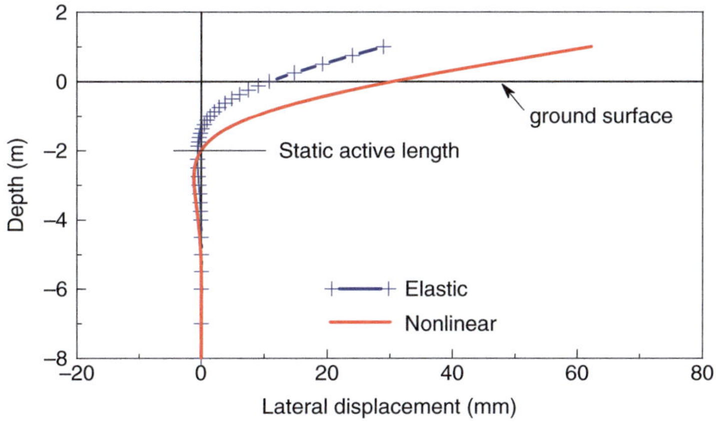


**Fig. 5.5** OpenSeesPL pile shaft deflected shape for elastic and inelastic modelling of pile-soil interaction

Another question is the effect of nonlinear soil behaviour on the active pile shaft length. This is illustrated in Fig. 5.5 which plots the deflected shape of a steel tube pile 273 mm in diameter, with a wall thickness of 9 mm, an embedded length of 8 m, and loaded laterally 1 m above the ground surface with a shear force of 140 kN. The soil is saturated clay with undrained shear strength of 80 kPa, the shear modulus is 10 MPa and Poisson's ratio is 0.49. The soil discretisation uses 20 node brick elements. The active length calculated from Eq. 5.4 is about 2 m. Figure 5.4 shows that at twice that depth there is some lateral displacement – about −0.5% of the lateral displacement of the pile shaft at the ground line for the elastic case and about −1.5% for the nonlinear case. Not surprisingly, it seems that the lateral deformation of the pile shaft extends to greater depths when nonlinear soil behaviour occurs. An additional nonlinear run of the model was made with the embedded length of the pile shaft reduced from the above 8 m to 4m in this case the lateral displacement at the ground line increased by 13%, and at 4 m depth by a similar proportion. From these results we conclude that when nonlinear soil behaviour occurs the active length of the pile shaft is about 2 or more times greater than that given by Eq. 5.4.

There are a number of factors that influence the output from the OpenSeesPL modelling. At low loads the modulus of the soil is the most important. At larger loads it is the undrained shear strength of the soil. In addition, Poisson's ratio, the value of $K_o$, and the shear strain at which the undrained shear strength is reached are of lesser significance. Also very important is any gap between the pile shaft and the soil – that is the unsupported length of pile shaft.

To obtain insight into the significance of gap formation we need to consider some of the field testing data.

## 5.5 Field Lateral Load Testing of Driven Piles in Residual Soil

Pile lateral load tests, both static and dynamic, were performed at a site in the Pine Hill subdivision at Albany, near Auckland, Sa'don (2011). The material at the site was classified as Auckland residual clay, a product of the in-situ weathering of Waitemata Group tertiary age sandstones and siltstones. The CPT recorded an average cone penetration resistance of about 1 MPa with friction ratio of 1.5–6%. One of the CPT profiles is illustrated in Fig. 5.6. Shear-wave velocities recorded from the seismic CPT was approximately 160 m/s and was fairly constant with depth. Undrained shear strength ($s_u$) values of about 100 kPa were estimated from the CPT results. In addition other in situ tests were done to determine the dynamic stiffness of the soils at the site using the WAK (wave-activated stiffness) and spectral analysis of surface waves (SASW) methods as well as seismic CPT testing. All of these methods lead to small strain shear moduli in the 30–40 MPa range. The pore pressure measurements from the CPT soundings indicate that the water table is at a depth of 5 m, however, in the winter rainy season in Auckland, and for much of the rest of the year, the fine grained residual soil is expected to be saturated to near the ground surface, Wesley (2010).

Four closed-ended steel pipe piles, having an outside diameter of 273 mm, wall thickness of 9.3 mm, and lengths of 7.5 m, were driven to depths of 6.5–7.0 m using a 3,000 kg drop hammer. The yield moment of the pile section is approximately 180 kNm.

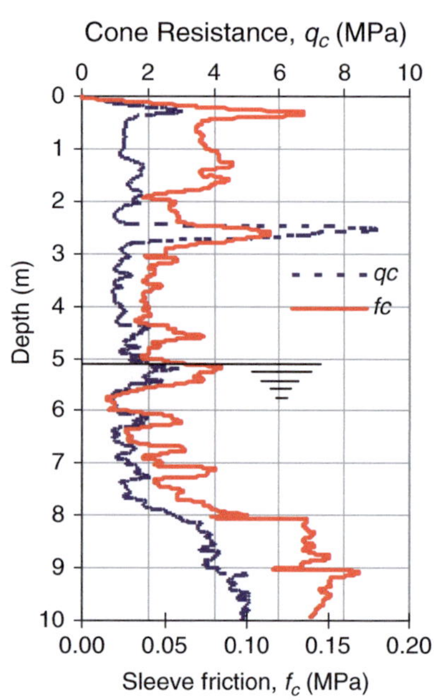


**Fig. 5.6** Representative CPT profile near one of the piles at the Albany site

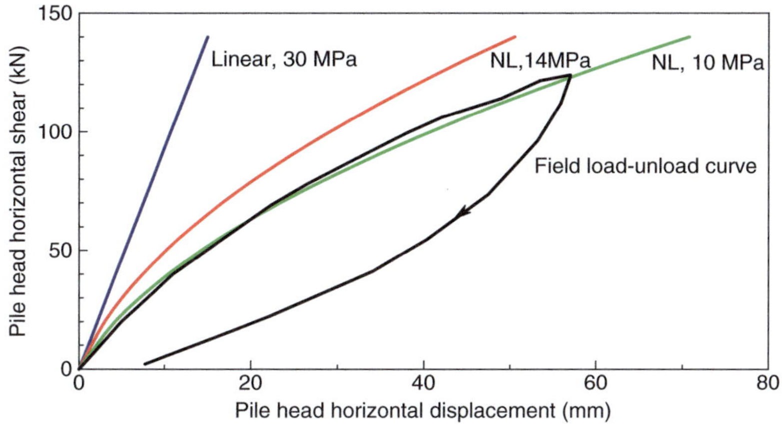


**Fig. 5.7** Comparison between measured static load-deformation behaviour of the pile and Davies and Budhu predictions with various soil operational shear modulus values (For the nonlinear cases the soil undrained shear strength = 80 kPa)

Static lateral loading was done on two of the piles by connecting a manually operated hydraulic jack between the parts of the pile shafts projecting above the ground surface. The piles were subject to a number of load-unload cycles with increasing maximum loads, but the maximum value was such that the yield moment of the pile shaft was not reached. The load-unload curve for the final cycle, up to a maximum lateral load to 125 kN, is shown in Fig. 5.7 along with data obtained with some Davies and Budhu calculations. Three things are apparent from these plots. First, the slope of the load deformation curve obtained by using the small strain shear modulus of the soil is applicable only at very small lateral loads and displacements. Second, a trial value of 14 MPa for the soil modulus is seen to give an improvement in the shape of the load-deformation curve, but still the calculated curve was too stiff. Finally, a gap 0.1 m deep was introduced into the Davies and Budhu equations, that is the effective ground line is lowered by 0.1 m, and a good match was achieved between the measured load deformation curve and the calculated values using a shear modulus of 10 MPa for the soil. Thus we conclude that the operational modulus of the soil is about one third to one quarter of the small strain modulus.

Finite element calculations in OpenSeesPL do not allow for a gap to open between the rear of the pile and the surrounding soil, rather it has the facility to prevent tensile stresses developing in the soil. When this is implemented the result is as shown in Fig. 5.8 indicating that the use of a small gap depth Davies and Budhu equations produces a similar result to the no-tension condition in OpenSeesPL.

In addition to the static lateral load testing the piles were subject to excitation with an eccentric mass shaker, of mass 600 kg, attached to the top of the pile shaft. Results for one pile are shown in Fig. 5.9 where hysteresis loops for a cyclic forcing

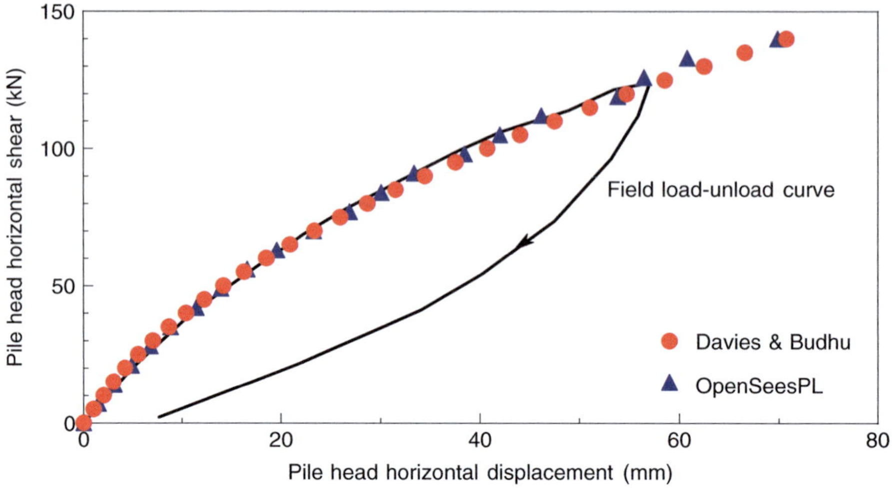


**Fig. 5.8** Comparison between the OpenSeesPL and Davies and Budhu predictions of the loading part of static load-deformation behaviour of the pile (Soil undrained shear strength = 80 kPa and operational shear modulus = 10 MPa)

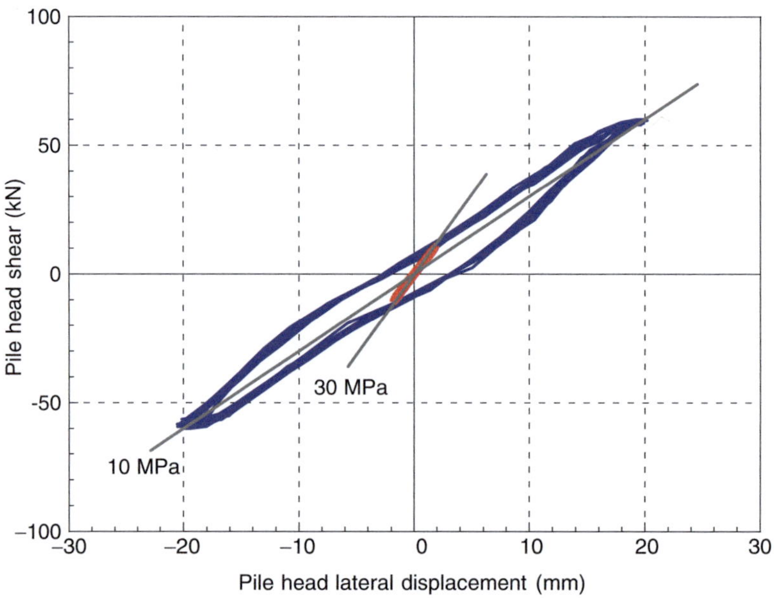


**Fig. 5.9** Response of a single pile driven into Auckland residual clay to excitation from an eccentric mass shaking machine at 5 Hz (amplitude 10 kN) and 6.5 Hz (amplitude 60 kN)

amplitude of 10 kN (frequency 5.0 Hz) as well as loops for a cyclic forcing amplitude of 60 kN (frequency 6.5 Hz) are plotted. These loops confirm again that the small strain shear modulus of the soil is operational only for small amplitude excitation. Also plotted in Fig. 5.9 is a line obtained from a soil shear modulus of 10 MPa and this time a gap depth of 0.2 m. This reveals, a potential disadvantage of eccentric mass shaking for estimating properties for earthquake design of foundations as very large numbers of cycles are required to get to the required excitation which may cause more cyclic softening of the system than earthquake excitation and so for Fig. 5.9a gap depth of 0.2 m was required, whereas for the monotonic loading in Fig. 5.8a gap depth of 0.1 m was satisfactory. Later in the programme of testing the piles we found that snap-back testing was a more satisfactory procedure, Pender et al. (2010).

## 5.6 Extension of the Davies and Budhu Method to Include Cyclic Loading

This brings us to the final part of our comparison of the Davies and Budhu equations and OpenSeesPL software. The question to be addressed in this section is the possibility of extending the Davies and Budhu equations to handle cyclic loading.

Looking at Figs. 5.3 and 5.7 it is clear that Eqs. 5.8 and 5.9 work well when applied to lateral loading from zero loads. Cyclic loading involves reversals of loading direction which is accompanied with a sudden increase in stiffness which occurs at each change in loading direction, which is each half cycle. We adopted a simple strategy for calculations following a turning point: (i) the data at the turning point was used as the starting point for the next half cycle, (ii) the calculation of incremental displacement was restarted from elastic conditions, (iii) the scale of the strain axis was doubled in the manner of Masing's rule, and (iv) the deformations calculated during the current half cycle were added to those at the end of the previous half cycle. The scale adjustment was achieved by noting that in Eq. 5.9 the expressions for $I_{uy}$ and $I_{\theta y}$ have the term $(h - 2.9\ k^{0.2})$ in the numerator. We halved this factor, that is applying Masing's rule, for all half cycles but the first loading from zero horizontal force; with this modification of Eq. 5.9 we obtain:

$$\begin{aligned} I_{uy\_cyclic} &= 1+\frac{0.5(h-2.9k^{0.2})}{10.5k^{0.45}} \\ I_{\theta y\_cyclic} &= 1+\frac{0.5(h-2.9k^{0.2})}{12.5k^{0.33}} \end{aligned} \tag{5.10}$$

A number of loops have been calculated and plotted in Fig. 5.10 which shows the shape of these to be as required. A backbone curve is marked on the diagram; this is simply the locus of the turning points and has no part in the calculation of the load deformation loops, although it does fall along the monotonic loading curve.

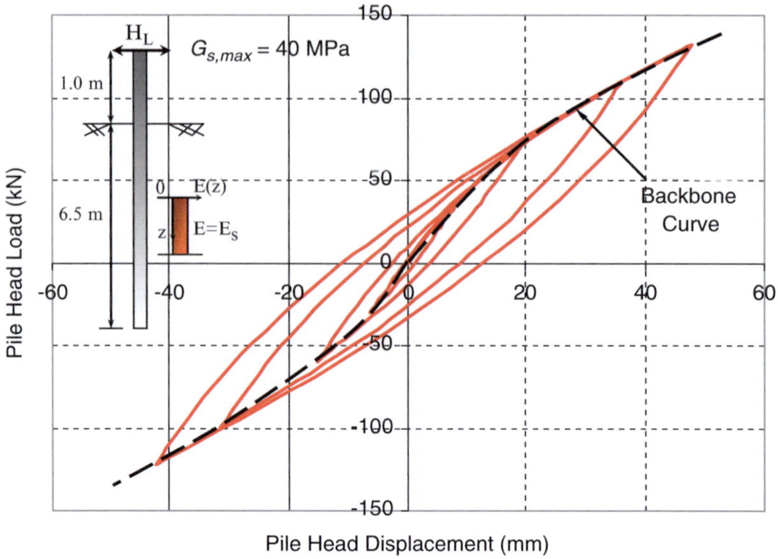


**Fig. 5.10** Hysteresis loops calculated with the cyclic extension to the Davies and Budhu equations

In Fig. 5.11 one of the test results from the steady state shaking machine response is plotted along with calculations based on the extended version of the Davies and Budhu equations. It is apparent that the loops are of similar size, although slightly fatter than those obtained from the shaking machine excitation. In Fig. 5.12 the same experimental data is compared with the loops obtained using OpenSeesPL. Once again the loops are a reasonable fit.

Figures 5.10, 5.11 and 5.12 all indicate that work is done during each of the calculated loading cycles. Although cyclic, these calculations are not dynamic so the work done during each loop is not a consequence of conventional velocity dependent damping. Even so it is of interest that the area enclosed by the calculated load-deformation loops in Figs. 5.11 and 5.12 are very similar to that of the loops measured during 6.5 Hz shaker excitation of the pile head. Perhaps this indicates that the observed damping during the shaker tests comes mainly from hysteretic deformation of the soil which is known not to be frequency dependent.

Although not presented here, the next stage in the work outlined in this chapter is the extension to modelling dynamic pile head excitation and the corresponding damping, in other words to extend the Davies and Budhu equations so that calculation of inertial interaction is possible.

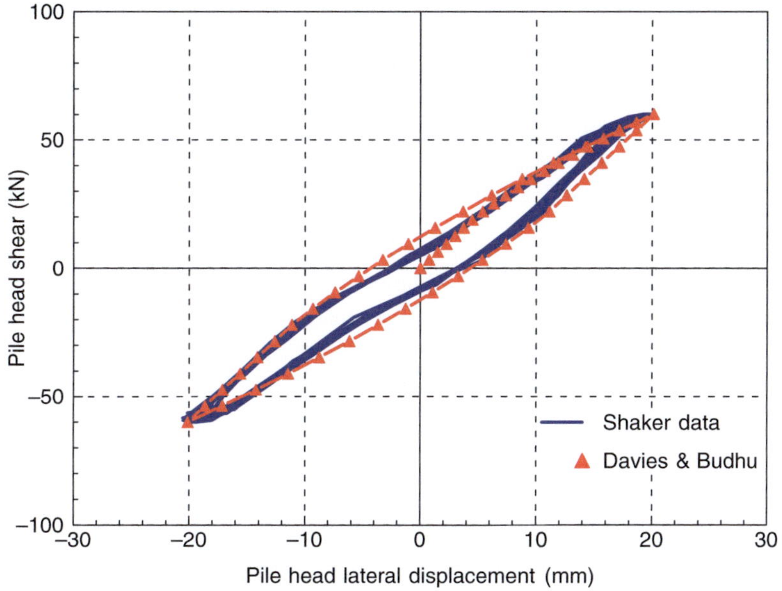


**Fig. 5.11** Comparison of the Budhu and Davies cyclic load deformation loops with those recorded during the dynamic excitation of the pile head. (Undrained shear strength = 80 kPa, soil shear modulus = 10 MPa, gap depth = 0.1 m)

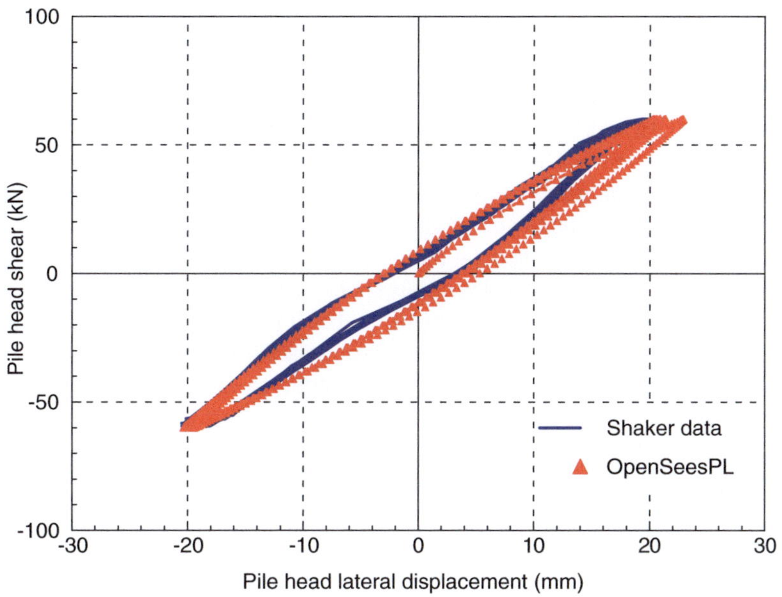


**Fig. 5.12** Comparison of the OpenSeesPL cyclic load deformation loops with those recorded during the dynamic excitation of the pile head. (Undrained shear strength = 80 kPa, soil shear modulus = 10 MPa, no-tension soil behaviour)

## 5.7 Conclusions

- The macro element for pile head lateral displacement and rotation developed by Davies and Budhu (1986) has been shown to be capable of modelling field test data on driven piles in a residual soil profile, and also to give similar displacements to those obtained from the OpenSeesPL 3D finite element software.
- Part of the reason for performing the field lateral load testing of driven piles in Auckland residual clay was to find how the operational modulus of the soil for this type of loading compares with the small strain soil modulus. The results indicate that the operational modulus of the soil is considerably less than the small strain modulus; values in the range of one third to one quarter appear to be appropriate.
- For an elastic soil layer having constant modulus with depth, four sets of expressions for pile head flexibility coefficients/pile head stiffnesses of long elastic piles, Kuhlemeyer (1979), Randolph (1981), Gazetas (1984 and 1991), and Davies and Budhu (1986), were found to be comparable.
- For a long elastic pile in an elastic soil OpenseesPL was found to duplicate the values for pile head flexibilities obtained from the four sets of expressions mentioned above.
- The Davies and Budhu nonlinear extension to the elastic pile in an elastic soil case was found to be closely modelled with the OpenSeesPL software.
- OpenSeesPL modelling of pile lateral loading indicates that when nonlinear soil behaviour occurs in the soil near the pile shaft the active length is rather longer than given by Eq. 5.4.
- The cyclic load extension of Davies and Budhu macro element was found to model the field test data well and also to match closely the calculated cyclic pushover results from OpenSeesPL.
- OpenSeesPL is an attractive 3D pile lateral loading software with capabilities well beyond what is displayed in this chapter. It is appealing because of the preprocessing, which makes data input and meshing straightforward, and the post-processing which provides comprehensive plotting facilities.
- The Davies and Budhu macro element, in that it produces results comparable to those obtained from OpenSeesPL, provides a useful tool for preliminary design and also a simple means of assessing output from sophisticated computer modelling.

**Acknowledgements** The second-named author, an EQC Research Fellow in Earthquake Engineering, acknowledges the support of the New Zealand Earthquake Commission (EQC).

The authors are grateful to the Ministry of Higher Education (MOHE), Malaysia and Universiti Malaysia Sarawak (UNIMAS), for the financial support granted for the doctoral studies of the third-named author.

## References

Comité Européen de Normalisation Eurocode 8, Part V: Geotechnical Design, General Rules. Draft 6, Jan 2003

Davies TG, Budhu M (1986) Nonlinear analysis of laterally loaded piles in heavily over consolidated clays. Geotechnique 36(4):527–538

Gazetas G (1984) Seismic response of single end-bearing piles. Soil Dyn Earthq Eng 3(2):82–93

Gazetas G (1991) Foundation vibrations. In: Fang H-Y (ed) Foundation engineering handbook, 2nd edn. Van Nostrand Reinhold, New York, 553–593

Kuhlemeyer RL (1979) Static and dynamic laterally loaded floating piles. J Geotech Eng Proc ASCE 105(GT2):289–304

Lu J, Yang Z, Elgamal A (2010) OpenSeesPL 3D lateral pile-ground interaction: User's manual. University of California, San Diego (http://neesforge.nees.org/projects/OpenSeesPL)

M.Sa'don (2011) Full scale static and dynamic loading of a single pile. PhD thesis, University of Auckland

Mazzoni S, McKenna F, Scott MH, Fenves GL (2009) Open system for earthquake engineering simulation user manual, University of California, Berkeley. (http://opensees.berkeley.edu/)

Pender MJ, Algie T, M.Sa'Don N, Orense RP (2010) Snap-back testing and estimation of parameters for nonlinear response of shallow and pile foundations at cohesive soil sites. In: Proceedings of the 5th international conference on earthquake geotechnical engineering, Santiago, Chile

Pranjoto S (2000) The effects of gapping on pile behaviour under lateral cyclic loading. PhD thesis, University of Auckland

Randolph MF (1981) Response of flexible piles to lateral loading. Geotechnique 31(2):247–259

Wesley LD (2010) Geotechnical engineering in residual soils. Wiley, New York

# Chapter 6
# Analysis of Piles in Liquefying Soils by the Pseudo-Static Approach

**M. Cubrinovski, Jennifer J.M. Haskell, and Brendon A. Bradley**

**Abstract** The pseudo-static method of analysis is a simplified design-oriented approach for analysis of seismic problems based on routine computations and conventional engineering models. The application of the method to analysis of piles in liquefying soils is burdened by significant uncertainties associated with soil liquefaction, soil-pile interaction in liquefying soils and the need to reduce a very complex dynamic problem to a simple equivalent static analogy. Hence, despite its simplicity, the application of the method is not straightforward and requires careful consideration of the uncertainties in the analysis. This chapter addresses some of the key issues that arise in the application of the pseudo-static analysis to piles in liquefying soils, and makes progress towards the development of a clear modelling (analysis) strategy that will permit a consistent and reliable use of the simplified pseudo-static analysis.

A comprehensive parametric study has been conducted in which a wide range of soil-pile systems, loading conditions and values for model parameters were considered. In the analyses, the pile responses during the strong ground shaking (cyclic phase) and post-liquefaction lateral spreading were considered by two separate pseudo-static analysis approaches. Results from the analyses are used to examine and quantify the sensitivity of the pile response to various model parameters, and to establish a fundamental link between the sensitivity of the response and

M. Cubrinovski (✉) • B.A. Bradley
Department of Civil and Natural Resources Engineering, University of Canterbury,
Private Bag 4800, Christchurch 8020, New Zealand
e-mail: misko.cubrinovski@canterbury.ac.nz

J.J.M. Haskell
Department of Engineering, Schofield Centre, University of Cambridge,
High Cross, Madingley Road, Cambridge CB3 0EL, UK

M.A. Sakr and A. Ansal (eds.), *Special Topics in Earthquake Geotechnical Engineering*, Geotechnical, Geological and Earthquake Engineering 16,
DOI 10.1007/978-94-007-2060-2_6, 

the mechanism of soil-pile interaction. On this basis, some general principles for conducting pseudo-static analysis of piles in liquefying soils can be established irrespective of the specific properties of the soil-pile system and loading conditions.

## 6.1 Introduction

Methods for assessment of the seismic performance of pile foundations in liquefying soils have evolved significantly over the past two decades. The 1995 Kobe earthquake, in particular, contributed to an improved understanding of the complex behaviour of piles in liquefying soils through evidence from well-documented case histories (JGS 1998) and extensive experimental studies that this event instigated (Cubrinovski et al. 2006; Tokimatsu and Suzuki 2009). On the analytical front, significant progress has been made across a very broad and diverse group of analysis methods ranging from simple design-oriented approaches to the most advanced (and complex) numerical procedures for dynamic analysis (O'Rourke et al. 1994; Tokimatsu and Asaka 1998; Yasuda and Berrill 2000; Finn and Thavaraj 2001; Cubrinovski et al. 2008).

Nominally, all these analysis methods have the same objective, to assess the seismic performance of the pile foundation and evaluate ground displacements, pile deformations and damage to piles. However, a close inspection of different methods reveals that they each focus on different aspects of the problem and provide a distinct contribution in the assessment of seismic performance (Cubrinovski and Bradley 2009). For example, Table 6.1 summarizes key features of three representative methods of analysis: (1) Pseudo-static analysis using a conventional beam-spring model (a simple design-oriented approach), (2) Seismic effective stress analysis (an advanced dynamic analysis incorporating effects of excess pore pressures and dynamic soil-pile-structure interaction), and (3) Probabilistic assessment within the Performance-Based Earthquake Engineering (PBEE) framework (a method rigorously quantifying the uncertainties and seismic risk). As outlined in Table 6.1, each of these methods of analysis provides a significant and different contribution in the assessment, and importantly, all methods have some shortcomings. In essence, these analysis methods are complementary in nature and it is envisioned that they will be all used in parallel in the future; hence they all require further development and improvement.

The pseudo-static method of analysis is a practical engineering approach based on routine computations and the use of relatively simple models. Application of this analysis method does not require excessive computational resources nor specialist knowledge, and it is thus a widely-adopted approach in current practice and seismic design codes. The application of the method to the analysis of piles in liquefying soils is not straightforward however, but rather is burdened by significant uncertainties associated with soil liquefaction during earthquakes, soil-pile interaction in liquefying soils and the need to reduce a very complex dynamic problem to a simple equivalent static analogy. Questions posed to the user are 'what stiffness and strength to adopt for the liquefied soil', 'how to combine oscillatory kinematic and inertial loads in a static analysis' and 'what is the sensitivity of the pile response to a certain model parameter', among others. This paper highlights the issues around the implementation of the

**Table 6.1** Methods for assessment of seismic performance of soil-structure systems: key features and contributions

| Method of assessment | Key features | Specific contributions in the assessment | Shortcomings |
|---|---|---|---|
| Pseudo-static analysis | • Simple to use<br>• Conventional data and engineering concepts | • Evaluates the response and damage level for the pile (parametric evaluation is needed)<br>• Enhances foundation design through better understanding of soil-pile interaction mechanism | • Gross approximation of dynamic loads and behaviour<br>• Aims at maximum response only |
| Seismic effective stress analysis | • Realistic simulation of ground response and seismic soil-foundation-structure interaction<br>• Complex numerical procedure | • Detailed assessment of seismic response of pile foundations including effects of liquefaction and SSI<br>• Considers inelastic behaviour of the entire soil-foundation-structure system<br>• Enhances communication of design between geotechnical and structural engineers | • Ignores uncertainties in the ground motion and numerical model<br>• High demands on the user |
| Probabilistic PBEE framework | • Considers 'all' earthquake scenarios<br>• Quantifies seismic risk | • Addresses uncertainties associated with ground motion and numerical model on a site specific basis<br>• Provides engineering measures (response and damage) and economic measures (losses) of performance<br>• Enhances communication of design and seismic risk outside profession | • Ignores details of the seismic response |

pseudo-static method of analysis for piles in liquefying soils and the effects of uncertainties in the analysis. It summarizes some of the key findings from a systematic analytical study and points towards methodology for the consistent and reliable use of the simplified pseudo-static analysis.

## 6.2 Soil-Pile Interaction in Liquefying Soils

### 6.2.1 Cyclic Phase

Soil-pile interaction in liquefying soils involves significant changes in soil stiffness, soil strength and lateral loads on the pile over a very short period of time during and immediately after the strong ground shaking. As illustrated in Fig. 6.1a, the excess

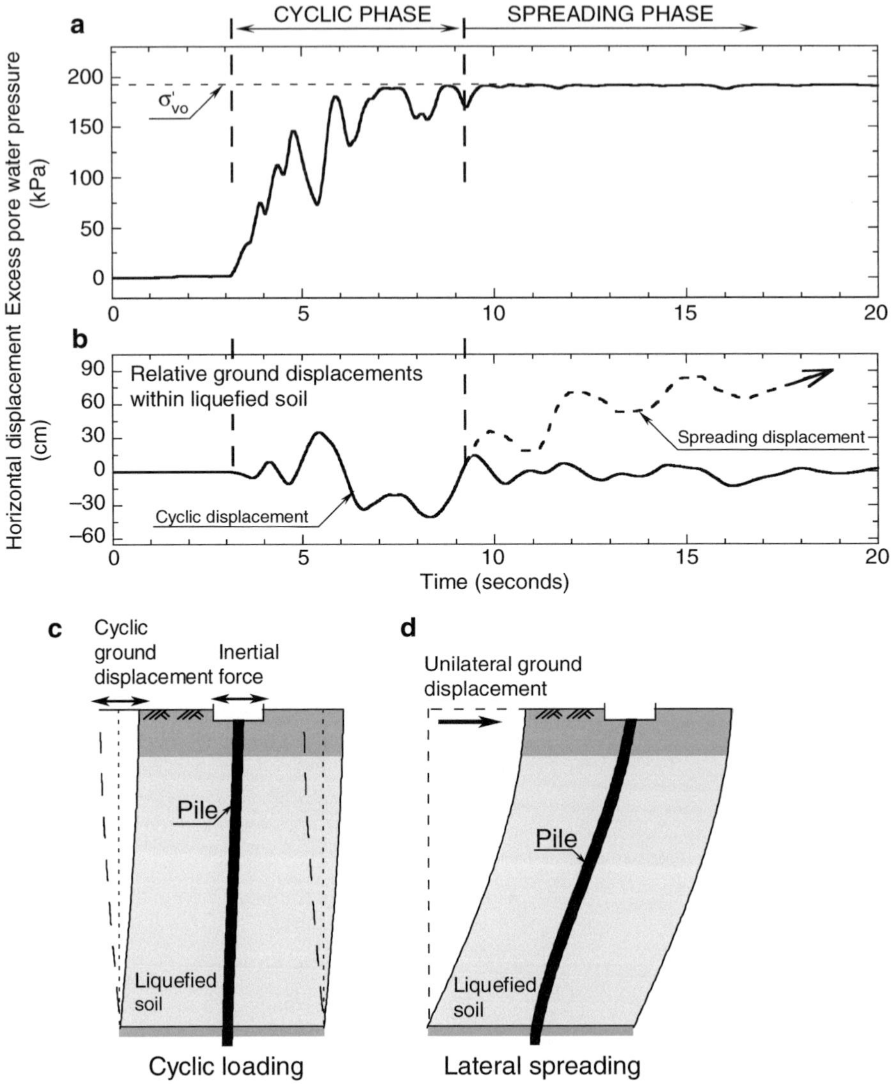


**Fig. 6.1** Illustration of ground response and soil-pile interaction in liquefying soils: (**a**) Excess pore water pressure; (**b**) Lateral ground displacement; (**c**) Loads on pile during strong ground shaking (cyclic phase) and (**d**) post-liquefaction lateral spreading

pore pressure may reach the level of the effective overburden stress in only a few seconds, and this is practically the time over which the soil stiffness reduces from its initial value to nearly zero. The intense reduction in stiffness and strength of the liquefying soil is accompanied by large lateral ground displacements, as illustrated with the solid line in Fig. 6.1b. Hence, during this phase of strong ground shaking (high accelerations) and development of liquefaction, the piles are subjected to significant kinematic loads due to lateral ground movement along with inertial loads from vibration of the superstructure (Fig. 6.1c). Both these loads are oscillatory in nature with magnitudes and spatial distribution dependent on a number of factors, including ground motion characteristics, soil density, the presence of a non-liquefied crust at the ground surface, predominant periods of the ground and superstructure, and the relative stiffness of the foundation soil and the pile, among others.

### *6.2.2 Lateral Spreading Phase*

In sloping ground or backfills behind retaining structures, liquefaction results in unilateral ground displacements or lateral spreading, as illustrated schematically with the dashed line in Fig. 6.1b. Lateral spreading typically results in large permanent ground displacements of up to several meters in the down-slope direction or towards waterways. There are many possible scenarios for the spatial and temporal distribution of lateral spreading displacements, depending on the stress-strain characteristics of soils, gravity-induced driving shear stresses and ground motion features. In general, lateral spreading may be initiated during the intense pore pressure build up and onset of liquefaction, however spreading displacements may continue well after the development of complete liquefaction and after the end of the strong shaking. One may argue that spreading is a post-liquefaction phenomenon or at least that a significant portion of the spreading displacements occurs after the foundation soils have liquefied. The spreading displacements may be one order of magnitude greater than the cyclic ground displacements, while the inertial loads during spreading are comparatively small. This is reflected in the schematic plot in Fig. 6.1d, where the inertial loads on the pile have been ignored. Thus, both the characteristics of the foundation soil and the lateral loads on piles are very different between the cyclic phase and the subsequent lateral spreading phase, and therefore, these two phases should be considered separately in the simplified pseudo-static analysis of piles.

## 6.3 Pseudo-Static Approach for Simplified Analysis

As a practical design-oriented approach, the pseudo-static analysis needs to be relatively simple, based on conventional geotechnical data and engineering concepts. In order to satisfy the objectives in the seismic performance assessment however, the pseudo-static analysis of piles also should: (a) capture the relevant deformational mechanism for piles in liquefying soils, (b) permit estimation of inelastic deformation and damage to piles, and (c) address the uncertainties associated with seismic behaviour of piles in liquefying soils. The adopted model in this study was based on this reasoning.

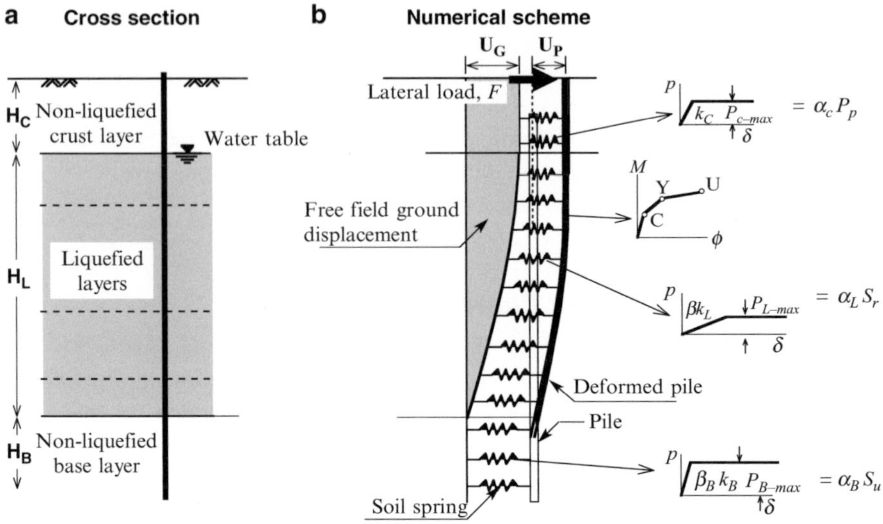


**Fig. 6.2** Beam-spring model for pseudo-static analysis of piles in liquefying soils: model parameters and characterization of nonlinear behaviour (Cubrinovski and Ishihara 2004; Cubrinovski et al. 2009a)

Although the pseudo-static analysis could in principle be applied to a pile group, it is often applied to a single-pile model. This approach was adopted in the analyses presented herein, as pile group effects introduce further complexities to the problem, beyond the scope of this study. A typical beam-spring model representing the soil-pile system in the simplified pseudo-static analysis is shown in Fig. 6.2 (Cubrinovski and Ishihara 2004; Cubrinovski et al. 2009a). The model can easily incorporate a stratified soil profile (multi-layer deposit) with liquefied layers of different thickness sandwiched between a crust of non-liquefiable soil at the ground surface and an underlying non-liquefiable base layer. Given that a key requirement of the analysis is to estimate the inelastic deformation and damage to the pile, the proposed model incorporates simple non-linear load-deformation relationships for the soil and the pile. The soil is represented by bilinear springs, the stiffness and strength of which can be degraded to account for effects of nonlinear behaviour and liquefaction. The pile is modelled using a series of beam elements with a tri-linear moment-curvature relationship. Parameters of the model are summarized in Fig. 6.2b for a typical three-layer configuration in which a liquefied layer is sandwiched between surface layer and base layer of non-liquefiable soils.

In the model, two equivalent static loads are applied to the pile: a lateral force at the pile head (*F*), representing the inertial load on the pile due to vibration of the superstructure, and a horizontal ground displacement ($U_G$) applied at the free end of the soil springs (for the liquefied layer and overlying crust), representing the kinematic load on the pile due to lateral ground movement (cyclic or spreading) in the free field. As indicated in Fig. 6.2, it has been assumed that the non-liquefied crust at the ground surface is carried along with the underlying liquefied soil and that it undergoes the same ground displacement as the top of the liquefied layer, $U_G$.

## 6.4 Model Parameters

The aim of the pseudo-static analysis is to estimate the maximum response of the pile induced by an earthquake, under the assumption that dynamic loads can be idealized as static actions. The key question in its implementation is thus how to select appropriate values for the soil stiffness, strength and lateral loads on the pile in the equivalent static analysis. In other words, what are the appropriate values for $\beta_L$, $p_{L\text{-}max}$, $U_G$ and $F$ in the model shown in Fig. 6.2? The following discussion demonstrates that this choice is not straightforward and that each of these parameters may vary within a wide range of values.

### *6.4.1 Crust Layer*

The lateral load from a crust of non-liquefied soil at the ground surface may often be the critical load for the integrity of the pile because of its potentially large magnitude and unfavourable position as a "top-heavy" load, acting above the unsupported portion of the pile embedded in liquefied soils.

The ultimate soil pressure from the surface layer per unit width of the pile can be estimated, for example, using a simplified expression such as, $p_{C\text{-}max} = \alpha_C\, p_p$, where $p_p(z)$ is the Rankine passive pressure while $\alpha_C$ is a scaling factor to account for the difference in the lateral pressure between a single pile and an equivalent wall. Figure 6.3 summarize values for $\alpha_C$ derived from experimental studies on piles which include benchmark lateral spreading experiments on full-size piles (Cubrinovski et al. 2006). In those experiments, the maximum lateral pressure on the single pile was found to be about 4.5 times the Rankine passive pressure. The very large values for the parameter $\alpha_C$ shown in Fig. 6.3 clearly indicate that excessive lateral loads can be applied from the crust layer to the pile. Notable also is a relatively significant variation in the values of $\alpha_C$ in the range between 3 and 5.

### *6.4.2 Liquefied Layer*

The stiffness degradation factor $\beta_L$, which specifies the reduction of stiffness due to liquefaction and nonlinear stress-strain behaviour ($\beta_L k_L$), is affected by a number of factors including the density of sand, excess pore pressure level, magnitude and rate of ground displacements, and drainage conditions. Typically, $\beta_L$ takes values in the range between 1/50 and 1/10 for cyclic liquefaction (Tokimatsu and Asaka 1998)

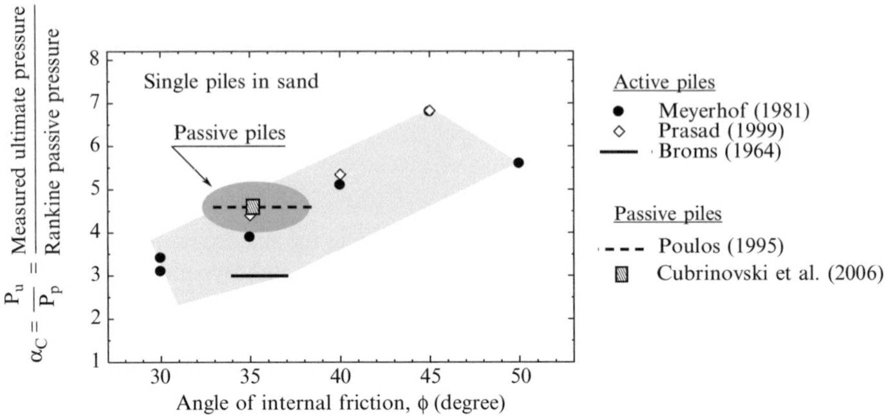


**Fig. 6.3** Ratio of ultimate pressure from the crust layer on a single pile and Rankine passive pressure obtained in experimental studies (Cubrinovski et al. 2006)

and between 1/1,000 and 1/50 in the case of lateral spreading (Yasuda and Berrill 2000; O'Rourke et al. 1994; Cubrinovski et al. 2006). In general, the value of $\beta_L$ should be related to the soil properties and anticipated ground deformation. For example, lower values of $\beta_L$ are expected for very loose soils because such soils are commonly associated with high and sustained excess pore water pressures and large ground deformation. While this sort of qualitative evaluation of $\beta_L$ should be considered, the quantification of the value for this parameter is very difficult and subjective because of the inherent uncertainties associated with properties of liquefying soils.

The residual strength of liquefied soils $S_r$ could be used in the evaluation of the ultimate pressure from the liquefied soil on the pile, e.g., $p_{L\text{-}max} = \alpha_L S_r$. Here, the residual strength $S_r$ can be estimated using an empirical correlation between the residual strength of liquefied soils and SPT blow count, such as that proposed by Seed and Harder (1991) or Olson and Stark (2002). Note that the former correlation assumes a constant residual strength for a given blow count while the latter one uses a normalized residual strength of the form $(S_r/\sigma'_{vo})$ and hence implicitly assumes that $S_r$ increases with depth for a given SPT blow count. Effects of strength normalization on the pile response are beyond the scope of this paper, but detailed analysis of these effects may be found in Cubrinovski et al. (2009b).

The shaded area in Fig. 6.4 shows the correlation between $S_r$ and the normalized SPT blow count for clean sand $(N_1)_{60cs}$ proposed by Seed and Harder (1991). A large scatter exists in this correlation indicating significant uncertainty in the value of $S_r$ for a given $(N_1)_{60cs}$ value. For example, for $(N_1)_{60cs} = 10$, the value of $S_r$ can be anywhere between 5 kPa (lower bound value) and 25 kPa (upper bound value). In addition, the multiplier $\alpha_L$ ($p_{L\text{-}max} = \alpha_L S_r$) is also unknown and subject to significant uncertainties. Note that $\alpha_L$ is different from the corresponding parameter $\alpha_C$ for the crust layer previously discussed, because the interaction and mobilization of pressure from surrounding soils on the pile is different for liquefied and non-liquefied soils.

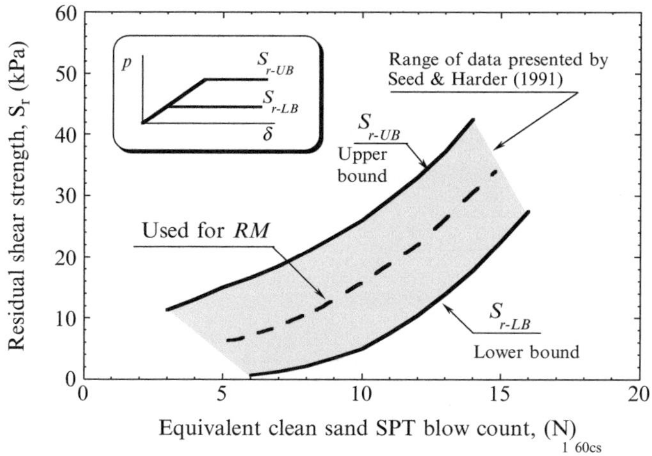


**Fig. 6.4** Residual shear strength of liquefied soils (After Seed and Harder 1991)

### *6.4.3 Other Parameters*

As indicated in Fig. 6.2, there are a number of other parameters of the soil-pile model that may influence the response of the soil-pile system, such as the parameters of the base layer or the initial (non-degraded) stiffness of soil springs. The latter is represented in the model by the subgrade reaction coefficient ($k_i$). A flow chart showing the various soil properties and relationships that are used to determine the model parameters for the soil springs is shown in Fig. 6.5. Note that both stiffness and strength properties in the model are derived based on the SPT blow count using conventional approaches (Rankine passive pressure theory, subgrade reaction coefficient) and empirical expressions, hence the selection of an appropriate representative blow count could be critical in the evaluation of the pile response.

### *6.4.4 Equivalent Static Loads*

The selection of appropriate equivalent static loads is probably the most difficult task in the pseudo-static analysis. This is because both input loads in the pseudo-static analysis ($U_G$ and $F$) are, in effect, estimates of the seismic response of the free field ground and superstructure respectively. Simplified methods for the evaluation of lateral displacements of liquefied and laterally spreading soils (e.g., Tokimatsu and Asaka 1998; Youd et al. 2002) and guidance how to combine kinematic and inertial loads on piles in liquefying soils (Boulanger et al. 2007; Tokimatsu and Suzuki 2009) are now available, however they all imply significant uncertainties in these loads.

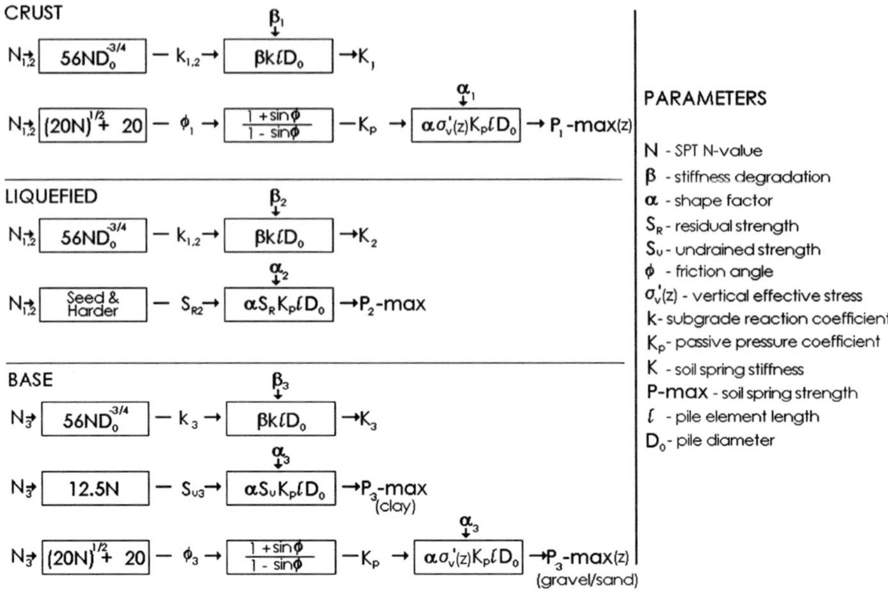


**Fig. 6.5** Flow chart illustrating the determination of soil spring parameters in the adopted model

### *6.4.5 Implementation of PSA*

The significant uncertainties associated with the model parameters in the pseudo-static analysis need to be considered in the assessment of the pile response. One of the first questions that needs to be answered is not 'what is the most appropriate value for a given model parameter', but rather 'how big is the effect of variation (uncertainty) in a given model parameter on the predicted pile response'. This will allow the user to focus their attention on critical uncertainties in the analysis and develop a suitable strategy for a robust implementation of the simplified pseudo-static analysis.

## 6.5 Concept of the Sensitivity Study

To investigate the sensitivity of the pile response to various model parameters and hence identify critical uncertainties in the pseudo-static analysis, a comprehensive parametric study has been conducted in which a wide range of soil-pile systems, loading conditions and values for model parameters were considered (Haskell 2009). The objectives of the analyses were to comparatively examine and quantify the sensitivity of the pile response to various model parameters, and to establish a

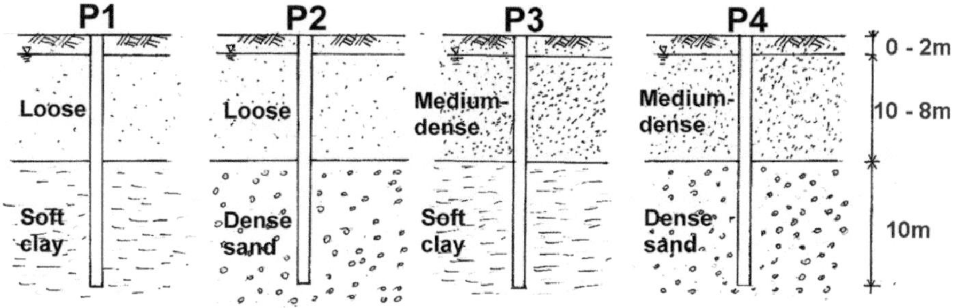


**Fig. 6.6** Soil profiles adopted in the parametric studies

**Table 6.2** Tri-linear moment-curvature relationships of piles used in the parametric studies

| | Units | S-Pile | M-Pile | F-Pile |
|---|---|---|---|---|
| $D$ | [mm] | 1,200 | 800 | 400 |
| $M_C$ | [kN-m] | 959 | 650 | 80.4 |
| $M_Y$ | [kN-m] | 1,970 | 1,240 | 126 |
| $M_U$ | [kN-m] | 2,470 | 1,420 | 133 |
| $\phi_C$ | [m$^{-1}$] | 0.00041 | 0.00109 | 0.00190 |
| $\phi_Y$ | [m$^{-1}$] | 0.00251 | 0.00609 | 0.01160 |
| $\phi_U$ | [m$^{-1}$] | 0.01040 | 0.01169 | 0.02450 |

fundamental link between the sensitivity of the response and the mechanism of soil-pile interaction. In this way, general principles for conducting pseudo-static analysis of piles in liquefying soils could be established irrespective of the specific properties of the soil-pile system or loading conditions.

Four different soil profiles, shown in Fig. 6.6 were adopted for the parametric study. Each profile essentially consists of two layers, a 10 m thick liquefiable sand layer overlying a 10 m thick non-liquefiable base layer. A loose sand with an SPT blow count of $N=5$ was adopted for the liquefiable soil in profiles *P1* and *P2*, whereas a medium-dense sand with $N=15$ was used for profiles *P3* and *P4*. Similarly, a soft clay base layer was adopted for profiles *P1* and *P3*, whereas a dense sand base layer was used in profiles *P2* and *P4*. The soil above the water table was assumed to act as a non-liquefiable crust at the ground surface, and five different scenarios were adopted for the location of the water table between $z=0$ and 2 m depth defining a crust with thickness of $H_C=0$, 0.5, 1.0, 1.5 and 2.0 m respectively. For all cases, three different piles were considered, with diameters of 400, 800 and 1,200 mm respectively, and tri-linear $M$-$\varphi$ relationships as summarized in Table 6.2.

As discussed earlier, the characteristics of liquefying soils and demands on piles are significantly different during the cyclic phase and lateral spreading phase of the response. For this reason and for clarity of the argument, three separate series of analyses were conducted, covering the following loading conditions: (1) Lateral Spreading Scenario, (2) Hypothetical Cyclic Scenario (without inertial force demand),

and (3) Cyclic Scenario. Even though the actual sequence of the phenomena has been practically reversed and the second scenario is purely hypothetical in nature (based on an unrealistic assumption), the results of the analyses will be presented in the abovementioned order because it will help the clarity of the presentation and allow for gradual introduction of complexities in the interpretation of results.

## 6.6 Lateral Spreading Pseudo-Static Analyses

As described earlier, there are significant uncertainties in the pseudo-static analysis (PSA) around the selection of appropriate values for model parameters. Thus, for a given soil-pile system and loading conditions, a relatively wide range of values could reasonably be used in the PSA for each parameter. Table 6.3 summarizes the ranges of values for all parameters of the pseudo-static model, for the four adopted soil profiles. The ranges of values for different parameters have been defined on the basis of well-documented case histories, evidence from experimental studies, and numerical analyses. Discussion on this can be found in Cubrinovski et al. 2006; Cubrinovski and Ishihara (2007) and Haskell (2009). Three values are listed in the table for each model parameter: a lower bound or minimum acceptable value (LB), a reference or 'mid-range' value, and an upper bound or maximum acceptable value (UB).

For each soil-pile system (e.g. *S-P1,* S-Pile in soil profile P1), an analysis was first conducted using the reference values for all model parameters (*RM* = Reference Model), thus establishing a reference pile response. Next, sensitivity analyses were carried out considering one model parameter at a time. For example, when examining the sensitivity of the pile response on $\beta_L$, an analysis was conducted in which the lower bound value of $\beta_L = \beta_{L\text{-}LB} = 0.001$ was used while all other parameters were set at their reference values. Another analysis was then conducted using instead the upper bound value for $\beta_L = \beta_{L\text{-}UB} = 0.02$, with all other parameters again being set at their reference values. From these two analyses, the sensitivity of the pile response to the $\beta_L$-value could be examined. The same procedure was followed and systematically applied to establish the sensitivity of the pile response to each of the model parameters.

In the lateral spreading analyses, seven different loading conditions (magnitudes of ground displacements) were applied, i.e. $U_G = 0.1$, 0.25, 0.50, 0.75, 1.0, 1.5 and 2.0 m respectively. Thus, for each case (soil-pile system and model parameter), the sensitivity was evaluated for seven different load levels. This approach was taken because the sensitivity of the response was expected to depend on the induced mechanism of soil-pile interaction which, in turn, depends on the induced pile response (displacement) and hence applied load to the pile (ground displacement).

**Table 6.3** Parameter variations considered in the sensitivity analyses

| | | Profile – P1 | | | Profile – P22 | | | Profile – P3 | | | Profile – P44 | | |
|---|---|---|---|---|---|---|---|---|---|---|---|---|---|
| | Units | LB | Ref | UB | LB | Ref | UB | LB | Ref | UB | LB | Ref | UB |
| $\beta_C$ | – | 0.3 | 1 | 1 | 0.3 | 1 | 1 | 0.3 | 1 | 1 | 0.3 | 1 | 1 |
| $\alpha_C$ | – | 3 | 4.5 | 5 | 3 | 4.5 | 5 | 3 | 4.5 | 5 | 3 | 4.5 | 5 |
| $\Phi_{C,L}$ | ° | 27 | 30 | 33 | 27 | 30 | 33 | 34 | 37 | 40 | 34 | 37 | 40 |
| $k_{C,L}$ | $MNm^{-3}$ | $168D^{-3/4}$ | $280D^{-3/4}$ | $392D^{-3/4}$ | $168D^{-3/4}$ | $280D^{-3/4}$ | $392D^{-3/4}$ | $504D^{-3/4}$ | $840D^{-3/4}$ | $1176D^{-3/4}$ | $504D^{-3/4}$ | $840D^{-3/4}$ | $1176D^{-3/4}$ |
| $N_{C,L}$ | – | 2 | 5 | 8 | 2 | 5 | 8 | 11 | 15 | 19 | 11 | 15 | 19 |
| $\beta_L$ | – | 0.001 | 0.01 | 0.02 | 0.001 | 0.01 | 0.02 | 0.001 | 0.01 | 0.02 | 0.001 | 0.01 | 0.02 |
| $\alpha_L$ | – | 1 | 3 | 6 | 1 | 3 | 6 | 1 | 3 | 6 | 1 | 3 | 6 |
| $S_r$ | kPa | 1 | 7 | 15 | 1 | 7 | 15 | 25 | 34 | 44 | 25 | 34 | 44 |
| $\beta_B$ | – | 0.3 | 1 | 1 | 0.3 | 1 | 1 | 0.3 | 1 | 1 | 0.3 | 1 | 1 |
| $k_B$ | $MNm^{-3}$ | $168D^{-3/4}$ | $280D^{-3/4}$ | $392D^{-3/4}$ | $840D^{-3/4}$ | $1400D^{-3/4}$ | $1960D^{-3/4}$ | $168D^{-3/4}$ | $280D^{-3/4}$ | $392D^{-3/4}$ | $840D^{-3/4}$ | $1400D^{-3/4}$ | $1960D^{-3/4}$ |
| $\alpha_B$ | – | 5 | 9 | 9 | 3 | 4.5 | 5 | 5 | 9 | 9 | 3 | 4.5 | 5 |
| $Su_B$ | kPa | 25 | 33 | 42 | 432 | 496 | 573 | 25 | 33 | 42 | 432 | 496 | 573 |
| $N_B$ | – | 2 | 5 | 8 | 20 | 25 | 30 | 2 | 5 | 8 | 20 | 25 | 30 |

In the above empirical terms, *D is given in cm*

### 6.6.1 Parametric Sensitivities

Results from the sensitivity analyses for *S-P1* (*S-Pile* in soil profile *P1*) with a crust of $H_C = 1.5$ m are presented in Fig. 6.7 In the form of 'tornado charts'. Here, the pile response is presented using the peak curvature (left-hand side plots in Fig. 6.7) and pile head displacement (right-hand side plots in Fig. 6.7) computed in PSA. The former illustrates the level of damage to the pile while the latter provides information on the relative displacement between the pile and the soil, and thus the respective soil-pile interaction mechanism. For example, the red bold line in Fig. 6.7 indicates that the pile displacement computed using the reference model parameters (RM) was $U_P = 0.21$ m in the analysis in which a lateral ground displacement of $U_G = 1.0$ m was applied. Hence, the relative displacement between the soil and the pile was $\delta = U_G - U_P = 0.79$ m at the pile head, indicating that the soil springs in the liquefied and crust layers yielded. This is illustrated schematically with the sketches in Fig. 6.7 where red indicates yielding in the soil and shows the depth along which the ultimate soil pressure was mobilized and applied to the pile.

Each bar in the tornado charts indicates the sensitivity of the pile response to the variation of a specific parameter. The left end of the bars shows the response computed when a lower bound value was used for a parameter of the liquefied or crust layer, while the right end of the bar was obtained in a respective analysis with an upper-bound value for the parameter. Note that the reverse is true for the base layer where the maximum response (right end of the bar) was obtained when lower bound values were used for the base layer. This simply reflects the fact that the liquefied layer and the crust provide driving forces to the pile deformation whereas the base layer provides a resisting force.

Note that in the tornado charts, model parameters for the surface layer, liquefied layer and base layer are ordered respectively from top to bottom for clearer illustration of the sensitivity of the pile response with regard to a particular soil layer. There are several key findings from the results presented in Fig. 6.7:

- The bars for parameters of the base layer ($\beta_B$, $k_B$, $\alpha_B$ and $S_{u\text{-}B}$) are relatively small in size showing that the pile response is less sensitive to the variation in the parameters of the base layer as compared to those of the crust or liquefied layer. Hence, when using PSA, parametric studies should focus on the parameters of the liquefied soil and non-liquefied layer at the ground surface.
- For small relative displacements $|U_G - U_P|$, the pile response is most sensitive to parameters affecting the stiffness of the soil ($\beta_L$). As the relative displacement increases and yielding in the soil occurs, the response becomes more sensitive to the strength of the soil (parameters $\alpha_L$ and $S_r$). This transition from stiffness to strength controlled pile response is schematically illustrated in Fig. 6.8 where the change in the size of the bar in the tornado chart is clearly related to the mobilized load (deformation) of the soil spring relative to the yield level.
- The response is sensitive to the SPT blow count reflecting its concurrent use for computing both the soil stiffness and the soil strength parameters.

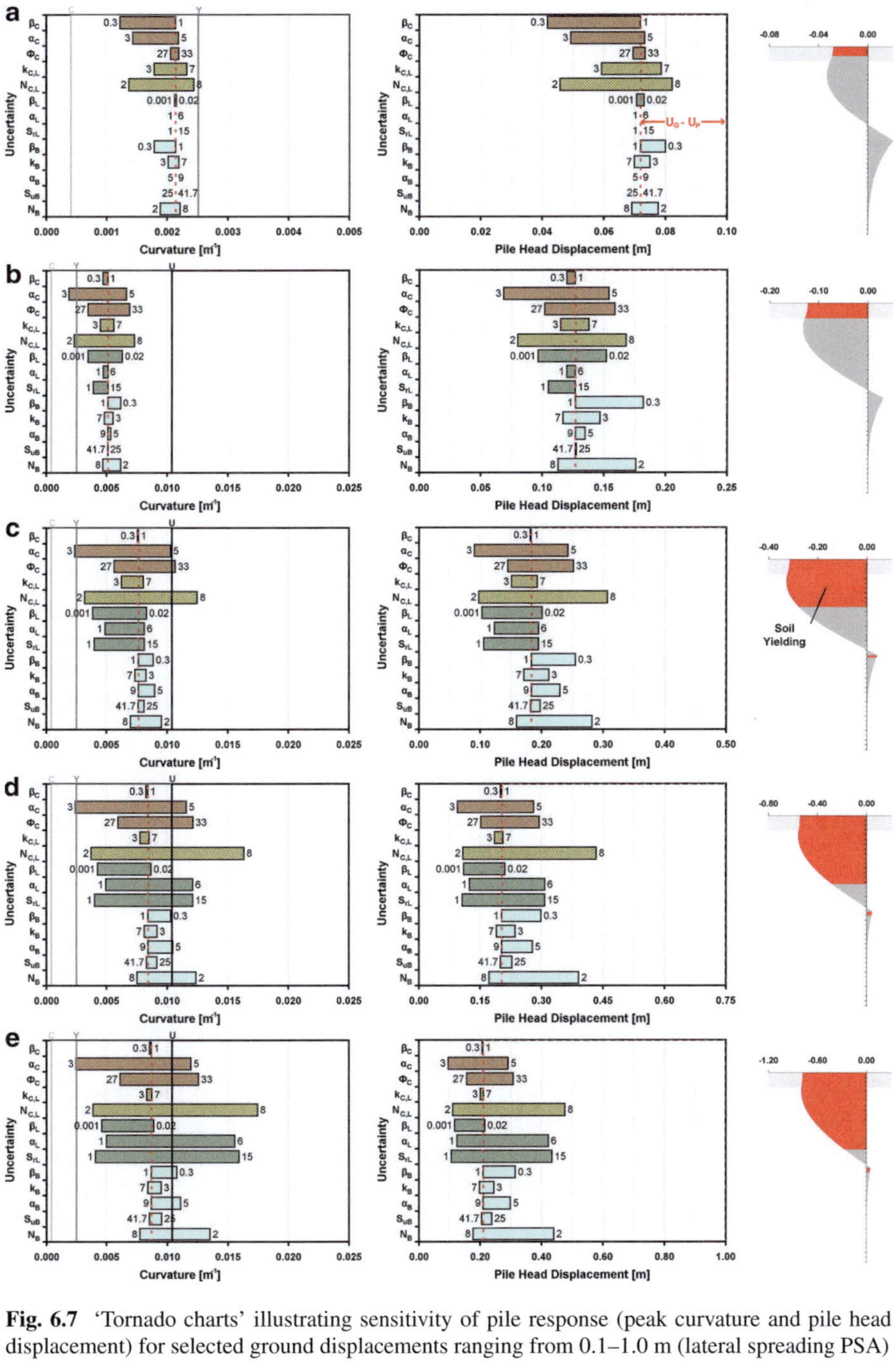


**Fig. 6.7** 'Tornado charts' illustrating sensitivity of pile response (peak curvature and pile head displacement) for selected ground displacements ranging from 0.1–1.0 m (lateral spreading PSA)

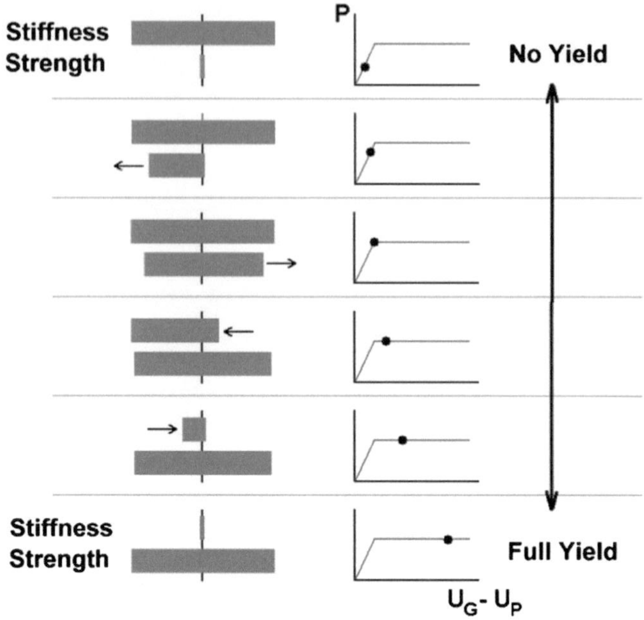


**Fig. 6.8** Conceptual illustration of transition from stiffness to strength controlled pile response and its relation to the mobilized load (deformation) of the soil spring relative to the yield level

Note that for this soil-pile system (*S-P1*), the displacement of the pile head was always less than the ground displacement, resulting in so-called stiff-pile behaviour in which the pile resists the lateral spreading displacements of the liquefied soil and non-liquefied crust.

### *6.6.2 Parameters with Relatively Small Influence on Pile Response*

As discussed earlier, the pile response is relatively insensitive to the properties of the base layer. The stiffness degradation of the crust soil, $\beta_C$, is also not considered a critical uncertainty as it only affects the pile response at very small (and not so relevant) ground displacements. Similarly, the uncertainties associated with the subgrade reaction coefficients ($k_i$) arising from the empirical relationship based on SPT blow counts generally have negligible effects on the pile response.

### *6.6.3 Critical Uncertainties*

It has already been noted that changes in the SPT blow counts of all soil layers have a significant influence on the pile response. This highlights the need for careful

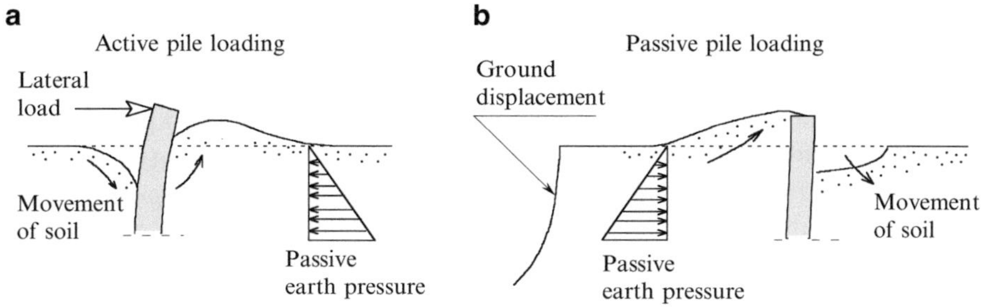


**Fig. 6.9** Schematic illustration of lateral loading of piles: (**a**) Active pile loading; (**b**) Passive pile loading

selection of the 'representative blow count' for each soil layer and the need for assessment of the effects of uncertainties in the SPT blow count on the pile response. Unlike the other model parameters, the SPT blow count simultaneously affects multiple soil properties (stiffness and strength of soils, as illustrated in Fig. 6.5), and therefore, it shows a relatively large influence on the pile response across various response levels and applied ground displacements.

The crust layer shape factor $\alpha_C$ is a key uncertainty because it affects the pile response over a wide range of ground displacements when varied between its lower and upper bounds of $\alpha_C=3$ and 5, as shown in Fig. 6.7. Note that in many guidelines and pseudo-static analysis procedures a value of $\alpha_C=3$ has been adopted, which is unconservative for piles in liquefying or laterally spreading soils. The value of $\alpha_C=3$ has been adopted based on the study of Broms (1964), in which active-pile-loading was considered. As illustrated in Fig. 6.9a, the crust layer resists the pile deformation (under active-pile-loading), hence use of a lower-bound value for $\alpha_C$ ($\alpha_C=\alpha_{C\text{-}LB}=3$) would be conservative for this loading condition. However, piles in liquefying and laterally spreading soils predominantly undergo passive-pile-loading in which the crust provides a driving force for the pile deformation (Fig. 6.9b). An upper-bound value for $\alpha_C$ ($\alpha_C=\alpha_{C\text{-}UB}$) would thus be the conservative choice for this loading mechanism. Recent experimental evidence from full-size tests on piles suggests that a value of $\alpha_C=4.5$ might be more appropriate for use in pseudo-static analysis of piles in liquefying or spreading soils (Cubrinovski et al. 2006).

There are inherent and very significant uncertainties associated with soil liquefaction during earthquakes and soil-pile interaction in liquefying soils. The extent of these uncertainties is reflected in the very wide ranges of values for the parameters of the liquefied soil in the pseudo-static model ($\beta_L$, $\alpha_L$, $S_r$), including the effects of significant scatter in the empirical relationships used for their evaluation. It was noted in the analysis of the results presented in Fig. 6.7 and schematic illustration in Fig. 6.8 that there is a clear link between the sensitivity in the pile response (or influence level of a given parameter) and the mechanism of soil-pile deformation. Thus, model parameters affecting the soil stiffness more strongly influence the pile response when the relative displacement between the soil and the pile $|U_G - U_P|$ is smaller than the yield displacement of the soil, $\Delta y$. Conversely, model parameters affecting soil strength (ultimate pressure from the soil on the pile) more strongly influence the pile response after soil yielding has been initiated or when $|U_G - U_P| > \Delta y$. In order to examine the relative importance of the parameters of the liquefied layer, results from the pseudo-static analyses were plotted in $\Delta\varphi/\varphi_{ref}$ against $(U_G - U_P)/\Delta y$ diagrams in Fig. 6.10a, b for $\beta_L$ and, $\alpha_L$ and $S_r$ respectively. Note that $\alpha_L$ and $S_r$ affect the strength of the soil (ultimate pressure on the pile) and therefore the results for both parameters were plotted together in Fig. 6.10b. Here, $\Delta\varphi = |\varphi_{UB} - \varphi_{LB}|$ represents the difference between the peak curvatures computed in analyses using the upper and lower bound values for a given parameter (essentially the size of the bar in the tornado charts), while $\varphi_{ref}$ is the curvature computed in the analysis using reference model parameters (RM). Hence, the ratio $\Delta\varphi/\varphi_{ref}$ is a measure for the sensitivity in the pile response (curvature), and the plots indicate how the sensitivity changes with the yield ratio $(U_G - U_P)/\Delta y$. Note that $(U_G - U_P)/\Delta y = 1.0$ indicates onset of yielding in the liquefied soil, at the pile head.

Figure 6.10a shows the sensitivity of the pile response to variations in the stiffness degradation factor $\beta_L$ for soil profiles with a 1.5 m thick crust, for all soil-pile systems considered (piles *S*, *M* or *F* embedded in profiles *P1*, *P2, P3* or *P4*). Here, different piles are represented by symbols of different colour, and the symbols shape indicates the soil profile. A very well defined relationship is seen irrespective of the wide range of soil and pile properties considered, thus clearly demonstrating the link between the influence levels of $\beta_L$ on the pile response (sensitivity) and loading (deformation) mechanism. Noting that a yield ratio of 1.0 corresponds to yielding of the soil (at the pile head at least), the plot shows that $\beta_L$ is most important (most strongly affects the pile response) prior to soil yielding, when soil stiffness 'controls' the pile response. The sensitivity of the response to variations in $\beta_L$ gradually decreases beyond yield ratios of 1.0, as more of the liquefied soil yields.

In reference to Fig. 6.10b, the results for $\alpha_L$ and $S_r$ also define a good relationship showing almost a linear increase in the influence of strength parameters ($\alpha_L$ and $S_r$) on the pile response with the yield ratio until a nearly stable maximum level has been reached at a yield ratio of about 3.0. Comparing Fig. 6.10a, b, it is apparent that the sensitivity of the pile response is greater for $\beta_L$ than for $\alpha_L$ or $S_r$ when $|U_G - U_P|/\Delta y \leq 1.0$, whereas the reverse is true for $|U_G - U_P|/\Delta y > 2.0$. For $|U_G - U_P|/\Delta y > 3.0$, the sensitivity of the pile response on $\alpha_L$ or $S_r$ is 3–4 times that of $\beta_L$.

The same results are re-plotted in Fig. 6.11a, b but now together with another set of results from analyses for profiles having no crust of non-liquefied soil at the ground surface, $H_C = 0$ m. These plots thus depict how the sensitivity of the pile response to $\beta_L$ (Fig. 6.11a), and $\alpha_L$ and $S_r$ (Fig. 6.11b) is affected by the presence and thickness of the crust. Clearly, the sensitivity of the pile response to the parameters of the liquefied layer ($\beta_L$, $\alpha_L$ and $S_r$) decreases with an increasing thickness of the crust. However, the trends previously established in relation to the stiffness and strength 'controlled' (dominated) response and mechanism of soil-pile deformation remain unchanged.

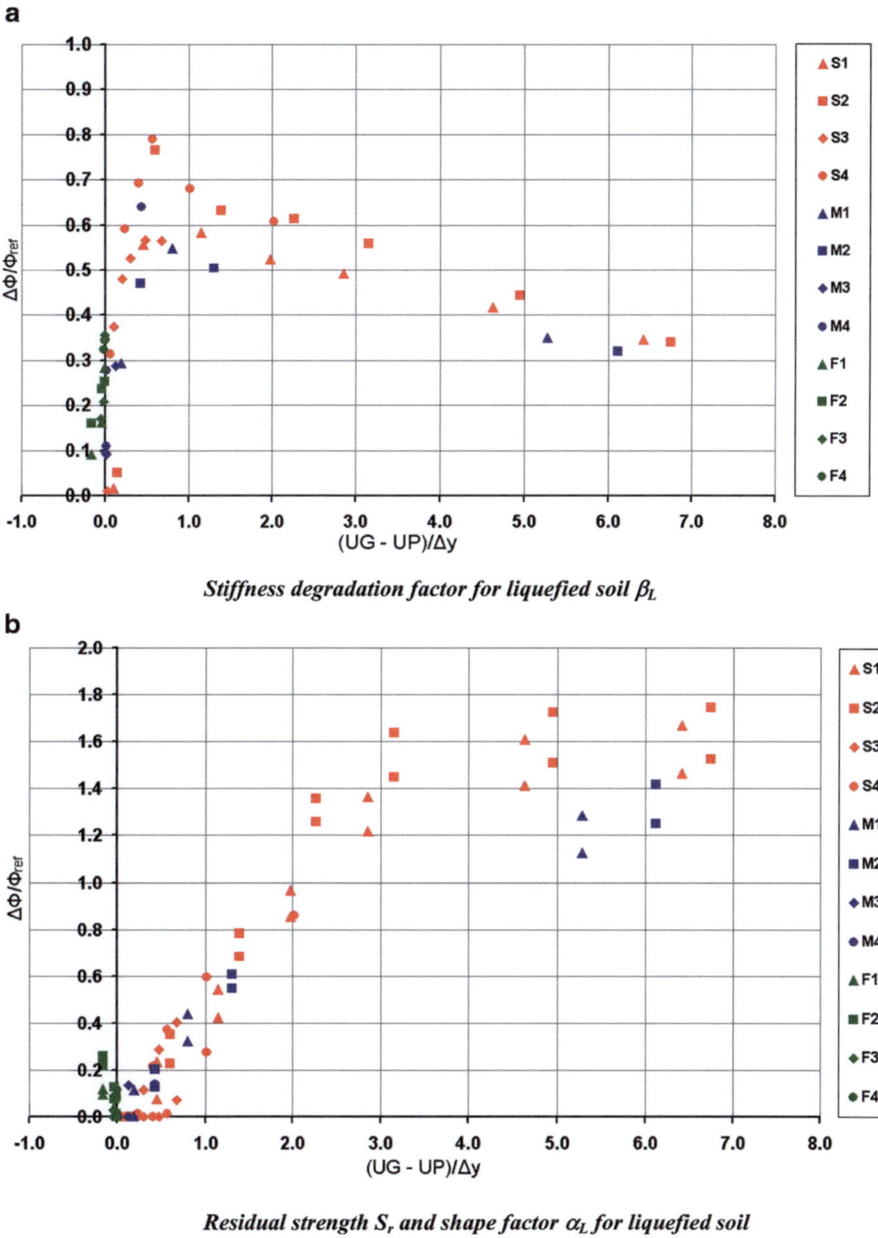


**Fig. 6.10** Sensitivity of pile response (peak pile curvature) to parameters of liquefied soil for soil profiles with a crust layer of $H_C = 1.5$ m in lateral spreading PSA

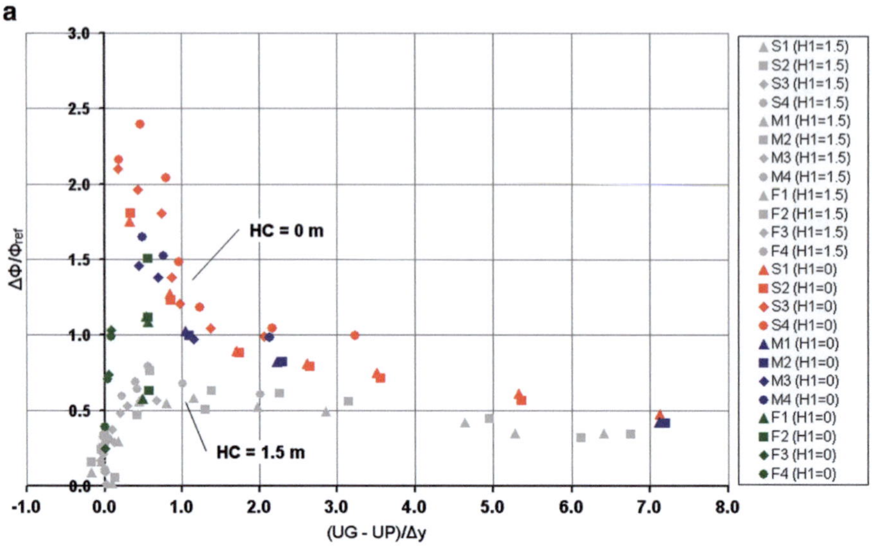


***Stiffness degradation factor of liquefied soil $\beta_L$***

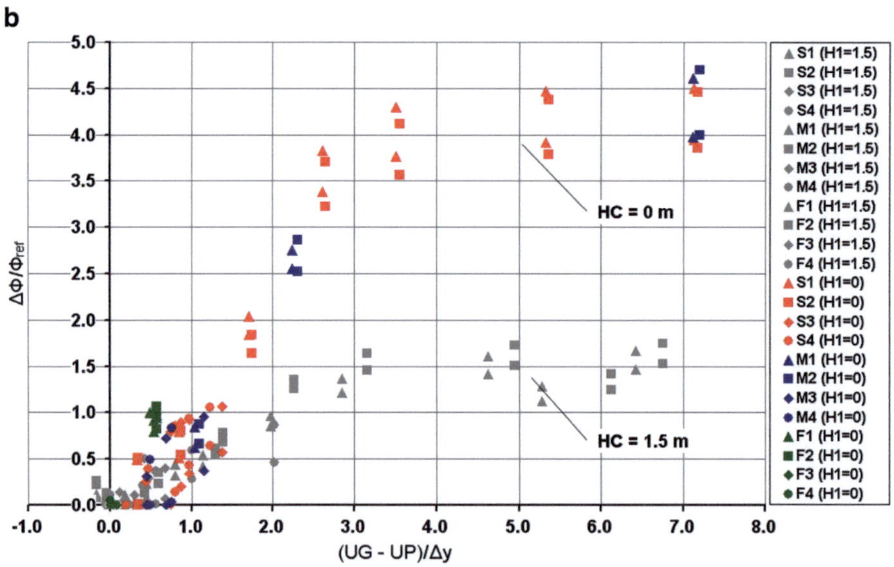


***Residual strength $S_r$ and shape factor $\alpha_L$ for liquefied soil***

**Fig. 6.11** Effect of crust thickness on sensitivity of pile response (peak pile curvature) to parameters of liquefied soil, in lateral spreading PSA

## 6.7 Hypothetical Cyclic Scenario

Using the same soil-pile systems (piles *S, M* or *F* embedded in soil profiles *P1, P2, P3* or *P4*) another series of PSA was carried out simulating a hypothetical cyclic scenario. In these analyses, model parameters and loading conditions considered were representative of the cyclic phase of the pile response during development and onset of liquefaction, except that inertial loads on the pile from a superstructure were ignored. In essence, these analyses were used to verify whether or not the findings obtained from the lateral spreading analyses are applicable to PSA evaluating the cyclic phase of the response. In the hypothetical cyclic scenario analyses, lower and upper bounds of $\beta_L=0.02$ and $\beta_L=0.10$ were adopted for the stiffness degradation factor for the liquefied soil, along with a reference value of $\beta_{L\text{-}ref}=0.05$. Note that this degradation in stiffness is much smaller $\beta_L$ than that adopted for lateral spreading reflecting the differences in excess pore pressures, and extent and severity of liquefaction between the cyclic phase and lateral spreading phase of the response. In the cyclic scenario PSA, ground displacements of 0.1, 0.2, 0.3 and 0.4 m were applied to the pile implying average shear stresses in the liquefied soil in the range between 1% and 4%.

Results from the hypothetical cyclic scenario PSA are presented in Fig. 6.12 for $\beta_L$, $\alpha_L$ and $S_r$, and two crust thicknesses, $H_C=0$ and 1.5 m respectively. By and large, the trends observed for all parameters with regard to the sensitivity, deformation mechanism and crust thickness effects were the same as those reported for lateral spreading. The relationship for $\alpha_L$ and $S_r$ shown in Fig. 6.12 (for $H_C=1.5$ m) is very similar to the corresponding relationship for lateral spreading shown in Fig. 6.10b. For the stiffness parameter $\beta_L$, on the other hand, the sensitivity of the response in the PSA simulating the cyclic phase was less than that observed in lateral spreading PSA, and was particularly small for the case with $H_C=1.5$ m across the whole range of response.

## 6.8 Cyclic Phase PSA: Combined Kinematic and Inertial Force Demands

One of the key issues in the PSA used for evaluating the peak response of the pile during the cyclic phase is how to combine the kinematic loads due to lateral ground displacements and the inertial load on the pile caused by vibration of the superstructure. The peak cyclic ground displacement and superstructure inertial force are transient conditions occurring momentarily during the course of strong shaking. They may or may not occur at the same instant, hence there is no clear and simple strategy how to combine these loads in the PSA. It has been suggested that the phasing of the kinematic and inertial demands varies, and depends primarily on the natural frequency of the superstructure and soil deposit (Tamura and Tokimatsu 2005).

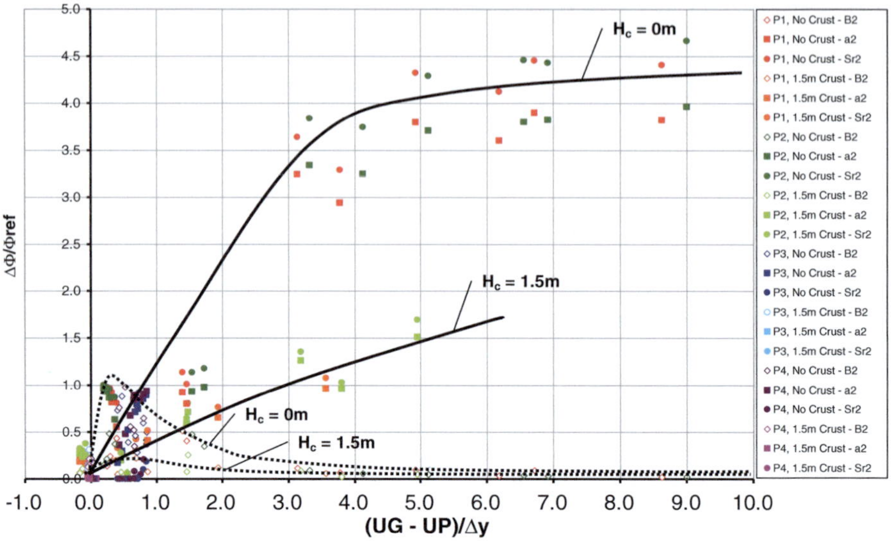


**Fig. 6.12** Sensitivity of pile response (peak pile curvature) to parameters of liquefied soil, for hypothetical cyclic scenario (ignored inertial loads) including effects of crust thickness ($H_C = 0$ m and $H_C = 1.5$ m); solid lines indicate sensitivity to $\alpha_L$ and $S_r$, while dashed lines are for $\beta_L$

Recently, Boulanger et al. (2007) suggested a simplified expression allowing for different combinations of kinematic and inertial loads on the pile while accounting for the period of the ground motion. As commonly acknowledged for pseudo-static approaches for analysis of seismic problems, the load combination producing the critical (peak) pile response in liquefying soils cannot be predicted with any high degree of certainty. The aim of the presented analyses herein was not to determine how best to approximate the critical pseudo-static demand on the pile, but rather to investigate the influence of this modelling decision (kinematic-inertial load combination) on the predicted pile response.

In this series of analyses, a horizontal force acting at the pile head representing the load on the pile from the superstructure was applied in addition to the lateral ground displacement. In total, 20 load combinations were considered for each soil-pile system: five inertial loads (lateral force at pile head) corresponding to horizontal accelerations of 0.1, 0.2, 0.3, 0.4 and 0.5 g, and four lateral ground displacements of 0.1, 0.2, 0.3 and 0.4 m at the ground surface. Depending on the base layer (soft clay or dense sand), different axial capacities were adopted for the piles resulting in substantially different inertial loads. For example, the inertial loads for case *S-P1* were in the range between 75 and 375 kN (for 0.1 and 0.5 g respectively) while the respective inertial loads for Case *S-P2* were 300 and 1,500 kN.

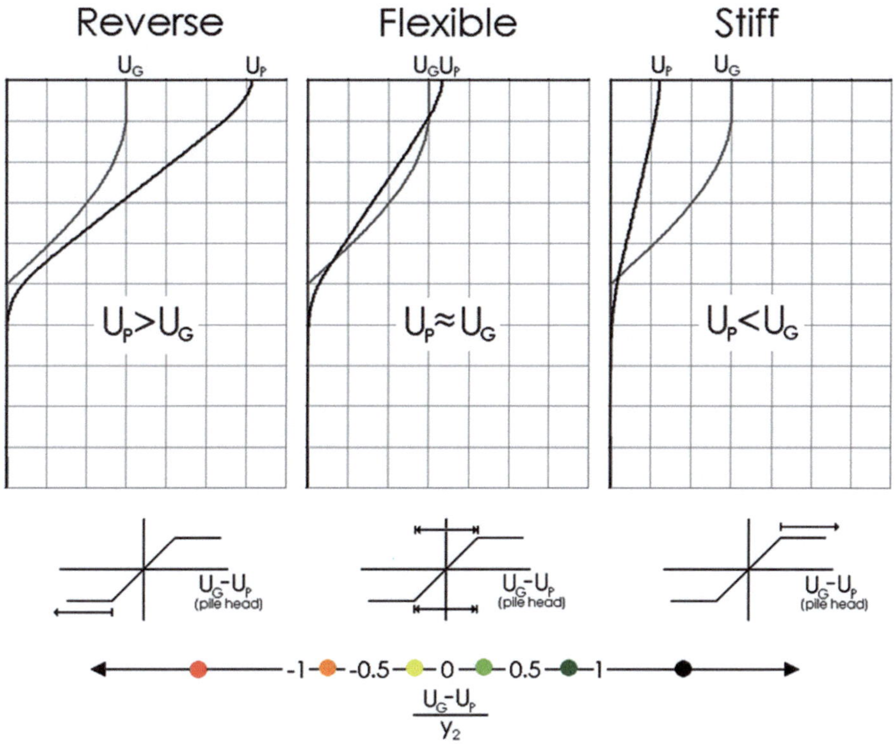


**Fig. 6.13** Schematic illustration of 'reverse', 'flexible-pile-behaviour' and 'stiff-pile-behaviour' based on relative displacements between the soil and the pile

## 6.8.1 *Parametric Sensitivities*

The introduction of an inertial force at the pile head, in addition to the kinematic soil demands adds another dimension to the already complex problem. For a given scenario, this force may change the fundamental mechanism of soil-pile interaction, increase the severity of the damage suffered by the pile, and alter the influence other parameters have on the predicted pile response. Detailed discussion on the combined kinematic and inertial effects on the pile response may be found in Haskell (2009). Herein the influence of the inertial force on the sensitivity of the pile response to parameters of the liquefied soil is examined in a fashion similar to that presented in the preceding sections. To simplify the problem, all cases that resulted in an unrealistic response (e.g. pile displacements significantly greater than the cyclic ground displacement) or unacceptable level of pile damage (well in excess of the ultimate level, $\varphi_u$) were not considered. In other words, in reference to the induced level of pile displacement and consequent soil-pile interaction mechanism, only the mechanism of 'stiff-pile-behaviour' illustrated in Fig. 6.13 where $U_G > U_P$

**Table 6.4** Acceleration levels associated with 'stiff-pile-behaviour' in cyclic PSA with combined kinematic and inertial force demands

| Crust Thickness, $H_C$ (m) | *S-Pile* | | | | *M-Pile* | | | | *F-Pile* | | | |
|---|---|---|---|---|---|---|---|---|---|---|---|---|
| | *S-P1* | *S-P2* | *S-P3* | *S-P4* | *M-P1* | *M-P2* | *M-P3* | *M-P4* | *F-P1* | *F-P2* | *F-P3* | *F-P4* |
| 0.0 | ≤0.5 g | ≤0.2 g[a] | ≤0.3 g | – | ≤0.5 g | ≤0.2 g | ≤0.3 g | – | – | – | – | – |
| 1.5 | – | ≤0.2 g | – | – | – | ≤0.2 g | – | – | – | – | – | – |

[a] "Stiff-pile-behaviour" was obtained for inertial loads corresponding to accelerations of less then 0.2 g

and $(U_G - U_P) > \Delta y$ is associated with sensitivity of the pile response to variation in the parameters of the liquefied soil. Note that, in the cyclic PSA with combined kinematic and inertial loads, the 'stiff-pile-behaviour' mechanism and hence parametric sensitivity was observed for only few of the examined soil-pile systems and loading conditions, as summarized in Table 6.4.

Results from analyses of the soil-pile system *S-P1* are shown in Fig. 6.14a, b for the parameters $\beta_L$ and $\alpha_L$ respectively. The application and increase in the inertial force reduces the relative soil-pile displacement making the response more flexible and hence decreasing the yield ratio. It is also apparent from the plots that an increase in the inertial force reduces the sensitivity of the pile response to the parameters of the liquefied soil. This effect on the sensitivity of the pile response to the liquefied soil parameters is analogous to that of a non-liquefied crust. Indeed, where the yield ratio is greater than one, equivalent crust and inertial forces have an identical effect on the sensitivity of the response to parameters of the liquefied soil. This outcome is intuitively expected, because the inertial force and the resultant lateral load from the crust act at nearly identical locations (at or near the pile head), and hence when the magnitudes of these two loads are similar, their effects on the pile response should also be similar. There is one important difference between these two loads however. Namely, whereas the size of the inertial force is predetermined as an input in the PSA, the magnitude of the lateral force from the crust on the pile depends on the computed pile response or relative displacement $(U_G - U_P)$. The general tendency in the effects of stiffness and strength parameters on the pile response and their relation to pre-yield or post-yield deformational behaviour was also evident for these (stiff-pile-behaviour) cases with combined inertial and kinematic loads. The sensitivity level of the response was also similar to that observed in lateral spreading PSA.

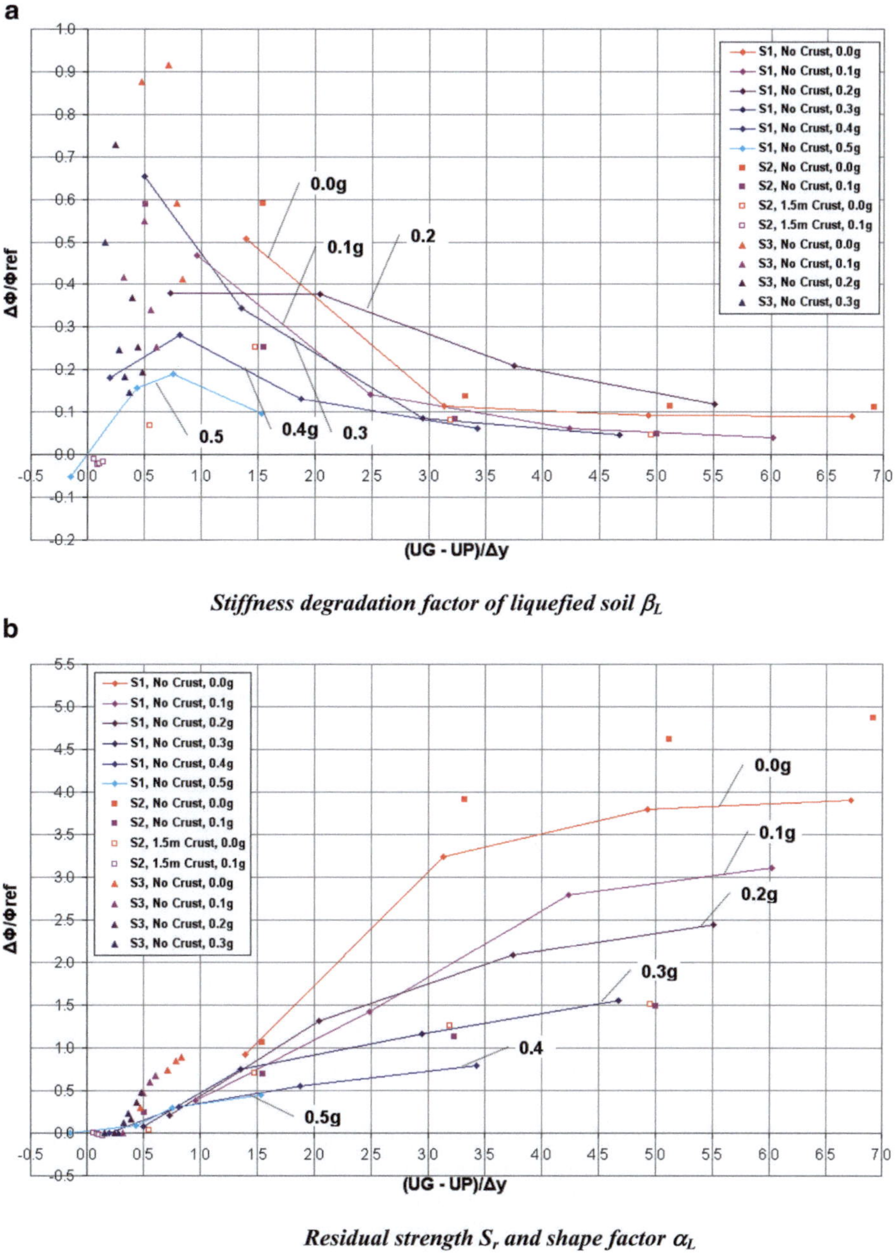


**Fig. 6.14** Sensitivity of pile response (peak pile curvature) to parameters of liquefied soil, for cyclic scenario (combined kinematic and inertial force demands)

## 6.9 Summary and Conclusions

Results from comprehensive series of parametric analyses have been used to examine and quantify the sensitivity of the pile response to various model parameters, and hence, to identify critical uncertainties in the pseudo-static analysis. Key findings from the study can be summarized as follows:

- Sensitivity of the pile response to parameters of the liquefied soil is clearly related the soil-pile interaction mechanism or load (deformation) of the soil spring relative to the yield level. For small relative displacements $|U_G - U_P|$, the pile response is most sensitive to the stiffness degradation factor ($\beta_L$). As the relative displacement increases and yielding in the soil occurs, the response becomes more sensitive to the strength of the soil (parameters $\alpha_L$ and $S_r$).
- The stiffness degradation factor ($\beta_L$) is most important and shows greater influence on the pile response than the strength parameters for yield ratios of up to 0.5–1.0. Its influence on the pile response gradually decreases as more of the liquefied soil yields.
- The influence of strength parameters on the pile response gradually increases with the relative displacement between the soil and the pile, and reaches the maximum level at yield ratios of about 3 or higher. For yield ratio of 1.0 (at the initiation of soil yielding) the effects of strength parameters ($\alpha_L$ and $S_r$) on the pile response are already at the same level or greater than those of $\beta_L$. At high yield ratios (≥3), the influence of the strength parameters on the pile response is substantially higher than that of $\beta_L$ (3–10 times greater sensitivity).
- Sensitivity of the pile response to the parameters of the liquefied layer ($\beta_L$, $\alpha_L$ and $S_r$) decreases with the thickness (or load contribution) of the crust.
- The above conclusions are applicable to pseudo-static analyses of piles for both cyclic ground displacements and lateral spreading displacements.
- For cyclic liquefaction with combined kinematic and inertial force demands, it was found that parametric sensitivity is an issue only for stiff-pile-behaviour where the pile resists the combined lateral loads and exhibits considerably smaller displacement than the applied ground displacement. For these cases, the effect of inertial load on the pile response is analogous to that of a non-liquefied crust; an increase in the inertial load decreases the sensitivity of the pile response to parameters of the liquefied soil.
- The shape factor $\alpha_C$ is a key uncertainty associated with the ultimate load on the pile from a crust of non-liquefied soil at the ground surface. The value of this factor needs to be selected in conjunction with the anticipated role of the crust layer in the loading mechanism (active-pile-loading vs. passive-pile-loading).
- The sensitivity of the pile response to parameters of the base layer and initial soil stiffness is negligibly small.
- The representative SPT blow count (soil characterization parameter) is a significant uncertainty in the analysis since it affects multiple model parameters (stiffness and strength) and shows relatively large influence on the pile response across various load (response) levels.

The above conclusions are generally applicable to pseudo-static methods for analysis of piles in liquefying soils even though specific parameters and details of the model may differ from those adopted in this study. The findings presented herein provide insight into the importance levels of different model parameters and their relation to the soil-pile deformation mechanism. Importantly, they provide basis for a more robust use of the simplified pseudo-static analysis and rigorous treatment of the uncertainties in the analysis.

**Acknowledgements** The authors would like to acknowledge the financial support provided by the Earthquake Commission (EQC), New Zealand.

## References

Boulanger R, Chang D, Brandenberg S, Armstrong R (2007) Seismic design of pile foundations for liquefaction effects. In: Pitilakis K (ed) Proceedings of the 4th international conference on Earthquake Geotechnical Engineering-invited lectures, pp 277–302

Broms B (1964) Lateral resistance of piles in cohesionless soils. ASCE J Soil Mech Found Eng 90(SM3):123–156

Cubrinovski M, Bradley BA (2009) Evaluation of seismic performance of geotechnical structures. In: Kokusho T, Tsukamoto Y, Yoshimine M (eds) Performance-based design in earthquake geotechnical engineering. Taylor & Francis Group, Boca Raton, pp 121–136

Cubrinovski M, Ishihara K (2004) Simplified method for analysis of piles undergoing lateral spreading in liquefied soils. Soils Found 44(25):119–133

Cubrinovski M, Ishihara K (2007) Simplified analysis of piles subjected to lateral spreading: parameters and uncertainties. In: Pitilakis K (ed) Proceedings of the 4th international conference on Earthquake Geotechnical Engineering-invited lectures, Paper 1385, pp 1–12

Cubrinovski M, Kokusho T, Ishihara K (2006) Interpretation from large-scale shake table tests on piles undergoing lateral spreading in liquefied soils. Soil Dyn Earthq Eng 26:275–286

Cubrinovski M, Uzuoka R, Sugita H, Tokimatsu K, Sato M, Ishihara K, Tsukamoto Y, Kamata T (2008) Prediction of pile response to lateral spreading by 3-D soil-water coupled dynamic analysis: shaking in the direction of ground flow. Soil Dyn Earthq Eng 28:421–435

Cubrinovski M, Ishihara K, Poulos H (2009a) Pseudo static analysis of piles subjected to lateral spreading. Special Issue, Bull NZ Soc Earthq Eng 42(1):28–38

Cubrinovski M, Haskell JJM, Bradley BA (2009b) The effect of shear strength normalisation on the response of piles in laterally spreading soils. In: Earthquake geotechnical engineering satellite conference 17th international conference on soil mechanics & geotechnical engineering, Alexandria, Egypt, 2–3 Oct 2009

Finn WDL, Thavaraj T (2001) Deep foundations in liquefied soils: case histories, centrifuge tests and methods of analysis. In: Proceedings of the 4th international conference on recent advances in geotechnical earthquake engineering and soil dynamics, San Diego, CA, CD-ROM, Paper SOAP-1

Haskell JJM (2009) Pseudo-static modelling of the response of piles in liquefying soils. Research report, University of Canterbury

Japanese Geotechnical Society (1998) Special issue on geotechnical aspects of the Jan 17 1995 Hyogoken-Nambu earthquake, soils and foundations

Meyerhof GG, Mathur SK, Valsangkar AJ (1981) Lateral resistance and deflection of rigid walls and piles in layered soils. Can Geotech J 18:159–170

O'Rourke TD, Meyersohn WD, Shiba Y, Chaudhuri D (1994) Evaluation of pile response to liquefaction-induced lateral spread, In: Proceedings of the 5th U.S.-Japan workshop on earthquake

resistant design of lifeline facilities and countermeasures against soil liquefaction, technical report NCEER-94-0026, pp 457–479
Olson SM, Stark TD (2002) Liquefied strength ratio from liquefaction flow case histories. Can Geotech J 39:629–647
Poulos HG, Chen LT, Hull TS (1995) Model tests on single piles subjected to lateral soil movement. Soils Found 35(4):85–92
Prasad YVSN, Chari TR (1999) Lateral capacity of model rigid piles in cohesionless soils. Soils Found 39(2):21–29
Seed RB, Harder LF (1991) SPT-based analysis of cyclic pore pressure generation and undrained residual strength. In: Bolton H (ed) Seed memorial symposium proceeding, vol 2. University of California, Berkeley, pp 351–376
Tamura S, Tokimatsu K (2005) Seismic earth pressure acting on embedded footing based on large-scale shaking table tests. ASCE Geotech Spec Publ 145:83–96
Tokimatsu K, Asaka Y (1998) Effects of liquefaction-induced ground displacements on pile performance in the 1995 Hyogoken-Nambu earthquake. Special issue of soils and foundations, pp 163–177, Sept 1998
Tokimatsu K, Suzuki H (2009) Seismic soil-pile interaction based on large shake table tests. In: Kokusho T, Tsukamoto Y, Yoshimine M (eds) Performance-based design in earthquake geotechnical engineering. Taylor & Francis Group, Boca Raton, pp 77–104
Yasuda S, Berrill JB (2000) Observation of the earthquake response of foundations in soil profiles containing saturated sands, GeoEng2000. In: Proceedings of the international conference on geotechnology and geological engineering, vol 1. Melbourne, pp 1441–1470
Youd TL, Hansen MC, Bartlett FS (2002) Revised multi-linear regression equations for prediction of lateral spread displacement. ASCE J Geotech Geoenviron Eng 128(12):1007–1017

# Chapter 7
# Experimental and Theoretical SFSI Studies in a Model Structure in Euroseistest

**Kyriazis Pitilakis and Vasiliki Terzi**

**Abstract** In the framework of a recent EU funded research project (EUROSEISRISK Seismic hazard assessment, site effects s soil-structure interaction studies in an instrumented basin), a bridge pier model was constructed, instrumented and tested in the EUROSEISTEST experimental site (http://euroseis.civil.auth.gr), located close to Thessaloniki in Greece. The prior aim is the experimental investigation of the dynamic characteristics of the model, the study of the soil-structure-interaction effects, and in particular the wave fields emanating from the oscillating structure to the surrounding ground; to accomplish this task a well-designed set of free-vibration tests were conducted. Experimental results were compared with careful 3D numerical simulations of the soil-foundation structure system, in the frequency and time domain. Several fundamental aspects of SSI are discussed and the available analytical impedance expressions are compared with the experimental and numerical results of the present study. The study of the dynamic behavior of a simple SDOF system consisting of a model bridge pier with surface foundation in real soft soil conditions and the numerical FE modeling of the experiments, enable us to enhance our knowledge on various soil-structure interaction aspects.

## 7.1 Introduction

In the framework of an EU research program entitled « Seismic hazard assessment, site effects and soil-structure interaction studies in an instrumented basin», (EUROSEISRISK project, EVG1-CT-2001-00040, http://euroseis.civil.auth.gr),

K. Pitilakis (✉) • V. Terzi
Department of Civil Engineering, Aristotle University, Thessaloniki, Greece
e-mail: kpitilak@civil.auth.gr

M.A. Sakr and A. Ansal (eds.), *Special Topics in Earthquake Geotechnical Engineering*, Geotechnical, Geological and Earthquake Engineering 16,
DOI 10.1007/978-94-007-2060-2_7, 

a bridge pier model was constructed and instrumented in the EUROSEISTEST experimental site, in view to conduct a set of free vibration tests on a model (scaled 1:3) SDOF structure. The aim of the tests was to study several fundamental aspects of the soil-structure interaction and the wave field created during the oscillation of a SDOF system, actually a bridge pier, in real soil conditions. In particular we investigated the dynamic characteristics of the soil-foundation-pier model system, under various excitation schemes and possible retrofitting. To accomplish the aforementioned goals the geotechnical conditions of the site have been well constrained through numerous field and laboratory tests, and the pier bridge, its foundation and the surrounding soil have been properly instrumented. Several experiments of static nature and free-vibration oscillations took place. In the paper we present the most important experimental results and the associated numerical modeling. We compare the computed and recorded dynamic characteristics of the system, in terms of eigen-frequencies and damping, discussing at the same time several aspects of soil-structure interaction phenomenon. Finally we present a hybrid method of evaluation of foundation impedance functions and the comparison of its results with available analytical expressions.

## 7.2 Bridge Pier Model, Instrumentation, Soil Conditions and Testing Program

The EUROSEISTEST experimental site (http://euroseis.civil.auth.gr), is located close to Thessaloniki in the epicentral area of the M = 6.4, 1978 Thessaloniki earthquake. The past decade various experiments took place in order to investigate the dynamic soil properties and the geometry of the valley (Raptakis et al. 2005; Manakou et al. 2010; Pitilakis et al. 1999) to study the complex site effects (Raptakis et al. 2004a, b) and the structural response of model structures under real soil conditions. Two prototype structures are established in the test site, a five-storey reinforced concrete building with infill masonry and a bridge pier model (Manos et al. 2005, 2008). In the present paper we will be concentrated ourselves on the pier bridge model.

### *7.2.1 The Model*

The reinforced concrete bridge pier model (scale 1:3) consists of a rectangular surface foundation (2.5 × 2.5 m), a column (20 × 50 cm) and a rectangular deck (2.00 × 3.90 m) (Fig. 7.1) (Manakou et al. 2010). Two axis of the pier vibration can be distinguished; a strong axis parallel to the face of 50 cm, axis XX in the present paper and the weak one (YY). The foundation is attached to the superstructure using eight auxiliary tendons, mainly to avoid accidental overturning of the deck, however

**Fig. 7.1** The bridge pier model (Terzi 2009)

their pre-stress may modify the fundamental period of the model structure. During the performance of the experiments various functional schemes of the tendons activation were adopted. The height of the structure is 4.60 m and the total mass of 30.71 tons including the added mass which is 3.35 ton.

### *7.2.2 The Soil Conditions*

The soil conditions of the test site have been thoroughly investigated through a comprehensive program of field geophysical, geotechnical surveys and laboratory tests (Raptakis et al. 2004a, b, 2005; Manakou et al. 2010; Pitilakis et al. 1999). Thus, the complete soil profile till the bedrock (−200 m) has been provided (Fig. 2.2), accompanied with full description of the dynamic soil characteristics in terms of $V_s(z)$ and G-γ-D curves for all soil layers (Figs. 7.2 and 7.3) (Manakou et al. 2010). The site is particularly interesting to study SSI effects, as the near surface soils are very soft with a mean shear wave velocity of about 140 m/s.

The experiments were conducted in two phases. The first one consists to the placement and connection of the prefabricated foundation slab, pier column and of its deck. In this stage we examine the contact pressures using load cells installed underneath the foundation slab. The second phase is the main experiment; it consists of several pull-out tests, producing different free vibration oscillations of the soil-foundation-bridge pier system. In the following paragraphs the results of this experiment are presented and discussed.

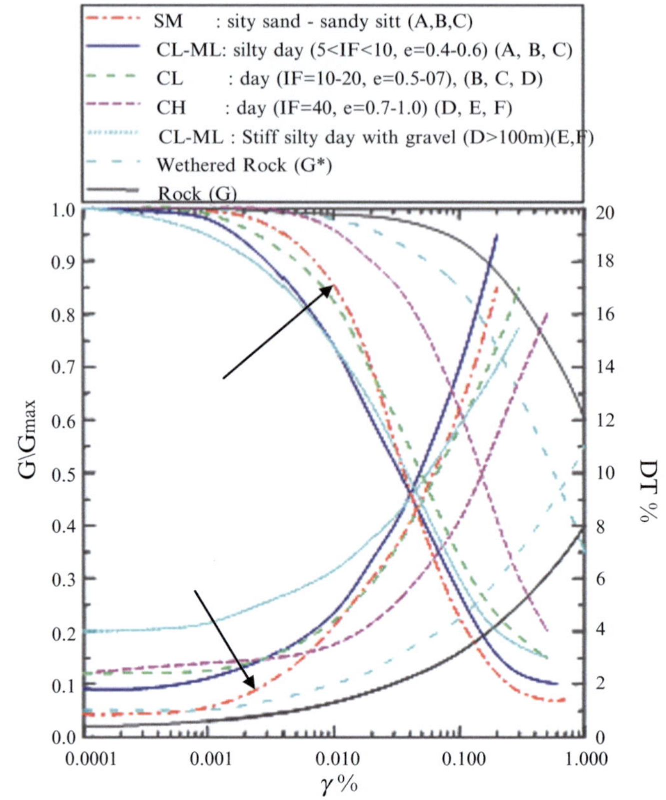


**Fig. 7.2** Mean G/Go-γ-D curves from resonant column and cyclic triaxial tests for all geotechnical formations in the N-S direction of the valley (A to G). The curves describe the shear modulus degradation with the shear strain and the respective internal damping increase (Raptakis et al. 2005; Manakou et al. 2010; Pitilakis et al. 1999). *Arrows* indicate the surface soft soil curves

## 7.2.3 Construction and Placement of the Model

The construction of the three main components of the bridge pier model including the foundation slab, the column (pier) and the deck, took place in the Laboratory of Experimental Mechanics of the Civil Engineering Department of Aristotle University of Thessaloniki (Manos et al. 2005). Then they were transported in the testing area and mounted in-place. The real contact stresses were evaluated with four load cells placed underneath the foundation (Figs. 7.4 and 7.5). A detailed description of the results of this stage may be found in (Manos et al. 2005).

Table 7.1 presents the recorded vertical forces of each load cell (Manos et al. 2005). The average force recorded during the placement of the foundation, the pier column and the deck is 2.17 kN.

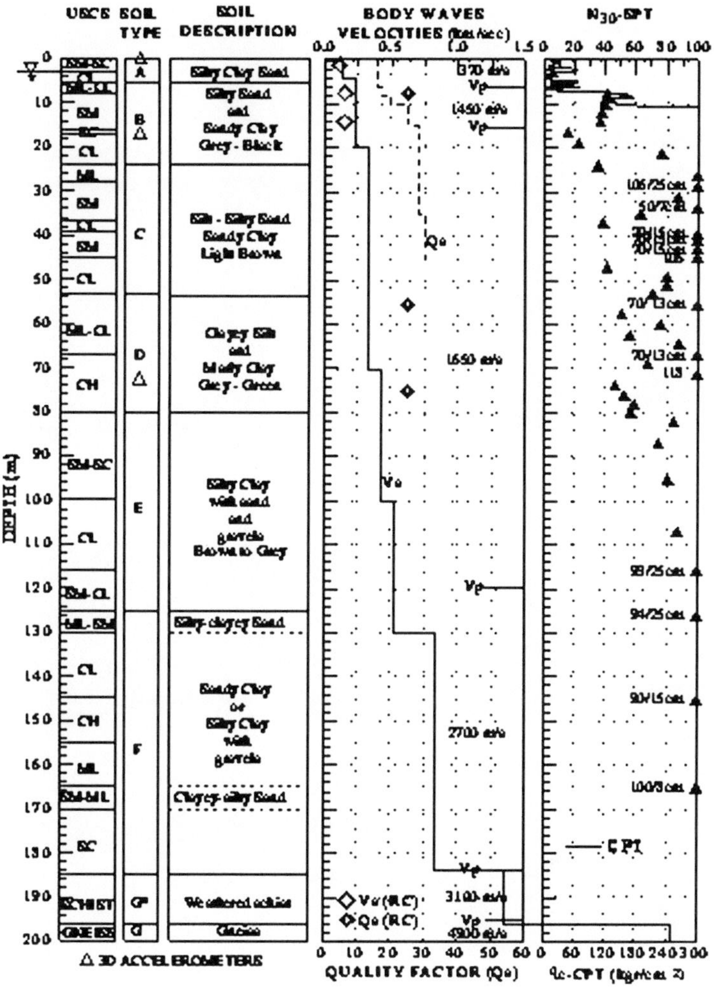


**Fig. 7.3** $V_s$ and $V_p$ profile and Q factors at the center of the valley (location S1, TST site) estimated using different in-situ and laboratory techniques for the main geological formations of the site (From A to G according to soil classification made by (Raptakis et al. 2004a, b, 2005; Manakou et al. 2010; Pitilakis et al. 1999))

## *7.2.4 Instrumentation Set Up and Testing Program*

In order to investigate the dynamic characteristics of the soil-foundation-bridge pier model system (eigenfrequencies and damping) and to study the SSI effects, we performed a set of free vibration tests (Manos et al. 2005, 2008; Terzi 2009; Terzi and Pitilakis 2009; Pitilakis 2009). Both the strong and weak axis of the bridge pier model is activated through tendons in each direction. Once the applied force had

**Fig. 7.4** Load cells (Manos et al. 2005)

**Fig. 7.5** Placement of the foundation slab (Manos et al. 2005)

**Table 7.1** Load cell recordings

| Load cell | Recorded forces [kN] |
|---|---|
| M1 | 1.609 |
| M2 | 1.662 |
| M3 | 1.915 |
| M4 | 3.493 |
| Average force | 2.17 |

reached the desired level (1.0–2.2 tons), the tendon was de-activated and the model performed a free vibration oscillation. (Fig. 7.6a, b).

The instrumentation of the ground surface with accelerometers and velocimeters is illustrated in Fig. 7.7. To monitor the wave field propagation emanating from the

**Fig. 7.6** (**a**), (**b**) Application of the force for the free vibration test on the strong and weak axis (Terzi 2009)

oscillating structure, and to study the SSI phenomenon, several tests are performed, varying the axis of the forced excitation, the mass of the deck and the activation of the tendons, connecting the foundation with the deck (Table 7.2). The total mass of the system is 30.71 ton including an added mass of 3.35 ton, which was placed on the deck.

## 7.3 Experimental Results

The experimental results of the free vibration tests are presented. The origin data refer to recorded ground motion, in terms of amplitude motion. The frequency content of the motion and its time decay is studied as well.

### *7.3.1 Recorded Ground Motion*

Figures 7.8, 7.9 and 7.10 illustrate free filed time history recordings for six different sets of free vibration pull out tests; three in each direction. It is observed that for the axis which is in-plane according to the direction of the oscillation, the surface ground motion is largely dominated by the vertical and the in-plane horizontal components. In the in-plane direction both vertical and in plane horizontal components follow the exponential decay shape of the free vibration oscillation of the system, whereas the out-of-plane horizontal components, having one order of magnitude

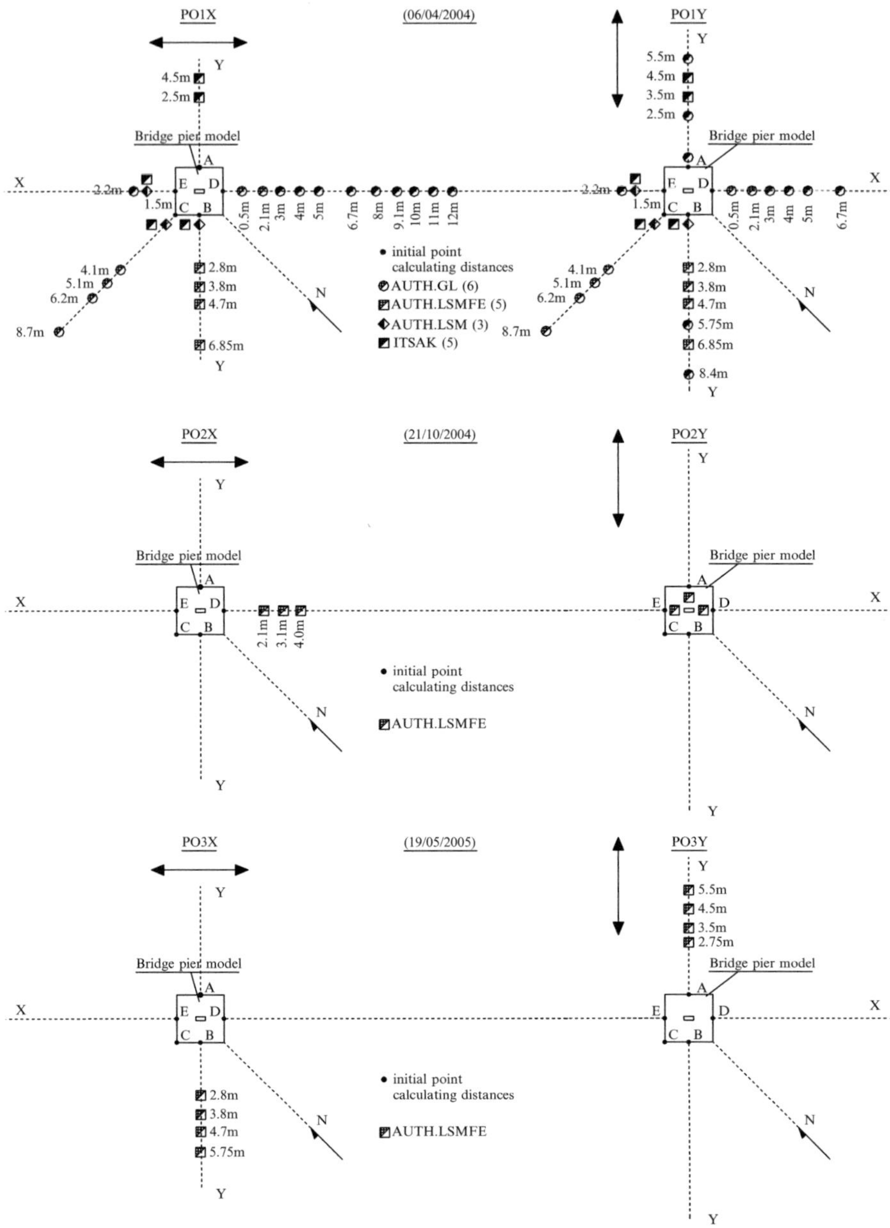


**Fig. 7.7** Free field instrumentation in three testing programs

lower amplitudes, include the contribution of frequencies different from the eigenfrequencies of the system. The in-plane horizontal components of the ground motion provide the most valuable information concerning the vibration characteristics of the soil-foundation-superstructure system.

**Table 7.2** Pull-out testing references and characteristics

| Date and name of experiment | Direction of pull-out test | Other characteristics | |
|---|---|---|---|
| | Axis XX, YY | Mass [ton] | Tendons |
| 06/04/2004_PO1 | Both | 27.36 | Yes |
| 21/10/2004_PO2 | Both | 30.71 | Yes |
| 19/05/2005_PO3 | Both | 30.71 | No |

## 7.3.2 Damping

Based on the exponential decay scheme of the ground motion recordings in different distance from the structure, we calculated the attenuation of the ground motion at different distances from the oscillating structure, applying the conventional definition of attenuation of free oscillating systems (Chopra 1995).

Table 7.3 presents the experimental attenuation for the in-plane horizontal components. All values refer to the free field ground motion recorded at the surface, except for the test PO2, which activated the weak axis of vibration (axis YY) and its values refer to the global damping for the whole soil-foundation-superstructure system. All other attenuation parameters given in Table 7.3 concern both material and geometrical soil damping.

The recorded maximum damping varies from 0.9% to 1.18%. For the soil material at the surface, and the strain ranges of this experiment ($\gamma < 10\text{-}5$), the material hysteretic damping given by the resonant column tests is between 1 and 2%, which is in general in good comparison with the values found from the free oscillating structure (Fig. 7.2). An important observation is that in all pullout tests the soil behavior remained practically in the elastic range (Chopra 1995). The global damping of the whole soil-foundation-superstructure system has a mean value equal to 0.92%.

## 7.3.3 Eigenfrequency Characteristics

The analysis of the recordings in the frequency domain provides valuable information of the frequency content of the motion. As the in plane horizontal component dominates the free surface motion, even in large distances from the oscillating model, it was decided to use this component to evaluate the eigenfrequencies of the model structure. Figure 7.10 presents for each experiment and for all locations, the experimental FFT (Fast Fourier Transform) (MATLAB 2004) of the in-plane horizontal component. Considering the free vibration oscillation, the frequency that dominates the motion is equal to the eigenfrequency of the soil-foundation-superstructure system.

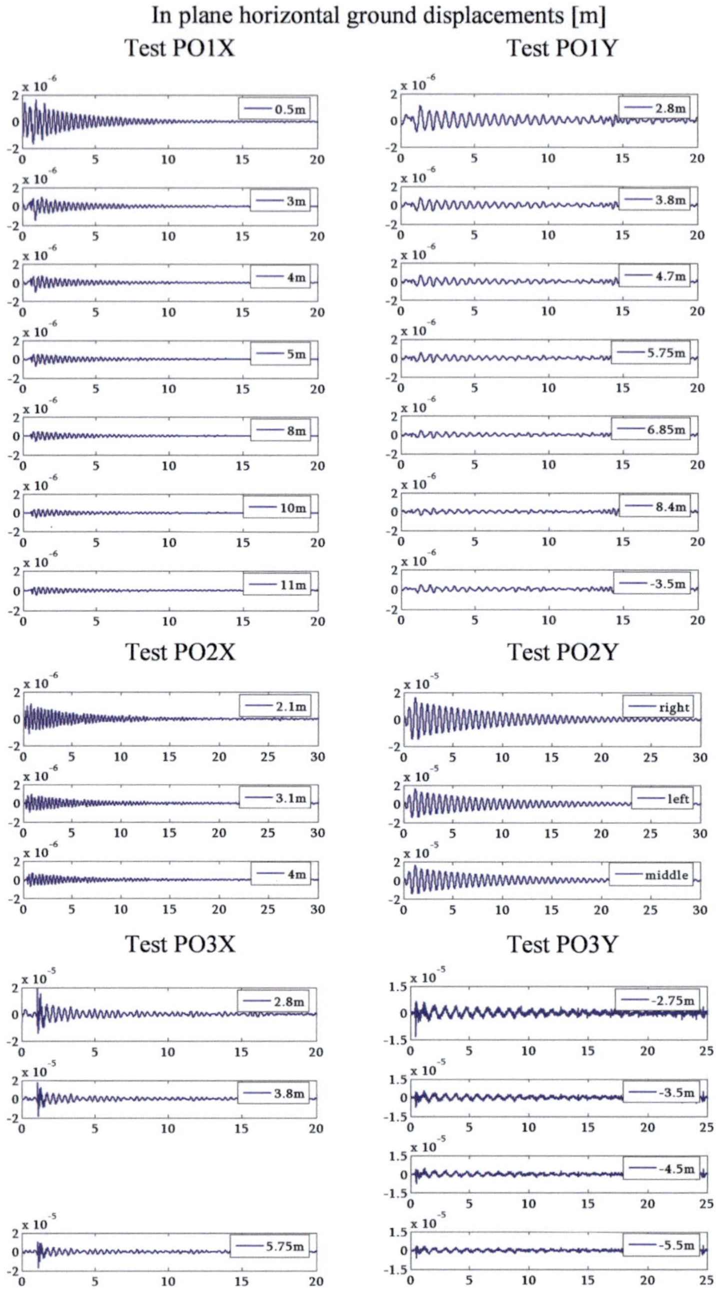


**Fig. 7.8** In-plane horizontal ground motion recordings at several distances from the oscillating structure (Terzi 2009)

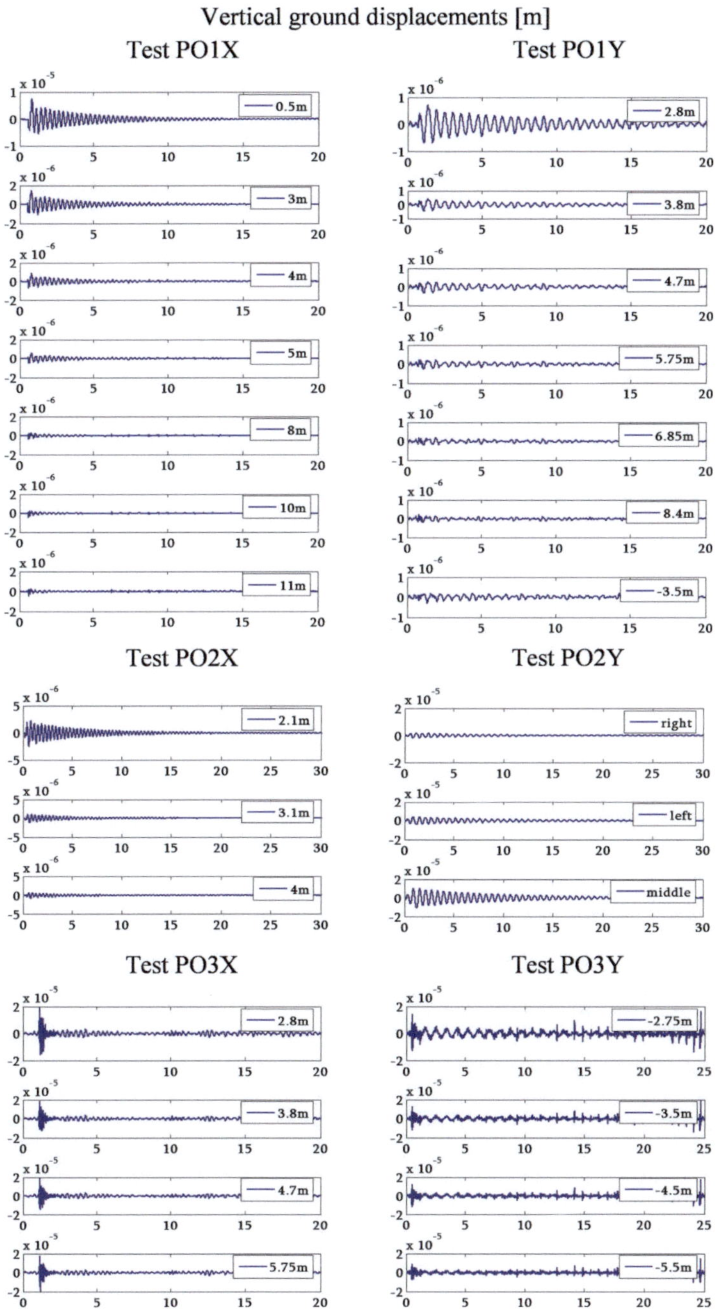


**Fig. 7.9** Vertical components of the recordings (Terzi 2009)

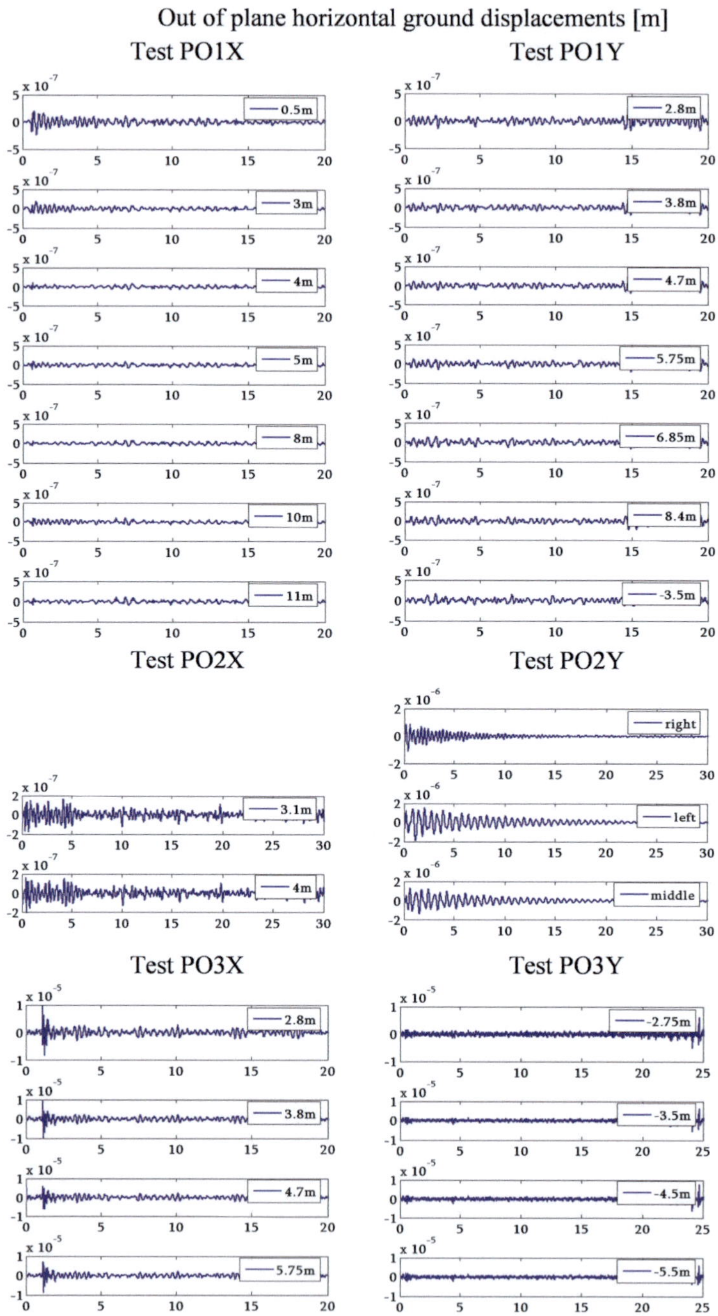


**Fig. 7.10** Out-of plane recordings (Terzi 2009)

**Table 7.3** Recorded damping at different distances

| | Free vibration in the strong axis XX | Location | Free vibration in the weak axis YY | | |
|---|---|---|---|---|---|
| Location | In plane component | Vertical component | | In plane component | Vertical component |
| PO1X | PO1Y | | | | |
| 0.5 m | 0.0118 | 0.0105 | 2.80 m | 0.0102 | 0.0097 |
| 3 m | 0.0108 | 00100 | 3.80 m | 0.0103 | 0.0100 |
| 4 m | 0.0103 | 0.0087 | 4.70 m | 0.0106 | 0.0099 |
| 5 m | 0.0099 | 0.0078 | 5.75 m | 0.0114 | 0.0114 |
| 8 m | 0.0096 | 0.0059 | 6.85 m | 0.0115 | 0.0128 |
| 10 m | 0.0090 | 0.0048 | 8.40 m | 0.0110 | 0.0146 |
| 11 m | 0.0090 | 0.0047 | –3.50 m | 0.0103 | 0.0060 |
| Average | 0.0100 | 0.0075 | Average | 0.0107 | 0.0106 |
| | PO2X | | | PO2Y | |
| 2.10 m | 0.0089 | 0.0093 | Middle | 0.0092 | 0.0090 |
| 3.00 m | 0.0107 | 0.0095 | Left | 0.0091 | 0.0089 |
| 4.00 m | 0.0107 | 0.0101 | Right | 0.0092 | 0.0093 |
| Average | 0.0101 | 0.0096 | Average | 0.0092 | 0.0091 |
| | PO3X | | | PO3Y | |
| 2.80 m | 0.0099 | – | 2.75 m | 0.0110 | 0.0121 |
| 3.80 m | 0.0094 | – | 3.50 m | 0.0103 | 0.0095 |
| 4.70 m | – | – | 4.50 m | 0.0093 | 0.0108 |
| 5.75 m | 0.0080 | – | 5.50 m | 0.0101 | 0.0100 |
| Average | 0.0091 | – | Average | 0.0102 | 0.0106 |

Changing the mass of the deck and activating the tendons it is possible to change the eigenfrequencies of the soil-foundation-structure model. During the experiment PO1 the tendons were activated. The eigenfrequency of the soil-foundation-pier system in the strong axis (XX) is equal to 3.3 Hz, whereas in the weak axis (YY) is only 1.83 Hz. In PO2 test the pier deck had an additional mass and therefore the total inertia of the system is increased and consequently the eigenfrequency in both directions has lower values than PO1 (2.78 Hz and 1.65 Hz for the strong and weak respectively). Finally in PO3 test the extra deck mass remained but the tendons are deactivated; hence the total stiffness of the system is decreased more and the eigenfrequencies of the system are now 2.48 Hz in the strong axis and 1.13 Hz in the weak axis respectively. Generally speaking the eigenfrequency in the strong axis varies from 2.48 to 3.3 Hz, while in the weak axis between 1.13 and 1.83 Hz.

The extra mass 12.24% on the deck of the bridge pier model decrease the eigenfrequency of the system in both axis by 15.8% in the strong and 10.0% in the weak one. On the other hand, the deactivation of the tendons decreases the eigenfrequancies by 10.8% in the strong and 31.50% in the weak axis. An interesting remark is that the oscillation in the strong axis is affected by the inertial alteration, whereas in the weak one by the stiffness alteration.

## 7.4 Numerical Modeling

For the numerical modeling we used the FE code ADINA (Automatic Dynamic Incremental Nonlinear Analysis 1999; Bathe 1996). Three different sets of numerical studies were conducted, each aiming at different target. The first one refers to the validation of the numerical code through the investigation of the generated wave field due to the harmonic oscillation of a surface foundation. In the second set we perform several numerical simulations of the experimental tests, while in the third one we calculate the foundation impedance functions via a hybrid method combining experimental and numerical results.

### *7.4.1 Validation of Numerical Code: Investigation of Generated Wave Field*

According to Chow (1986), the results of the numerical methods must agree with the analytical results that refer to Lamb's problem (Lamb 1904). In the first stage we respected this principle. The results of the analysis are present below.

#### 7.4.1.1 Description of Studied Case and Finite Element Model

The wave field is generated applying a vertical harmonic load on the surface of a plan, rigid, rectangular foundation (2B = 2.5 m) in the half space. The amplitude of the force is equal to unity and its circular frequency 108 rad/s. The elastic soil properties correspond to the first soft soil layer in the EUROSEISTEST experimental site (Table 7.4).

The finite element models for the foundation and the soil are given in Figs. 7.11 and 7.12. The dimensions of the elements respect the principle that the minimum length should be smaller than 1/8 of the wave length (Kuhlemeyer and Lysmer 1973). To avoid undesirable reflections on the lateral boundaries, we extend laterally the mesh, which final has 161.616 elements and 171.419 nodes. The dimensions of the mesh are 70 m (vertical direction) × 70 m (both horizontal directions) (Fig. 7.13).

**Table 7.4** Surface soil properties

| G [kPa] | ρ [t/m³] | ν | $V_P$ [m/s] | $V_S$ [m/s] | $V_{Rayleigh}$ [m/s] |
|---|---|---|---|---|---|
| 37361.25 | 2.05 | 0.3333 | 270 | 135 | 125.89 |

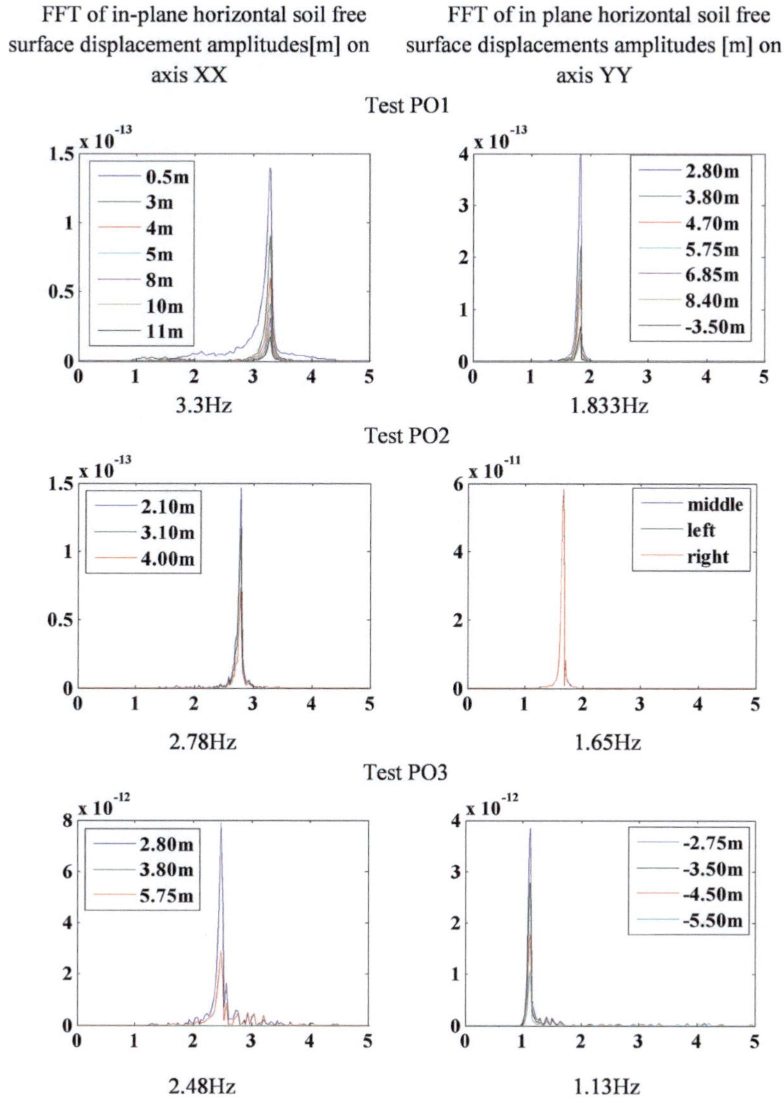


**Fig. 7.11** FFT of the horizontal in plane records at different distances from the foundation (Terzi 2009)

#### 7.4.1.2 Orbits of Motion

The orbits of motion were investigated in the 3D domain along the three axis according to Fig. 7.10. In Fig. 7.12 we present the ground motion in the XZ plane and the XX direction. The vertical motion being in the direction of the applied force it is obviously greater that the horizontal. The orbits have a characteristic elliptical shape

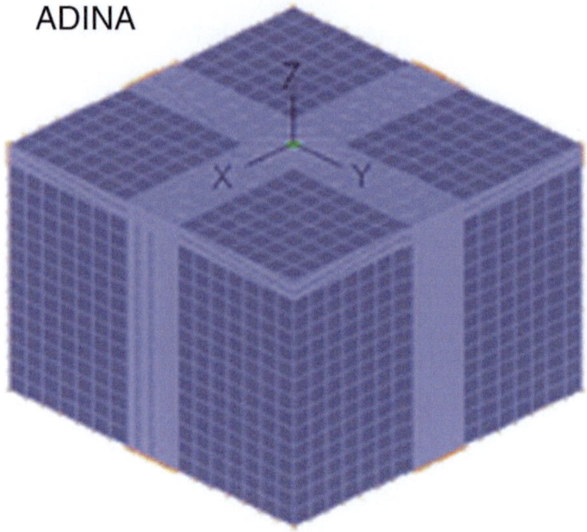


**Fig. 7.12** 3D finite element model (Terzi 2009)

**Fig. 7.13** Finite element foundation slab model (Terzi 2009)

with the main axis inclination varying according to the contributing part of S, P and surface waves. The ground motion on the free surface is the result of various wave forms. In particular the motion on the vertical axis, which is perpendicular to the direction of wave propagation, is composed by shear waves, whereas on the horizontal axis, which is parallel to the direction of wave propagation, the wave filed is composed of primary waves. The composition of the actions of the two body waves is limited until the arrival of surface waves, which can be considered responsible for the inclination of the elliptical axis.

In Fig. 7.15 the computed time histories of the free field particle motion in XX are depicted. The amplitude scale is kept constant based on the maximum vertical displacement that corresponds to the foundation's acme. The analysis is performed with zero material damping, and hence the decay of the amplitude is attributed only to the radiation damping. Another interesting observation concerns the time arrival of the generated wave field in each point away from the foundation.

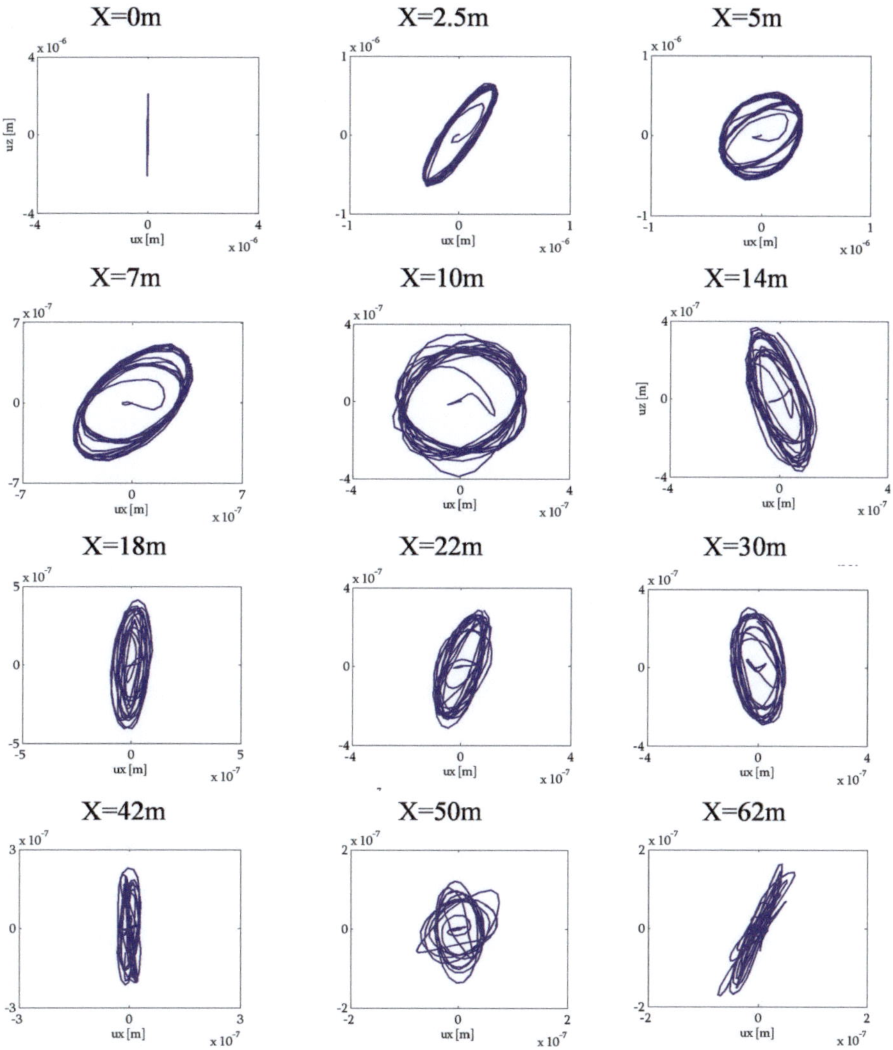


**Fig. 7.14** Orbits of motion in XZ plane and XX direction (Terzi 2009)

#### 7.4.1.3 Velocity of Wave Propagation

Two types of waves are generated: body P and S waves as well as surface waves. In Figs. 7.14, 7.15 and 7.16 the arrival of each type of body wave is depicted at distances of 2.5, 3.125, 3.75, 4.375 and 5.0 m away from the foundation. The primary waves, due to their greater velocity are the first that arrive at each point and they are responsible for the initiation of the excitation of each point. The secondary waves with smaller velocity arrive next and their arrival can be distinguished by the

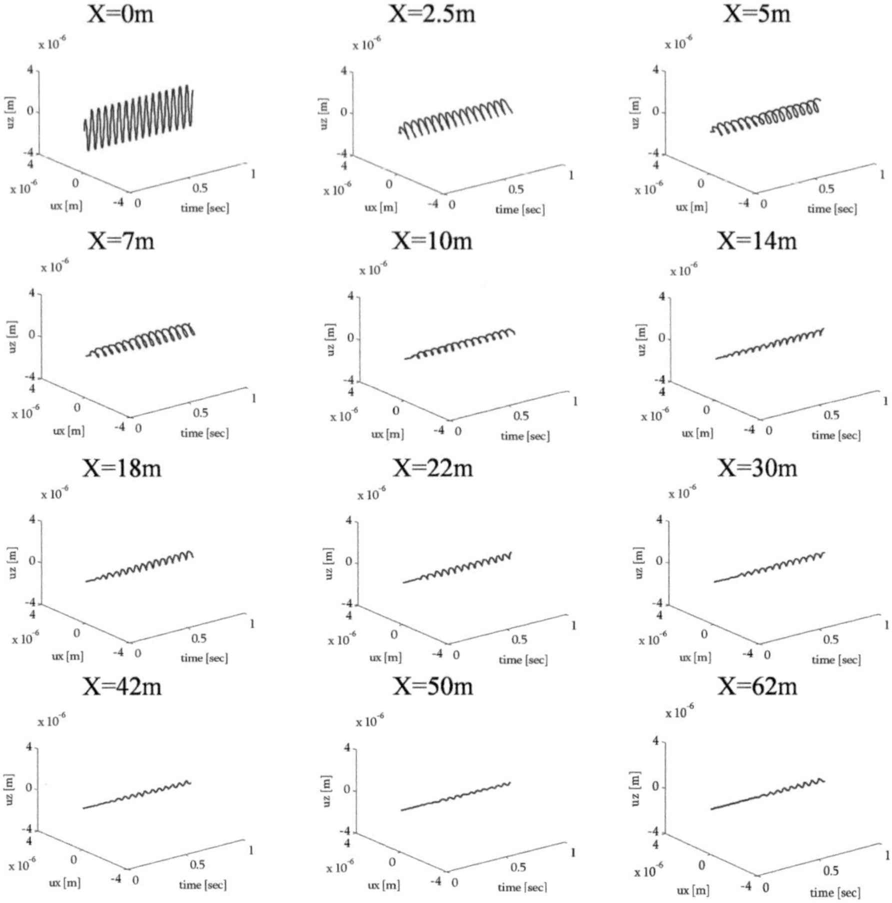


**Fig. 7.15** Time histories of orbits of motion in XZ plane and XX direction (Terzi 2009)

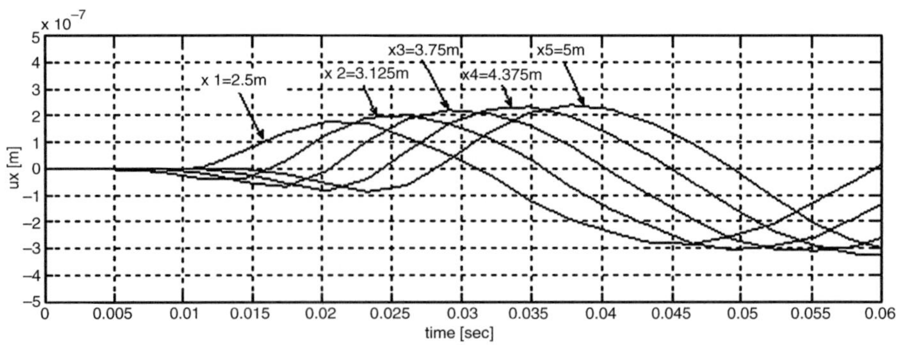


**Fig. 7.16** Primary wave's forms at various distances from the foundation (Terzi 2009)

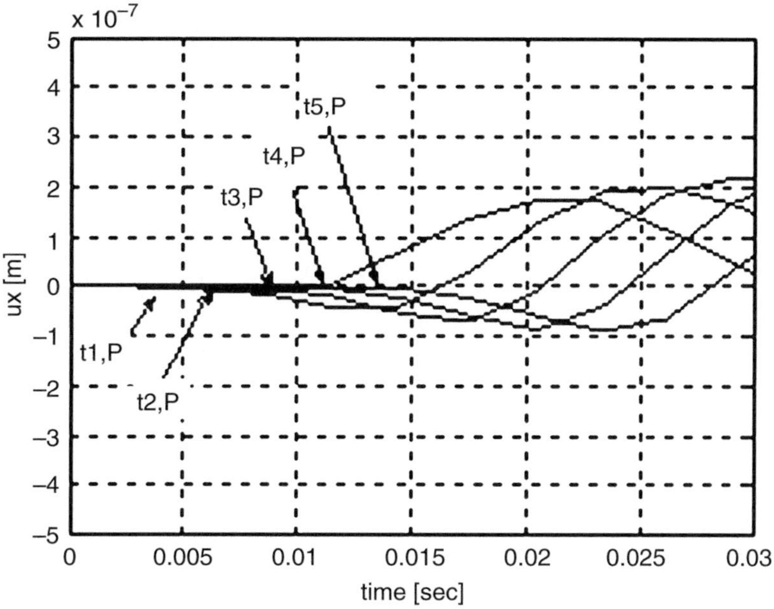


**Fig. 7.17** Detection of the first arrivals of primary waves (Terzi 2009)

increase of motion amplitude. The small difference between the velocity of the secondary (or shear) and surface Rayleigh waves, makes difficult any effort to distinguish the first R arrivals and hence their propagation velocity.

Using the time arrivals of body waves at different distances from the vibrating foundation, the evaluation of wave velocities is straightforward. The points in Fig. 7.17 correspond to the pairs of arrival time and distance, whereas the thick black line corresponds to the medium line adopting the linear model (MATLAB 2004). The propagation velocity of the P waves based on the numerical simulation is equal to 271.1 m/s, very close to the experimental velocity (270 m/s) adopted in the analysis. For the S waves, the numerical value is equal to 135.2 m/s, compared to the experimental one adopted in the model (135 m/s). The excellent comparison is a direct validation and consistency of the numerical code and modeling characteristics (Figs. 7.18, 7.19 and 7.20).

#### 7.4.1.4 Spatial Decay of Amplitude of Free Field Motion

We examined next the spatial decay of the amplitude of the free field ground motion. Taking into account the harmonic nature of the oscillation and the absence of material or modal damping, the maximum amplitude of the motion of each node corresponds to the quadratic root of the sum of the quadratic powers of each component. In Fig. 7.21

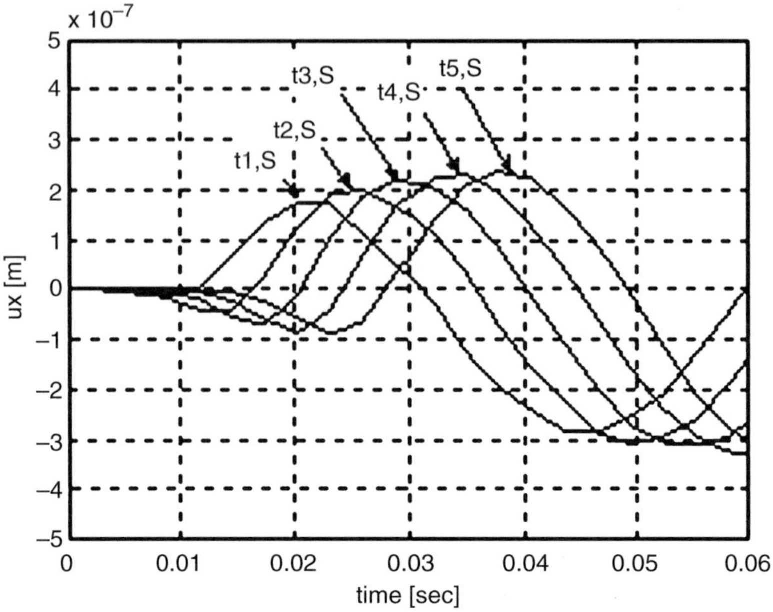


**Fig. 7.18** Detection of the first arrivals of secondary waves (Terzi 2009)

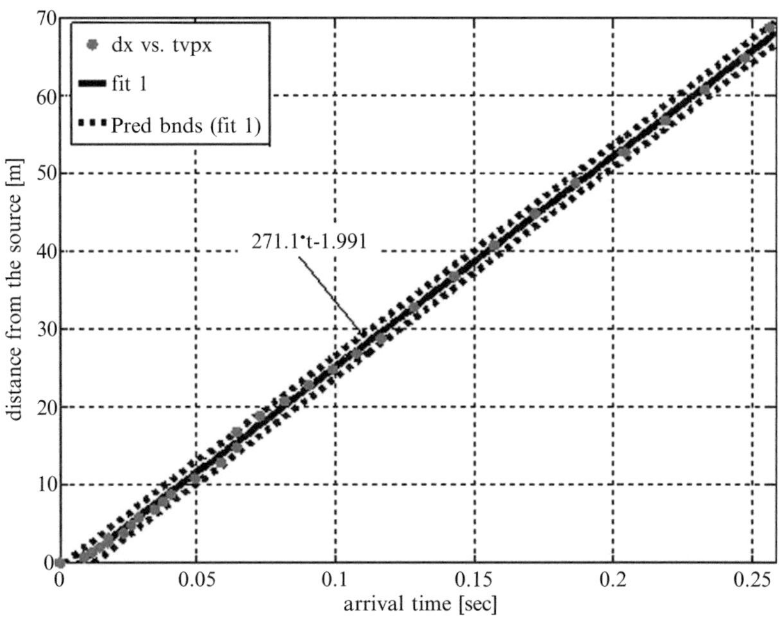


**Fig. 7.19** Evaluation of the P wave velocity based on time arrivals (Terzi 2009)

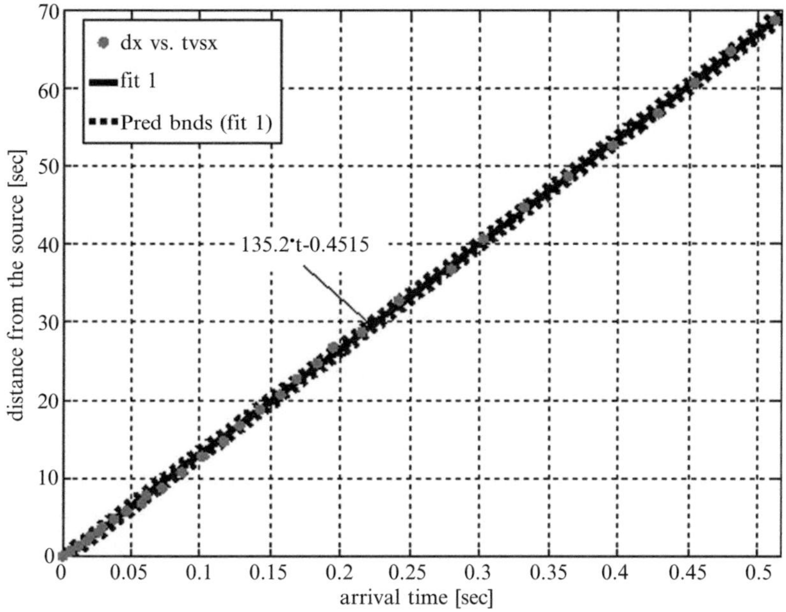


**Fig. 7.20** Evaluation of S wave velocity based on time arrivals (Terzi 2009)

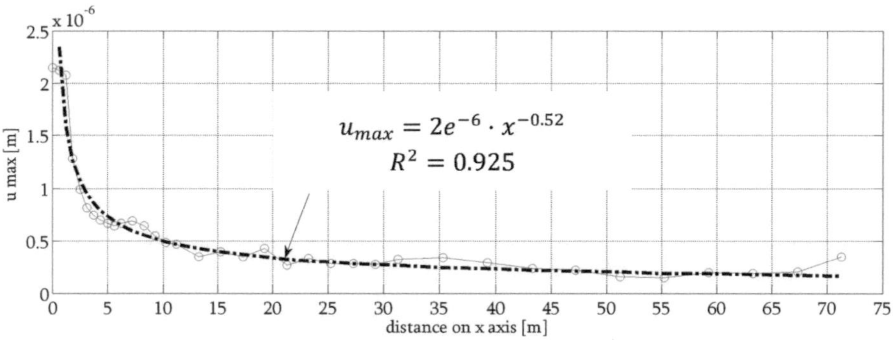


**Fig. 7.21** Spatial decay of maximum motion amplitude for the whole duration of the phenomenon (Terzi 2009)

we present the spatial decay of the maximum amplitudes of motion, considering the whole duration of the phenomenon. Therefore, all the types of generated waves contribute to the motion. The computed mean exponential curve has an exponential coefficient equal to −0.52 very close to the theoretical value −0.5 of the decay of surface waves (Graff 1975). In Fig. 7.22 we present again the spatial decay of the maximum amplitudes of motion, but in this case considering only in the time win-

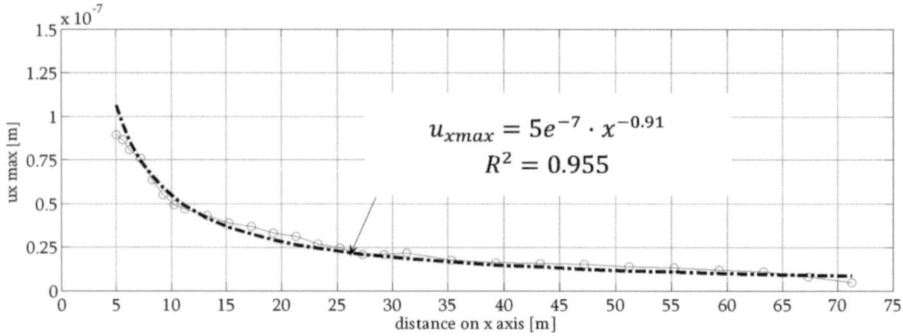


**Fig. 7.22** Spatial decay of maximum motion amplitude for the first arrival time window of the body waves (Terzi 2009)

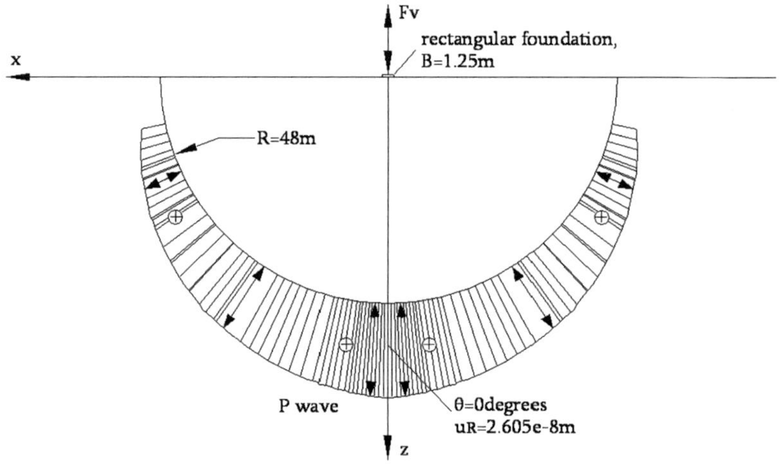


**Fig. 7.23** Maximum amplitude of the computed radial P-wave amplitudes in the XZ plane (Terzi 2009)

dow until the first arrivals of the waves. We remark that now the exponential coefficient takes a value of –0.91, which is close to the theoretical value −1.0 of the body waves (Graff 1975).

#### 7.4.1.5 Wave Field in Large Distances from the Source

The next stage of the validation concerns the investigation of the generated wave field at long distances from the foundation. In Figs. 7.23 and 7.24 we present the computed radial and tangential components of motion in the XZ plane, at 48 m from the foundation centre. The numerical results are then compared to the accurate

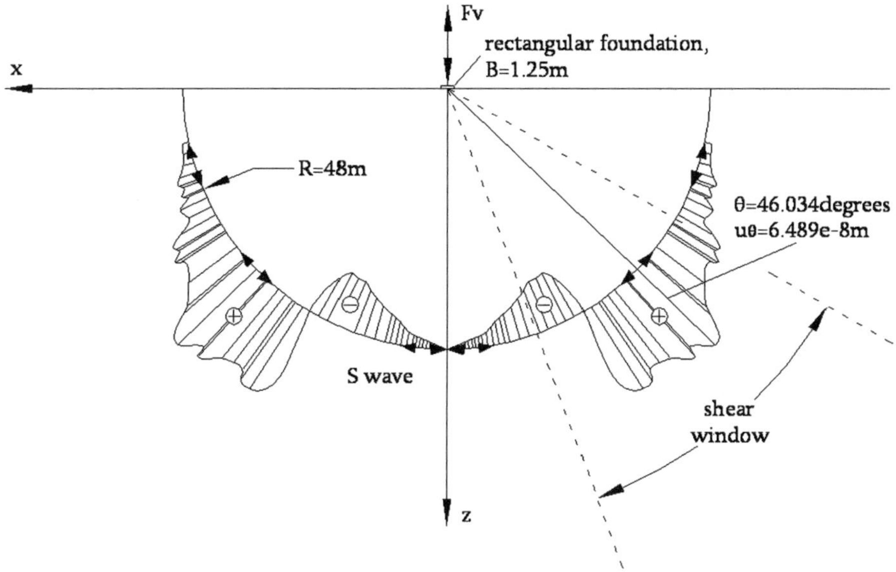


**Fig. 7.24** Maximum amplitude of the computed tangential S wave amplitudes in the XZ plane (Terzi 2009)

analytical solution for the half space provided by Miller and Pursey (1954, 1955) in Fig. 7.23. The comparison between analytical and numerical results is quite satisfactory, taking into account the inherent limitations of the finite element modeling.

Regarding Rayleigh waves their presence is identified in the total duration of the phenomenon, taking into account the time needed for the decay of body waves. The horizontal and vertical components of the R-waves are presented graphically in Fig. 7.24 at 51.25 m distance from the foundation, which may be considered as far field. The numerical results are then compared with analytical ones, provided again by Miller and Pursey (1954, 1955) and displaced graphically in Fig. 7.25. The comparison between the forms of the analytical and numerical results is quite good.

## *7.4.2 Comparison of Numerical and Experimental Results*

The comparisons of the numerical and experimental results have been performed in three stages. First type we compare the static contact pressures after the placement of the foundation slab, the pier and the deck. Then we perform FE modal analysis of the soil-structure system, in order to evaluate the eigenfrequencies of the system. Finally we analyzed the soil-structure systems in the time domain. The aim of these analyses is to validate the numerical model against the experimental results, and to gain extra information regarding the SSI phenomenon through the numerical analysis of the complete system.

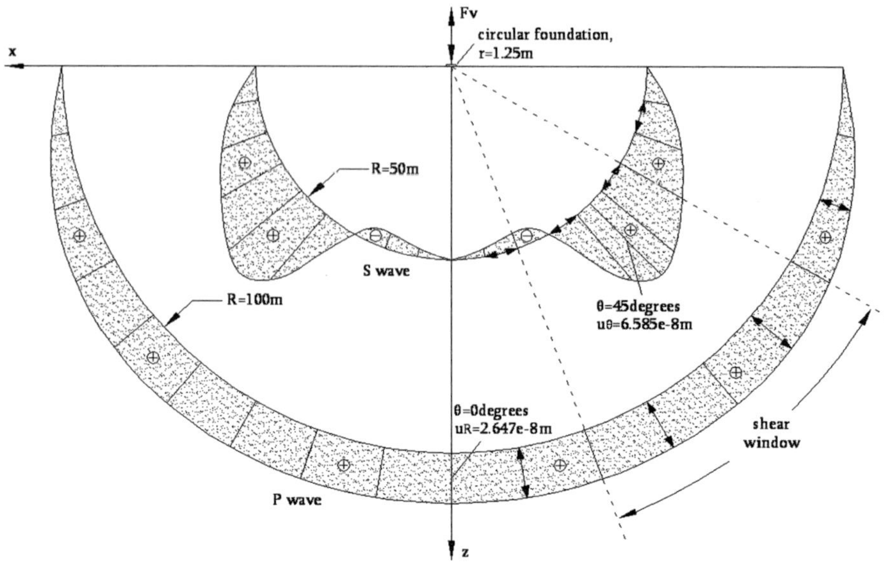


**Fig. 7.25** Theoretical maximum amplitudes of radial (P-wave) and tangential (S-wave) amplitudes in the XZ plane (Terzi 2009)

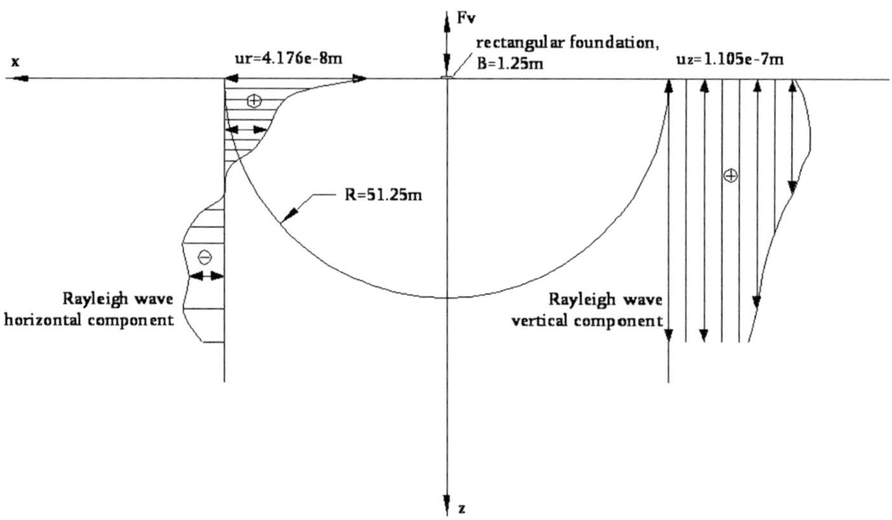


**Fig. 7.26** Computed maximum amplitudes of horizontal and vertical displacements of the Rayleigh waves in the XZ plane (Terzi 2009)

The 3D finite element model we used is presented in Fig. 7.26. Considering the low intensity of the induced vibration, only the first soft soil layer of the soil profile has been modeled, with solid elements having adequate dynamic soil properties (Figs. 7.2, 7.3 and Table 7.4). The lateral viscous boundaries are located far enough

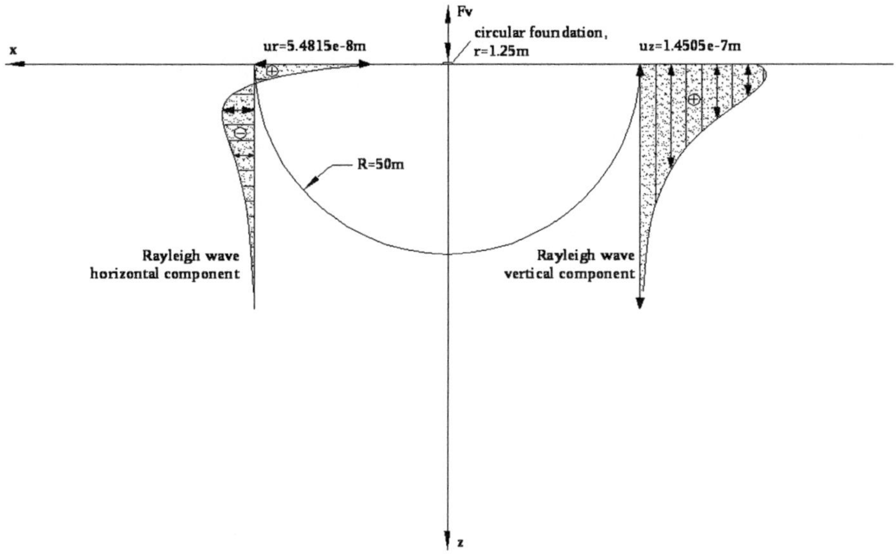


**Fig. 7.27** Theoretical maximum amplitudes of the horizontal and vertical displacements of the Rayleigh waves in XZ plane (Terzi 2009)

from the vibrating structure in order to avoid effects of reflected waves (Kuhlemeyer and Lysmer 1973). The dimensions of the mesh are 8.25×8.25×4 m with 42,519 elements and 47,943 nodes.

The average forces acting at the load cells are calculated as the sum of the products of the stresses in the load cells area. During the placement of the foundation slab, the average computed force in each load cell is 1.95 kN whereas the corresponding experimental value is 2.13 kN. The difference (8.6%) between the numerical and the experimental results is quite satisfactory, considering the inherent differences between the ideally symmetrical numerical simulation of the application of the dead loads and the actual placement of the pre-constructed foundation slab. When the pier and the deck are placed, the difference between the computed and measured forces is reduced to 3.2% (2.10 kN against 2.17 kN respectively). Figure 7.27a, b and c present (a) the mesh of the foundation surface with the area of the load cells, (b) the developed stresses in each cell and (c) the zoom in one load cell, all after the placement of the deck.

So the static modeling of the system is considered very satisfactory.

#### 7.4.2.1 Modal Analysis

During the free vibrations tests the two main translational modes of vibration of the bridge pier model were excited. The first three eigenmodes of the full model correspond to the eigenmodes of the structure itself. These three modes for the model

**Fig. 7.28** Modeling of load cells and soil (Terzi 2009)

**Table 7.5** Comparison of computed and recorded eigenfrequncies

| Case | Date of experiment | Eigenmodes | Eigenfrequencies, f [Hz] FEM | Experiment | Δf |
|---|---|---|---|---|---|
| PO1 | 06/04/2004 | Translational on weak axis | 1.822 | 1.833 | 0.60% |
| | | Torsional | 2.705 | – | – |
| | | Translational on strong axis | 3.288 | 3.3 | 0.36% |
| PO2 | 21/10/2004 | Translational on weak axis | 1.651 | 1.65 | 0.06% |
| | | Torsional | 2.710 | – | – |
| | | Translational on strong axis | 2.902 | 2.78 | 4.39% |
| PO3 | 19/05/2005 | Translational on weak axis | 1.185 | 1.13 | 4.91% |
| | | Torsional | 2.625 | – | – |
| | | Translational on strong axis | 2.484 | 2.48 | 0.15% |

structure with and without the tendons are given in Fig. 7.28. The first eigenmode corresponds to the translation in the weak axis of vibration (YY), whereas the other to two to the translation on the strong axis and the torsion. The added mass on the deck didn't affect the order of the eigenmodes. The activation of the tendons affects mostly the translational mode on strong axis and the torsional one. In particular the activation of the tendons results to the appearance of the torsional mode in the second place and of the translational on the strong axis in the third place.

Table 7.5 summarizes the computed eigenfrequencies and the corresponding translational experimental ones.

The maximum difference between numerical and measured values is only 4.91%, while in most cases is less than 1%, which is more than satisfactory, proving the successful numerical modeling of the experiments.

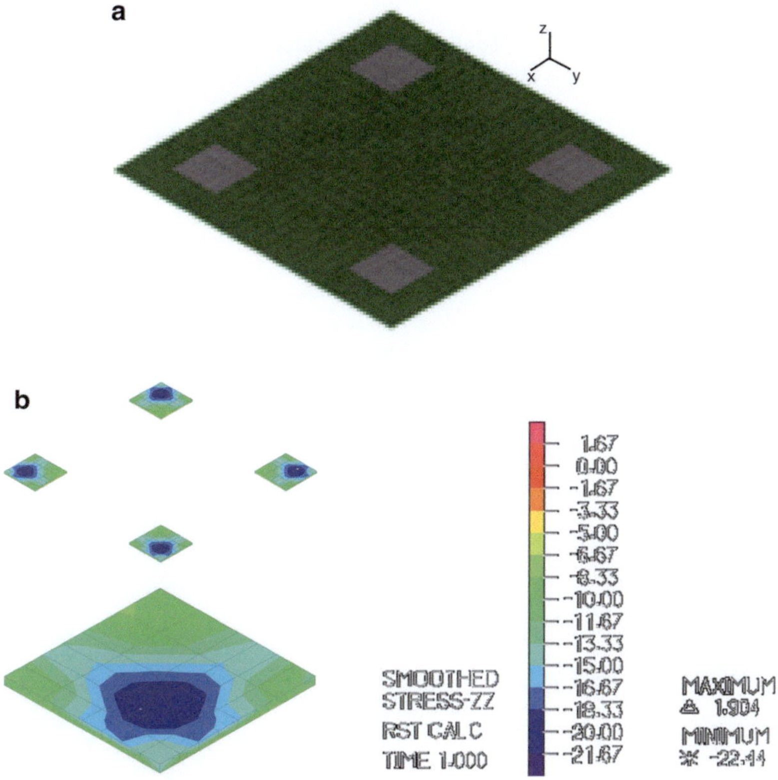


**Fig. 7.29** (**a**) Developed pressures load cells. Deck placement (Terzi 2009), (**b**) Developed pressures on one load cell. Deck placement (Terzi 2009)

### 7.4.2.2 Analysis in the Time Domain

We performed several analyses changing the inertial and stiffness characteristics of the model. In order to compare both time histories and eigenfrequencies we selected and present herein the tests PO2, where the instruments are placed on the foundation slab (Fig. 7.28) and hence the recorded signals provide direct information regarding the eigenfrequency characteristics of the soil-foundation-pier bridge system. Of course due to the low intensity of the applied forces the whole system remains in the elastic range. We used a Rayleigh damping and viscous lateral frontiers (ADINA 1999), while the dimensions of the elements are kept as low as possible (Kuhlemeyer and Lysmer 1973; Ewing et al. 1957). The time step used in the time integration is also kept very low ($\Delta T < T/20$). In Fig. 7.29 we compare the computed time histories in the three stations situated on the foundation slab with the measured ones.

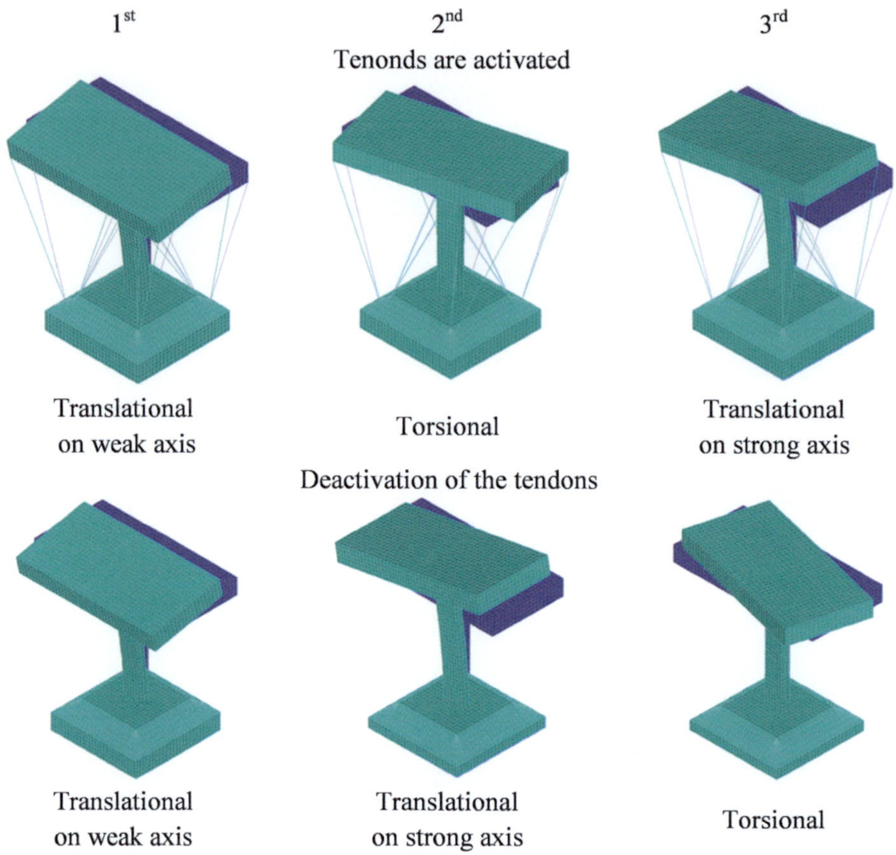


**Fig. 7.30** Eigenmodes of the bridge pier model with and without the tendons (Terzi 2009)

The pull out force is applied parallel to the weak axis of symmetry. Therefore the motion of the model is dominated by the in-plane horizontal motion $u_y$, parallel to the weak axis YY. It is observed that both in-plane and out-of plane motion follows the exponential decay scheme.

The comparison between the experimental and numerical results is successful, with the exception the out of plane component of the "left" instrument, which is probably due to possible malfunction of this particular component of the instrument.

In order to investigate the contribution of frequency sources, other than the oscillation of the bridge pier model itself, the Short Time Fourier Transform (STFT) is applied in all of the components of motion. We observe that the experimental frequency 1.65 Hz is exactly the same with the dominating numerical frequency (Fig. 7.30). An interesting observation of the numerical modeling concerns the identification of a second important frequency equal to 2.78 Hz,

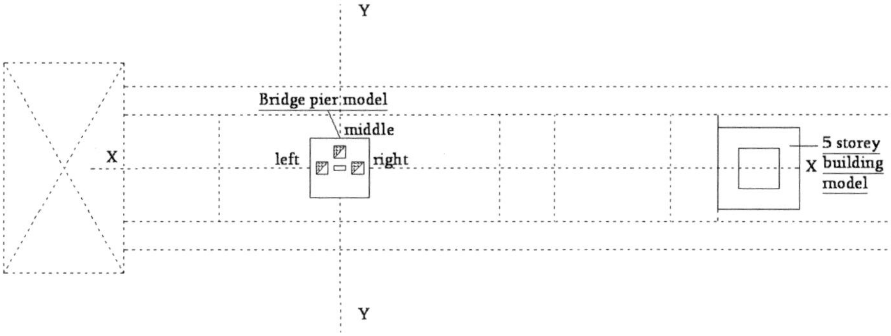


**Fig. 7.31** Instrumentation of foundation slab. Test PO2Y (Terzi 2009)

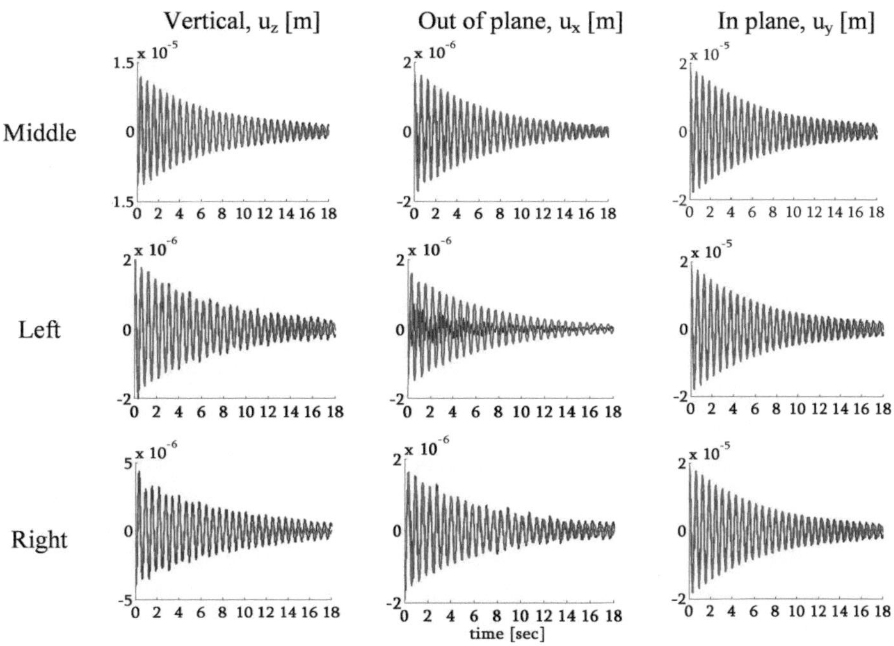


**Fig. 7.32** Computed (*green*) and measured (*blue*) time histories in three stations on the foundation slab (Terzi 2009)

which refers to the eigenfrequency of the system on axis XX (strong axis). Therefore, the system translates also in the axis perpendicular to that of the main vibration. The contribution to the total motion is limited in terms of time and amplitude, but it could be recognized via STFT (MATLAB 2004) (Figs. 7.31, 7.32 and 7.33).

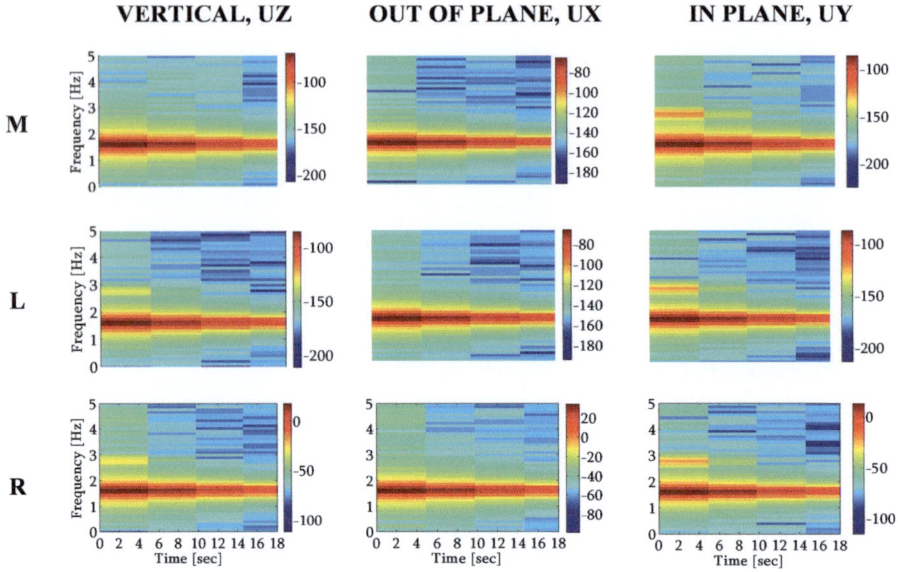


Fig. 7.33 Short Time Fourier Transform of the computed motions (Terzi 2009)

#### 7.4.2.3 SSI Effects

The presence and importance of the SFSI effects in these low intensity experiments, are examined comparing the modal analysis characteristics of the previously validate full soil-foundation- structure model, with the case where the pier bridge model is founded on no-deformable base (fixed base conditions). The contribution of the soil flexibility and the foundation compliance is highlighted in the following Table 7.6. As it is expected the fixed based foundation leads to a less compliant system and therefore, the eigenfrequencies are affected in all vibration modes. In particular in case of compliant foundation it is observed that the SFSI, decreases the eigenfrequency up to 11%, in both translational modes. The stiffer the original structure, the higher is the importance of soil flexibility. In case of really strong seismic excitation this differences will be certainly much higher. The torsional mode of vibration is in general less affected compared to the translational modes in both strong and weak axis.

## 7.5 Evaluation of Foundation Impedance Functions

The experimental evaluation of impedance foundation functions of real full scale or model structures is a complicated problem, mainly because the experimental records can never describe all the details of the complex response of the structure, and they cannot provide all necessary information (Luco et al. 1986; Lin AN 1982; Crouse et al. 1984). In this work we propose to evaluate the foundation impedance functions of the EUROSEISTEST model applying a hybrid method that combines

**Table 7.6** Soil-foundation-structure interaction effects

| Test characteristics | | | | Eigenfrequencies, f [Hz] | | |
|---|---|---|---|---|---|---|
| Case | Added mass | Tendons | Eigenmodes | Compliant foundation | Fixed base conditions | Δf |
| PO1 | No | On | Translational in weak axis | 1.822 | 2.013 | −10.53% |
| | | | Torsional | 2.705 | 2.741 | −1.30% |
| | | | Translational in strong axis | 3.288 | 3.647 | −10.92% |
| PO2 | Yes | On | Translational in weak axis | 1.651 | 1.690 | −2.35% |
| | | | Torsional | 2.710 | 2.718 | −0.30% |
| | | | Translational in strong axis | 2.902 | 3.124 | −7.64% |
| PO3 | Yes | Off | Translational in weak axis | 1.185 | 1.202 | −1.42% |
| | | | Torsional | 2.625 | 2.632 | −0.27% |
| | | | Translational in strong axis | 2.484 | 2.640 | −6.29% |

**Table 7.7** Basic characteristics of the four case studies

| Case study | Date of experiment | Axis of vibration | Eigen frequency f [Hz] | Tendons state | ω [rad/s] | $\alpha_o$ |
|---|---|---|---|---|---|---|
| 1st | 06/04/2004 | XX (in plane) | 3.234 | Active | 20.320 | 0.188 |
| 2nd | 21/10/2004 | XX (in plane) | 2.841 | Active | 17.851 | 0.165 |
| 3rd | 21/10/2004 | YY (out of plane) | 1.651 | Active | 10.374 | 0.096 |
| 4th | 19/05/2005 | XX (in plane) | 2.450 | Inactive | 15.394 | 0.143 |

experimental and numerical results. Since the comparison between the two methods is successful, all extra information provided by the numerical analysis, and missing by the experimental procedure, like stresses and displacements in the time domain, in various points of the model, shall be used in order to study SSI effects and evaluate the impedance functions of the foundation.

For this purpose we used the results of four experiments that were modeled and analyzed in time domain (Table 7.7). Since the excitation of the bridge pier model refers to a free vibration oscillation, the eigenfrequency of the system is the frequency of excitation. In three out of the four case studies the load is applied in the XX axis. The last column of Table 7.7 provides the normalized frequency α0, which is usually used in the graphical representation of foundation impedance functions.

In Fig. 7.34 the two excitation systems are depicted. The horizontal force applied at the deck of the bridge pier model is transferred at the interface between the column and the foundation as a pair of horizontal force and a moment. Therefore, the motion of the foundation is a combination of translation parallel to the axis of vibration and rocking around the aforementioned axis. Thus, three types of foundation impedance functions can be evaluated, horizontal translation, rotation and coupling of the two types of movements.

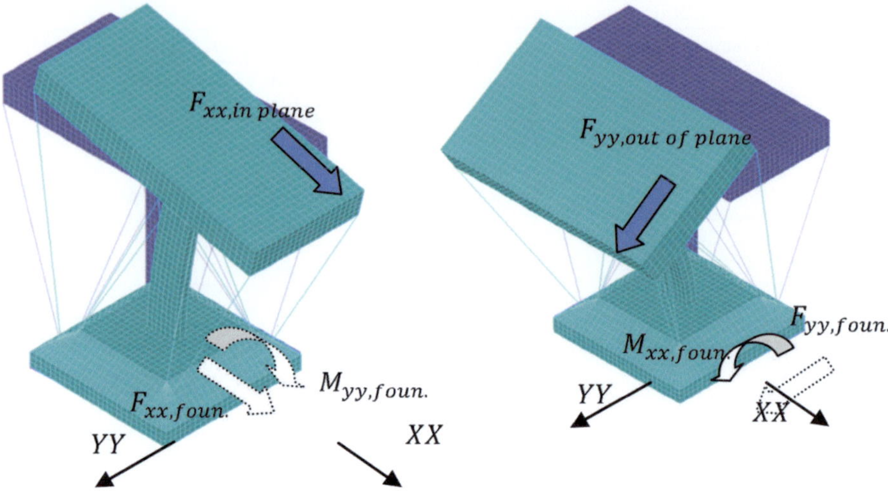


**Fig. 7.34** In plane and out of plane excitation (Terzi 2009)

The equations describing the real and imaginary part of the motion, accounting also the exponential decay scheme, are given below (Terzi 2009): Equation 7.1a, b provides the real and imaginary parts for the rotational mode, Eq. 7.2a, b for the translational motion, and Eq. 7.3a, b the coupling of the two types of motion. The coefficient $\zeta_1$ corresponds to the decay of the load, whereas the coefficient $\zeta_2$ to the decay of the motion. In our case the two coefficients take the same value and therefore, the impedance functions do not vary to time and have a unique value for each frequency of excitation.

| Degree of freedom | Kreal | |
|---|---|---|
| Rotation | $\dfrac{M_o \cdot e^{-\zeta_1 \cdot \omega \cdot t}}{u_{rot,0} \cdot e^{-\zeta_2 \cdot \omega \cdot t}} \cdot cos\ \varphi$ | (7.1) |
| Translation | $\dfrac{F_o \cdot e^{-\zeta_1 \cdot \omega \cdot t}}{u_{hori,0} \cdot e^{-\zeta_2 \cdot \omega \cdot t}} \cdot cos\ \varphi$ | (7.2) |
| Coupling of rotation and translation | $\dfrac{M_o \cdot o^{-\zeta_1 \cdot \omega \cdot t}}{u_{hori,0} \cdot e^{-\zeta_2 \cdot \omega \cdot t}} \cdot cos\ \varphi = \dfrac{F_o \cdot o^{-\zeta_1 \cdot \omega \cdot t}}{\varphi_{rot,0} \cdot e^{-\zeta_2 \cdot \omega \cdot t}} \cdot cos\ \varphi$ | (7.3) |

| Degree of freedom | Kimag | |
|---|---|---|
| Rotation | $-i \cdot \dfrac{M_o \cdot e^{-\zeta_1 \cdot \omega \cdot t}}{u_{rot,0} \cdot e^{-\zeta_2 \cdot \omega \cdot t}} \cdot sin\ \varphi$ | (7.4) |
| Translation | $\dfrac{F_o \cdot e^{-\zeta_1 \cdot \omega \cdot t}}{u_{hori,0} \cdot e^{-\zeta_2 \cdot \omega \cdot t}} \cdot cos\ \varphi$ | (7.5) |
| Coupling of rotation and translation | $-i \cdot \dfrac{M_o \cdot e^{-\zeta_1 \cdot \omega \cdot t}}{u_{hori,0} \cdot e^{-\zeta_2 \cdot \omega \cdot t}} \cdot sin\ \varphi = -i \cdot \dfrac{F_o \cdot e^{-\zeta_1 \cdot \omega \cdot t}}{\varphi_{rot,0} \cdot e^{-\zeta_2 \cdot \omega \cdot t}} \cdot sin\ \varphi$ | (7.6) |

**Table 7.8** Static foundation impedance functions

| Case study | $\alpha_o$ | $K_{h,static}$ [kN/m] | $K_{r,static}$ [kNm/rad] | $K_{hr,static}$ [kN/rad] |
|---|---|---|---|---|
| 3rd | 0.096 | 177,308 | 243,687 | 79,921 |
| 4th | 0.143 | 357,901 | 495,609 | 140,823 |
| 2nd | 0.165 | 292,677 | 362,418 | 119,994 |
| 1st | 0.188 | 294,584 | 379,003 | 120,381 |

**Table 7.9** Dynamic foundation impedance functions

| Horizontal translational degree of freedom | | | | | | |
|---|---|---|---|---|---|---|
| Case study | $\alpha_o$ | $K_{h,static}$ [kN/m] | $K_{h1}$ | $K_{h2}$ | $K_{h1}/K_{h,static}$ | $K_{h2}/K_{h,static}$ |
| 3rd | 0.096 | 177,308 | 176,695 | 14,732 | 0.9965 | 0.0831 |
| 4th | 0.143 | 357,901 | 353,932 | 39,090 | 0.9889 | 0.1092 |
| 2nd | 0.165 | 292,677 | 289,682 | 41,764 | 0.9898 | 0.1427 |
| 1st | 0.188 | 294,584 | 289,932 | 42,874 | 0.9842 | 0.1455 |
| Rotational degree of freedom | | | | | | |
| Case study | $\alpha_o$ | $K_{r,static}$ [kNm/rad] | $K_{r1}$ | $K_{r2}$ | $K_{r1}/K_{r,static}$ | $K_{r2}/K_{r,static}$ |
| 3rd | 0.096 | 243,687 | 243,521 | 8,969 | 0.9993 | 0.0368 |
| 4th | 0.143 | 495,609 | 495,066 | 23,193 | 0.9989 | 0.0468 |
| 2nd | 0.165 | 362,418 | 361,846 | 20,351 | 0.9984 | 0.0562 |
| 1st | 0.188 | 379,003 | 378,207 | 24,547 | 0.9979 | 0.0648 |
| Coupling degree of freedom | | | | | | |
| Case study | $\alpha_o$ | $Kh_{r,static}$ [kN/rad] | $K_{hr1}$ | $K_{hr2}$ | $K_{hr1}/K_{hr,static}$ | $K_{hr2}/K_{hr,static}$ |
| 3rd | 0.096 | 243,687 | 243,521 | 8,969 | 0.9993 | 0.0368 |
| 4th | 0.143 | 495,609 | 495,066 | 23,193 | 0.9989 | 0.0468 |
| 2nd | 0.165 | 362,418 | 361,846 | 20,351 | 0.9984 | 0.0562 |
| 1st | 0.188 | 379,003 | 378,207 | 24,547 | 0.9979 | 0.0648 |

Taking all necessary parameters in Eqs. 7.1–7.3 form the numerical modeling and analyses of the respective tests (Table 7.7), we were able to calculate the values of the static and dynamic impedance factors presented in Tables 7.8 and 7.9 respectively.

The estimated values of the static and dynamic impedance functions are then compared with analytical impedance functions available in the literature. Table 7.10 summarizes the analytical expressions used in the present comparison. In all cases the soil parameters used for the analytical calculation conform to the soil properties of the upper layer of EUROSEISTEST site, whereas the dimension of the circular foundation are compatible, for each degree of freedom, with the foundation of the model.

Figure 7.35 presents the comparison of these results with analytical solutions for rectangular foundation on halfspace. The comparison is within acceptable limits for the translational and coupled mode, while in the rotational motion we observe rather important divergences. Considering the real part, we remark that our estimation leads to a stiffer behavior, certainly due to the activated tendons, whereas the imaginary part has generally higher values due to the combined action of radial and material damping.

**Table 7.10** Analytical expressions of foundation impedance functions used in the comparison

| Foundation shape | Soil profile | Analytical expression provided by |
|---|---|---|
| Rectangular | Half space | Dominguez and Roesset (1978) |
| | | Wong and Luco (1976) |
| | | Rucker (1982) |
| | | Schmid et al. (1988) |
| | Soil layer over half space | Wong and Luco (1985) |
| Circular | Half space | Veletsos and Wei (1971) |
| | Soil layer over rigid bedrock | Kausel (1974) |
| | | Luco and Westmann (1971) |
| Arbitrary | Half space | Dobry and Gazetas (1986) |

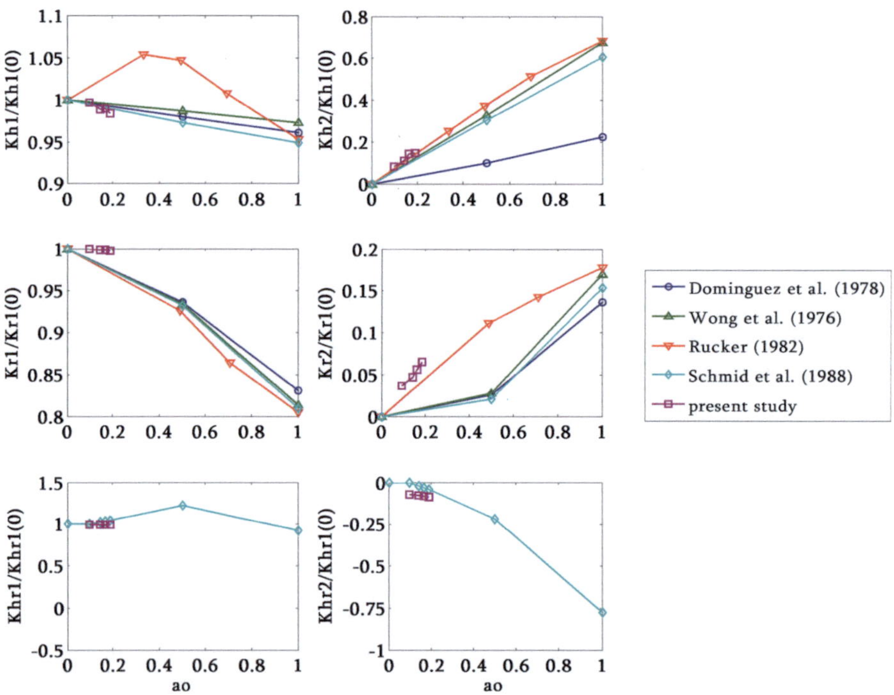


**Fig. 7.35** Comparison of impedances issued from this study with analytical expressions for rectangular foundation on halfspace (Terzi 2009)

In Fig. 7.36 we compare our results with analytical expression provided by Wong and Luco (1985), for rectangular foundation on a soil layer over halfspace. The comparison for the translational and coupled motion is fairly satisfactory. The real part of the rotational motion matches very well between the two approaches, whereas in the imaginary part we observe higher differences but within the acceptable limits.

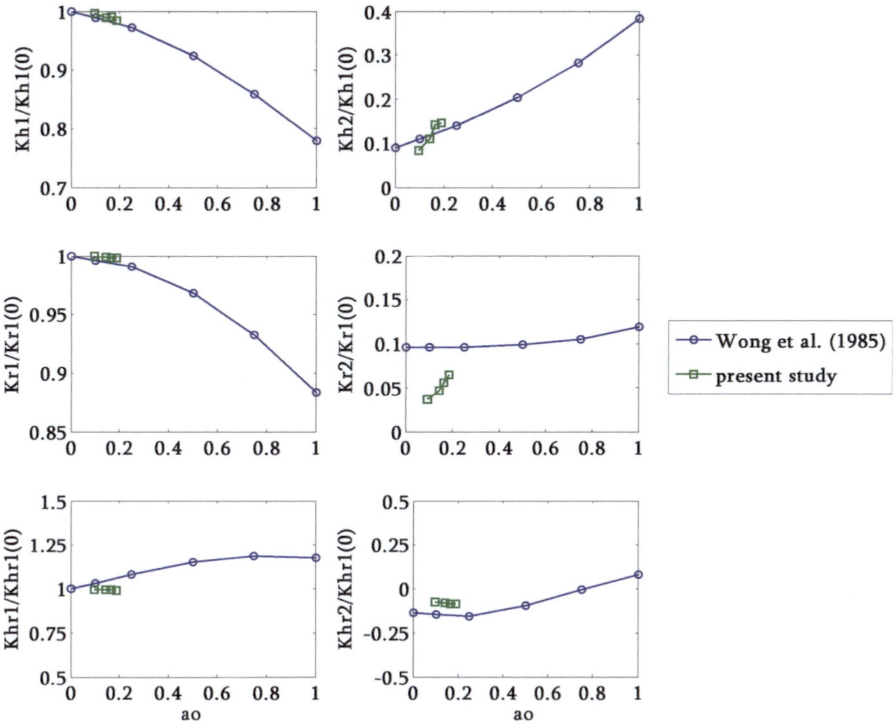


**Fig. 7.36** Comparison of impedance issued from this study with analytical expression for rectangular foundation on soil layer over halfspace (Terzi 2009)

The main diverges are observed in the imaginary part of the rotation motion. The aforementioned differences can be attributed to the fact the finite element soil model is fixed at the depth of 4 m, whereas the analytical expression refers to a soil layer over halfspace. Therefore the participation of the radiation damping is expected to be larger in the analytical expression. Finally the small differences observed in the coupling term can be attributed to the soil stratification assumption, and the role of the tendons.

In Fig. 7.37 we compare our results with the classical analytical expression by Veletsos and Wei (1971), for circular foundation on halfspace. Referring to the real part of the horizontal translational and the rotational motion, the observed differences may be attributed to the action of the tendons, and the soil-foundation interface conditions. In particular in our numerical model we assume full contact condition between the soil and the foundation slab, whereas the analytical study assumes relaxed conditions. Furthermore, the higher values of the imaginary part concerning the aforementioned degrees of freedom can be attributed to the combined effects of the radiation and material damping. The small differences concerning the coupling term may be considered negligible.

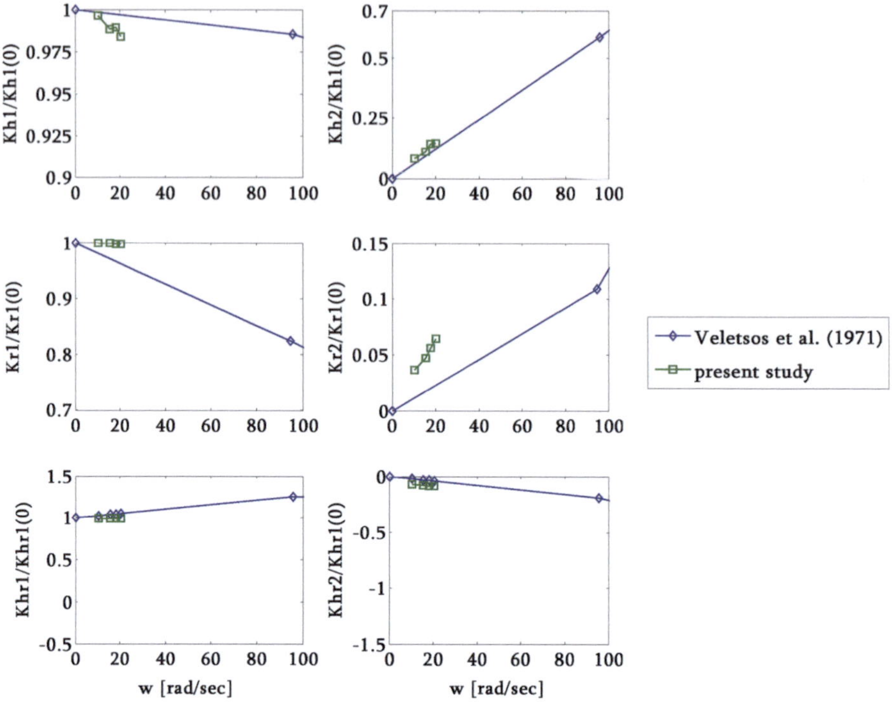


**Fig. 7.37** Comparison of impedance issued from this study with analytical expression for circular foundation on halfspace (Terzi 2009)

The next comparison is between our study and the analytical expressions by Kausel (1974) and Luco and Westmann (1971), for circular foundation on soil layer over rigid bedrock. Three different analytical cases are considered according to the ratio of the depth H of the soil layer and the radius r of the foundation. The differences are in general important for small values of the H/r ratio. The analytical expressions fail in estimating the measured values, following our hybrid procedure. For the real part of the horizontal translational and the rotational motion the differences may be attributed to the action of the tendons and the assumption in soil stratification. Furthermore, the lower values of the imaginary part, always for the aforementioned degrees of freedom, can be attributed to the combination of the action of radiation and material damping in a limited depth of soil layer assumed by the finite element model, while the analytical study takes into account a damping coefficient of 0.05, which is considerably larger than the measured one in these tests (Fig. 7.38).

Finally in Fig. 7.39 we compare our results with a well known analytical expression proposed by Dobry and Gazetas (1986), for arbitrary shaped foundation on halfspace. In the real part of the translational motion the hybrid method gives lower values, which can be attributed to the limited soil depth assumed in the numerical model. On the contrary for the imaginary part the comparison is perfect, probably

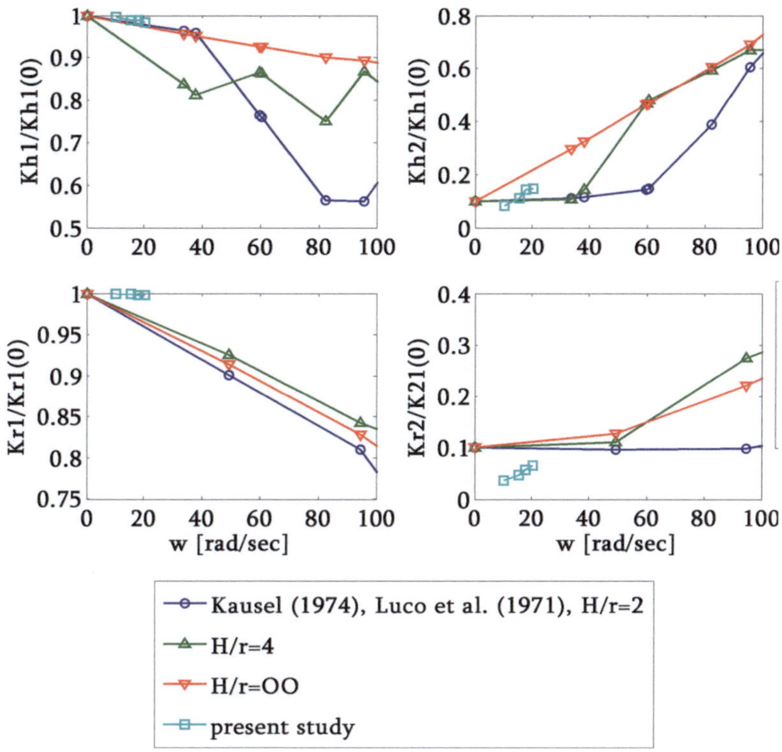


**Fig. 7.38** Comparison of impedances for this study and analytical expressions for circular foundation on soil layer over rigid bedrock (Terzi 2009)

because the effects of the radiation and material damping in our analysis gives similar results to the radiation damping in halfspace assumed by the analytical study. The characteristic point concerning the analytical results referring to the rotation degree of freedom and particular in the real part is the different values even under the prism of same foundation dimensions in both axes. Our approach gives higher values, which can be attributed to the action of the tendons. Furthermore, higher values are displaced by the hybrid method in terms of the imaginary part. Thus, the rotational degree of freedom is more affected by the action of damping, either radiation or material one.

## 7.6 Conclusion

We presented and discussed a free vibration experiment in a SDOF model structure, (scaled 1:3), constructed in the EUROSEISTEST experimental test site (http://euroseis.civil.auth.gr). The aim was four-fold: (a) to model accurately the experiment and

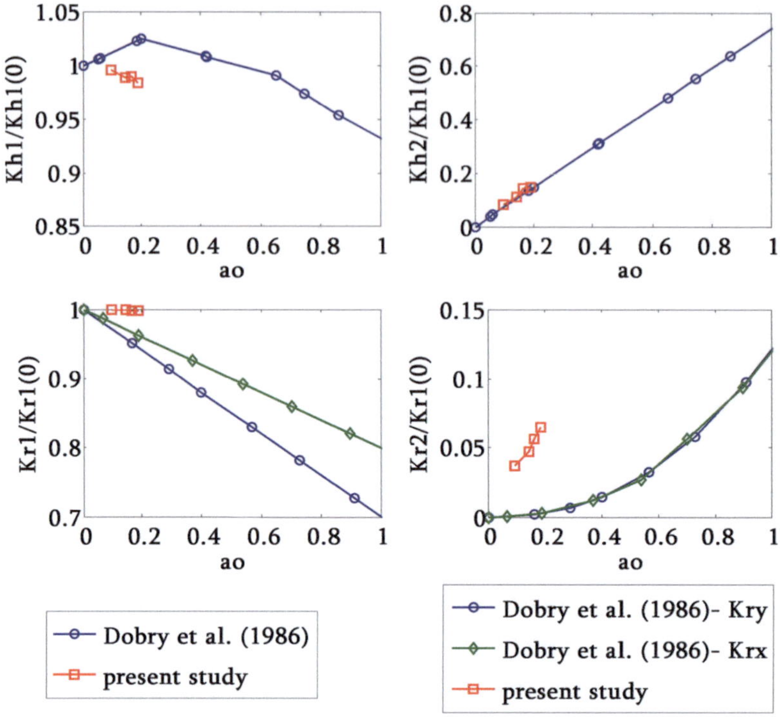


**Fig. 7.39** Comparison of impedances for this study and analytical expressions for arbitrary shaped foundation on halfspace (Terzi 2009)

the results with a 3D numerical model, (b) to study in frequency and time domain the wave field generated by the pull-out tests and propagated in the ground during the free oscillation of the structure, (c) to study several important issues regarding the numerical modeling of SFSI, and (d) to estimate the impedance factors in this specific case, and to compare them with available analytical solutions. The main results are summarized as follows:

The experiments have been accurately simulated with a 3D finite element model, both in time and frequency domain. The validation of the numerical modeling has been made according to Lamb's problem. The wave field generated has been correctly reproduced at different distances, and the body and surface waves have been well identified and compared to theoretical solution

The attenuation of ground motion of body and surface waves, in terms of material and radiation damping has been correctly reproduced from the exponential amplitude decay of the recorded ground motion at different distances from the model. Its very low values (<1%) prove that the structure and the soil remain practically in the linear elastic range.

The generated wave field along the axis of the excitation is dominated by the in plane horizontal and the vertical components of the ground motion, whereas in the axis perpendicular to the excitation, the motion is dominated only by the in plane horizontal component.

The initial geostatic conditions and the foundation-soil contact stresses have been accurately simulated.

The modal analysis captured successfully three pairs of eigenfrequncies in the translational axis and provided valuable information concerning the torsional mode of vibration, which is seriously affected by the presence of the diagonal tendons offering extra stiffness to the model.

The activation of the tendons affects the order of appearance of the torsional and translational strong axis eigenmode.

Regarding the effect of the soil flexibility in eigenfrequency characteristics, it is demonstrated that the torsional mode of the vibration is practically unaffected, contrary to the translational mode, which in both the weak and strong axis, decreased of about 11% compared to the fixed base case.

To evaluate the foundation impedance functions of the foundation model, a hybrid approach has been developed, which combines experimental and numerical results. The prototype of the method is attributed to the use of free field recordings as validation, to the exponentially decay of the time histories and to the fact that the same structure under small changes in inertial and stiffness characteristics produced various eigenfrequency configurations.

The hybrid method results in terms of translational, torsional and coupled impedance factors are compared with analytical solutions, for rectangular, circular and arbitrary shaped foundation over halfspace and various soil profiles. The comparison of the real and imaginary part is not always successful for all three modes. The main reason may be attributed to the idealized scheme for the foundation and the soil used in the analytical solutions, contrary to our method, which considers closely the real soil and foundation conditions.

**Acknowledgments** The research presented in this paper has been performed in the framework of the EU research program "Seismic hazard assessment, site effects and soil-structure interaction studies in an instrumented basin", (EUROSEISRISK, EVG1-CT-2001-00040). Profound acknowledgments are due to professor George Manos who provided the model set up, and performed the free vibration tests. We also acknowledge the crucial contribution of assistant professor Dimitrios Raptakis and Dr. Maria Manakou, who were deployed the free field ground motion instrumentation and helped us in getting good quality records. The support of the Hellenic State Scholarships Foundation to the second author is also acknowledged.

## References

ADINA (1999) Theory and modeling guide, vol I, ADINA. Report ARD 99–7

Bathe KJ (1996) Finite element procedures. Prentice Hall, Upper Saddle River, p 07458

Chopra AK (1995) Dynamics of structures-theory and applications to earthquake engineering. Prentice Hall International Inc, Upper Saddle River

Chow YK (1986) Simplified analysis of dynamic response of rigid foundations of arbitrary shape. Earthq Eng Struct Dyn 14:643–653

Crouse CB, Liang GC, Martin GR (1984) Experimental study of soil-structure interaction at an accelerograph station. Bull Seismol Soc Am 74:1995–2013

Dobry R, Gazetas G (1986) Dynamic response of arbitrarily shaped foundations. J Geotech Eng 112:109–135

Dominguez J, Roesset JM (1978) Dynamic stiffness of rectangular foundations. Research Report R78-20 MIT

Ewing W, Jardetzky WS, Press F (1957) Elastic waves in layered media. McGraw-Hill, New York

Graff KF (1975) Wave motion in elastic solids. Dover Publications INC, New York

Kausel E (1974) Forced vibrations of circular foundations on layered media. Research Report R74-11, MIT

Kuhlemeyer LR, Lysmer J (1973) Finite element method accuracy for wave propagation problems. J Soil Mech Found Div Proc Am Soc Civ Eng 99:421–427

Lamb H (1904) On the propagation of tremors over the surface of an elastic solid. Philos Trans R Soc Lond A203:1–42

Lin AN (1982) Experimental observations of the effect of foundation embedment on structural response. EERL 82–01, Earthquake Engineering Research Laboratory, California Institute of Technology, Pasadena

Luco JE, Westmann RA (1971) Dynamic response of circular footings. J Eng Mech Div Am Soc Civ Eng 97:1381

Luco JE, Wong HL, Trifunac MD (1986) Soil-structure interaction effects on forced vibration tests. Report 86–05, University of Southern California, Department of Civil Engineering. Los Angeles

Manakou M, Raptakis D, Chavez-Garcia FJ, Apostolidis P, Pitilakis K (2010) 3D soil structure of the Mygdonian basin for site response analysis. Soil Dyn Earthq Eng 30:1198–1211. doi:10.1016/j.soildyn.2010.04.027

Manos GC et al (2005) Final report for W5, Structural behavior and SSI effects studies. EUROSEIS-RISK, Seismic hazard assessment, site effects and soil structure interaction studies in an instrumented basin, Contract: EVGI-CT-2001-00040, 2005

Manos GC, Kourtides V, Sextos A (2008) Model bridge pier-foundation-soil interaction implementing in-situ/shear stack testing and numerical simulation. In: 14th World conference on earthquake engineering, Beijing

MATLAB (2004) Signal processing toolbox, user's guide. Version 7.0.0.19920 (R14)

Miller GF, Pursey H (1954) The field and radiation impedance of mechanical radiators on the free surface of a semi-infinite isotropic solid. Proc Roy Soc Lond A Math Phys Sci 223:521–541

Miller GF, Pursey H (1955) On the partition of energy between elastic waves in a semi-infinite solid. Proc Roy Soc Lond A Math Phys Sci 223:55–69

Pitilakis KD (2009) Experimental and theoretical SSI studies in a model structure in Euroseistest. In: Earthquake geotechnical engineering satellite conference, XVIIth international conference on soil mechanics and geotechnical engineering, 2–3 Oct 2009, Alexandria

Pitilakis K, Raptakis D, Lontzetidis K, Tika-Vassilikou Th, Jongmans D (1999) Geotechnical and geophysical description of EURO-SEISTEST, using field, laboratory tests and moderate strong motion recordings. J Earthq Eng 3:381–409

Raptakis D, Makra K, Anastasiadis A, Pitilakis K (2004a) Complex site effects in Thessaloniki (Greece) – I: soil structure and confrontation of observations with 1D analysis. Bull Earthq Eng 3:271–290

Raptakis D, Makra K, Anastasiadis A, Pitilakis K (2004b) Complex site effects in Thessaloniki (Greece) – II: 2D SH modelling and engineering insights. Bull Earthq Eng 3:301–327

Raptakis D, Manakou M, Chavez-Garcia FJ, Makra K, Pitilakis K (2005) 3D configuration of Mygdonian basin and preliminary estimate of its site response. Soil Dyn Earthq Eng 25:871–887

Rucker W (1982) Dynamic behavior of rigid foundations of arbitrary shape on a halfspace. Earthq Eng Struct Dyn 10:675–690

Schmid G, Willms G, Huh Y, Gibhardt M (1988) Ein progammsystem zur berechnung von Bauwerk-Boden-Wechsel-wirkungs-problemen mit der Randelementmethode. SFB 151 – Berichte Nr. 12, Dezember, RUB

Terzi VG (2009) Experimental and theoretical investigation of specific fundamental issues concerning the dynamic soil-foundation-superstructure interaction. PhD thesis (in greek)

Terzi VG, Pitilakis KD (2009) Experimental and numerical analysis of the dynamic characteristics of a bridge pier model. In: COMPDYN 2009, Rhodes, 22–24 June 2009

Veletsos AS, Wei YT (1971) Lateral and rocking vibrations of footings. J Soil Mech Found Div Proc Am Soc Civ Eng 97:1227–1248

Wong HL, Luco JE (1976) Dynamic response of rigid foundations of arbitrary shape. Earthq Eng Struct Dyn 4:579–587

Wong HL, Luco JE (1985) Tables of impedance functions for square foundations on layered media. Soil Dyn Earthq Eng 4:64–81

# Chapter 8
# Soil-Structure Interaction for Seismic Improvement of Noto Cathedral (Italy)

**Michele Maugeri, Glenda Abate, and Maria Rossella Massimino**

**Abstract** Noto Cathedral, one of the most famous example of Baroque architecture in Italy, was damaged by the December 13, 1990 earthquake ($M_L = 5.4$) and because of the damage it partially collapsed on May 13, 1996. After the collapse, accurate investigations were performed on the structure and on the subsoil, in order to rebuild and improve this very significant religious heritage building.

The present paper deals with a pseudo-static and two 2-D FEM analyses of soil-structure interaction. These analyses were carried out for the maximum credible scenario earthquake bigger than the 1990 earthquake. Pseudo-static analysis as well as transient dynamic analysis by FEM modelling was performed. The bearing capacity of the foundation-soil system, as well as foundation displacements and rotations were investigated.

## 8.1 Introduction

Noto Cathedral, one of the most famous examples of Baroque architecture in Italy, was damaged by the December 13, 1990 earthquake ($M_L = 5.4$; Rovelli et al. 1991). In particular, some columns of the right side of the main nave were severely fessured; due to the degeneration of these cracks, in 1996 the cathedral suffered a major partial collapse, which involved the columns of the right side of the main nave, the roof of the main nave and of the right nave and part of the dome (Fig. 8.1).

M. Maugeri (✉) • G. Abate • M.R. Massimino
Department of Civil and Environmental Engineering,
University of Catania, Viale Andrea Doria 6, 95125 Catania, Italy
e-mail: mmaugeri@dica.unict.it; glenda.abate@dica.unict.it; mmassimi@dica.unict.it

M.A. Sakr and A. Ansal (eds.), *Special Topics in Earthquake Geotechnical Engineering*, Geotechnical, Geological and Earthquake Engineering 16,
DOI 10.1007/978-94-007-2060-2_8, 

**Fig. 8.1** The collapse of the main and right naves and of the dome after the damage caused by the 1990 South-East Sicilian earthquake

Accurate investigations were performed on the structure and on the subsoil, in order to rebuild and improve this very significant religious heritage building (Maugeri et al. 2006). The structural investigation (Binda et al. 2001) was mainly devoted to the characterization of masonry properties, as well as to the analysis of the masonry construction technique. The soil foundation investigation (Cavallaro et al. 2003) was devoted to estimate soil static and dynamic properties. The results of the investigations were utilised to perform an accurate geotechnical model for the numerical simulation of the seismic behaviour of the soil underneath the cathedral (Massimino and Maugeri 2003a). A pseudo-static analysis (Massimino and

Maugeri 2003b), a numerical analysis of soil-foundation interaction (Cavallaro et al. 2003) and dynamic numerical analyses of (SSI) soil-structure interaction (di Prisco et al. 2006; Abate et al. 2009) were performed.

The pseudo-static analysis was carried out for the design spectrum (Massimino and Maugeri 2003a), derived from a previous analysis on seismic hazard of Noto (Campoccia and Massimino 2003). The design spectrum is referred to the maximum credible scenario earthquake, the 1,693 Val di Noto earthquake (estimated magnitude $M_L = 7.0 \div 7.3$; Priolo 1999, 2000).

The numerical analyses were carried out for the synthetic accelerograms evaluated on the basis of source mechanism modelling by Bottari et al. (2003).

As regards the reconstruction phase, not only the superstructure but also the foundation was subjected to remedial works.

By the pseudo-static analysis, the bearing capacity of the foundation-soil system, as well as foundation displacements and rotations were investigated taking into account the inertial effects on the foundation soil as described by the European Seismic Code EC8 (2003). The dynamic SSI analysis allowed us to investigate the effectiveness of the main remedial works, taking into account the effect of the soil deformability on the behaviour of the over structure.

## 8.2 Dynamic Geotechnical Characterization

After the 1996 collapse, accurate investigations were performed, in order to find the reasons for the collapse as well as to plan the seismic improvement and the reconstruction of the cathedral.

The geotechnical characterization of the foundation soil of the Noto Cathedral regarded an area of 3,200 $m^2$ (40 m × 80 m), with a maximum depth of 81 m. Twelve boreholes, including Standard Penetration Tests, three Down-Holes, four Cross-Holes and four Menard Pressiometer Tests were performed in-situ (Fig. 8.2).

Different laboratory tests were also performed on undisturbed samples: three Odometer Tests, eight Direct Shear Tests, nine (CD, CU and UU) Triaxial Tests, six Cyclic Loading Torsional Shear Tests (CLTST; $v = 0.1 \div 0.01\%$/min), six Resonant Cyclic Tests (RCT; $v = 10 \div 2{,}000\%$/min) and three Cyclic Triaxial Tests (CLTx; $v = 1, 0.1, 0.01\%$/min).

Soil non linearity was evaluated by means of CLTST and RCT (Cavallaro et al. 2003). Figures 8.3 and 8.4 show respectively the normalized shear modulus $G/G_0$ versus the shear strain $\gamma$ and the damping ratio D versus the normalized shear modulus $G/G_0$, obtained from RCT tests.

The comparison among the SPT, CLTST and RCT results allowed to propose $G_0$ versus z profiles for all the 12 boreholes (Fig. 8.5), as reported in Massimino and Maugeri (2003a).

According to the $G_0$ versus z profiles an average $G_0$ equal to 127,500 kPa is considered in the present paper, as well as an operative value of $D = 10\%$ was chosen for an operative shear strain $\gamma = 0.1\%$.

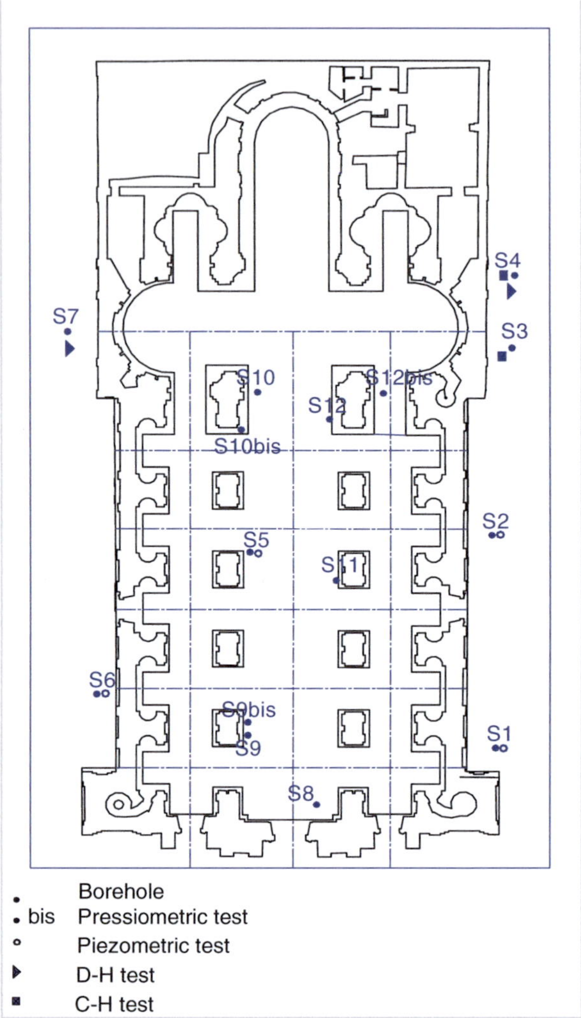


**Fig. 8.2** Location of the boreholes and in-situ tests performed after the 1996 partial collapse of Noto Cathedral (After Massimino and Maugeri 2003a)

## 8.3 Seismic Action and Design Spectrum

The seismic action was defined by the Italian Seismic Regulation on the basis of an exceedance less than 10% in a period of 50 years. According to the Italian Seismic Regulation the seismic action for the city of Noto was $a_g = 0.07$ g at the time of the reconstruction design of the Cathedral.

Because of the importance of the religious building, a deterministic evaluation of the seismic action was performed for the scenario January 11, 1,693 earthquake ($M_L = 7.0 \div 7.3$; Priolo 1999, 2000), which was strongly bigger than the 1990 earthquake.

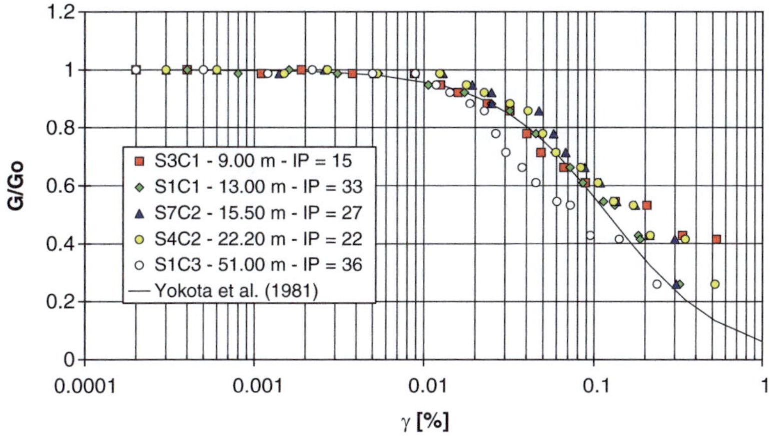


**Fig. 8.3** Degradation of normalized shear modulus $G/G_0$ with shear strain $\gamma$ (After Cavallaro et al. 2003)

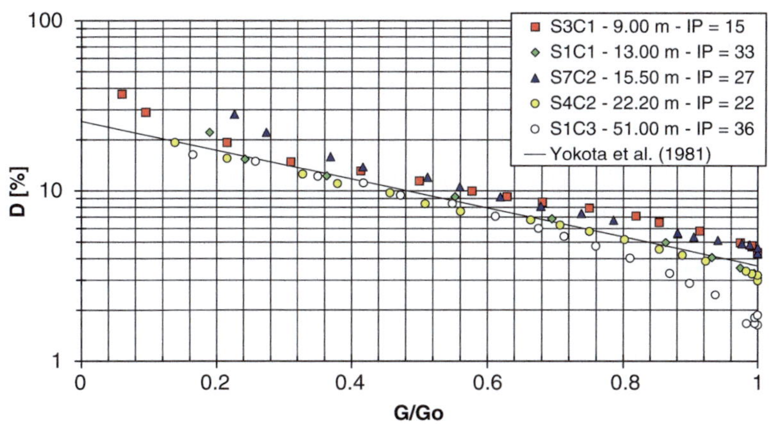


**Fig. 8.4** Increasing of damping ratio D with degradation of normalized shear modulus $G/G_0$ (After Cavallaro et al. 2003)

Synthetic accelerograms were evaluated by means of three source rupture models at the Ibleo-Maltese Fault: (i) unilateral failure S-N; (ii) unilateral failure N-S; (iii) bilateral failure (Fig. 8.6). The synthetic accelerograms were computed at the location of the Cathedral at three different depths: 0, 30, and 70 m (Bottari et al. 2003).

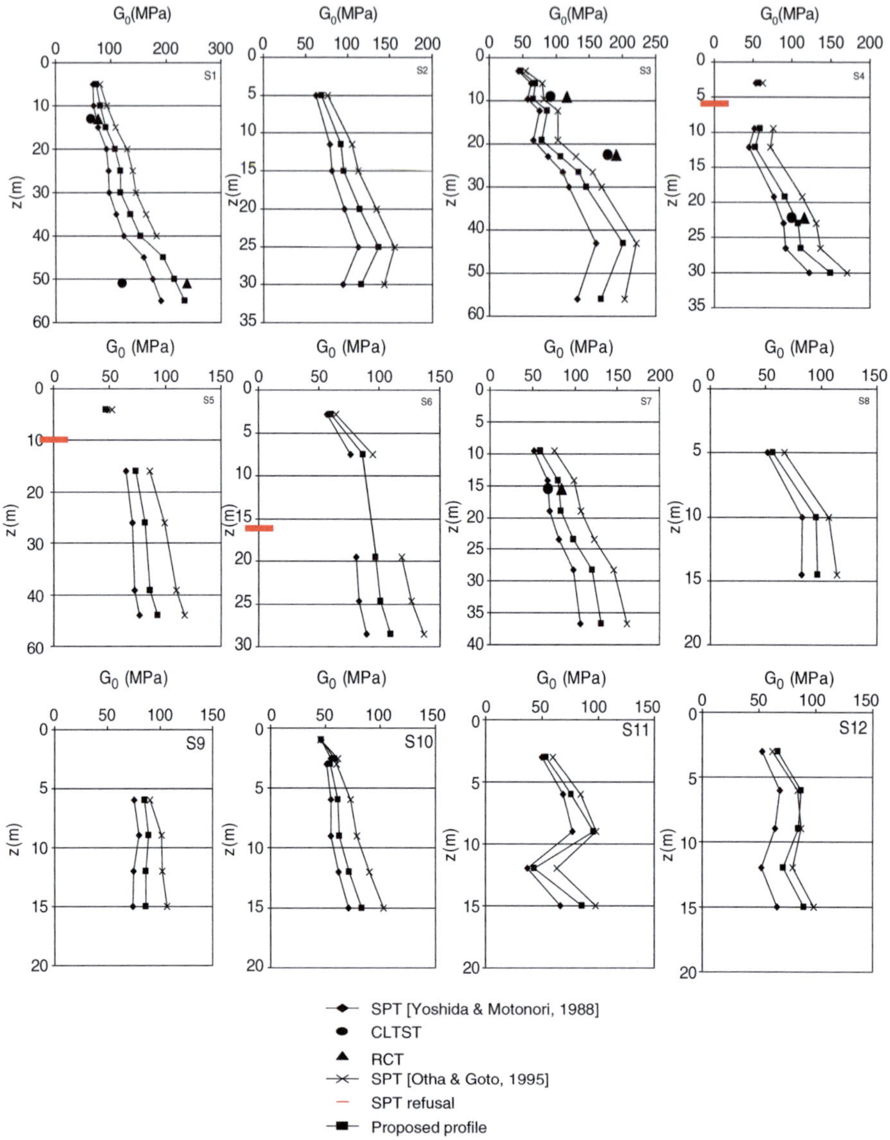


**Fig. 8.5** Shear modulus profiles by empirical correlations with SPT results for Noto Catherdal foundation soil; comparison with laboratory tests results and designed profiles (After Massimino and Maugeri 2003a)

According to the geotechnical investigations, the bedrock level for the foundation soil of Noto Cathedral is 45–50 m below the soil surface, so in order to perform DSSI analyses, the synthetic accelerograms at 30 m below the soil surface were considered. As a benchmark, the behaviour of the structure was evaluated considering

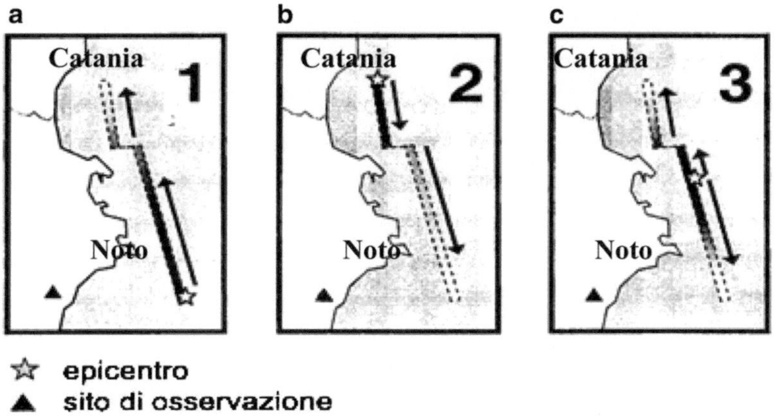


**Fig. 8.6** Rupture propagation models for the Ibleo-Maltese fault: (**a**) propagation from South to North; (**b**) propagation from North to South; (**c**) bilateral propagation (After Bottari et al. 2003)

**Table 8.1** Main characteristics of the input synthetic accelerograms for z = 30 m (After Massimino and Maugeri 2003a)

| Name | Model | Direction | PGA (m/s$^2$) | $T_0$ (s) |
|---|---|---|---|---|
| ACC1EW-30 | 1 | E-W | −1.48800 | 0.33 |
| ACC1NS-30 | 1 | N-S | −0.39971 | 0.20 |
| ACC2EW-30 | 2 | E-W | +0.98752 | 0.23 |
| ACC2NS-30 | 2 | N-S | −0.40684 | 0.21 |
| ACC3EW-30 | 3 | E-W | −1.28410 | 0.25 |
| ACC3NS-30 | 3 | N-S | +1.07720 | 0.36 |

$T_0$ = predominant period

a fixed-base (without soil underneath) utilising the synthetic accelerograms at the soil surface. Thus, considering the two horizontal component of motion (E-W and N-S) the analyses were performed utilising six synthetic accelerograms at 30 m below the soil surface and six synthetic accelerograms on the soil surface.

The main properties of the input synthetic accelerograms are reported in Table 8.1 and Table 8.2, respectively; in Fig. 8.7 two accelerograms, at 30 m below the soil surface (Fig. 8.7a) and at 0 m on the soil surface (Fig. 8.7b), are shown.

In order to evaluate a design spectrum for the seismic improvement of the Cathedral, a 1-D site response analysis was performed for each shear modulus profile of Fig. 8.5 and taking into account the soil non linearity laws reported in Figs. 8.3 and 8.4. The evaluated elastic response spectra (Massimino and Maugeri 2003a) were then transformed in inelastic spectra according to Giuffrè and Giannini (1982), referring to a structural ductility estimated equal to 2 for Noto Cathedral (Fig. 8.8).

So, the medium spectrum was computed and it was compared with the anelastic spectrum suggested by EC8-Part1 (2003) for soil type C, with refer to a structural damping ratio D = 10%. By this comparison the design spectrum shown in Fig. 8.9 were proposed.

Table 8.2 Main characteristics of the input synthetic accelerograms for z=0 m (After Massimino and Maugeri 2003a)

| Name | Model | Direction | PGA (m/s²) | $T_0$ (s) |
|---|---|---|---|---|
| ACC1EW-0 | 1 | E-W | −2.17740 | 0.29 |
| ACC1NS-0 | 1 | N-S | −0.43126 | 0.21 |
| ACC2EW-0 | 2 | E-W | +1.14580 | 0.23 |
| ACC2NS-0 | 2 | N-S | −0.54016 | 0.24 |
| ACC3EW-0 | 3 | E-W | −1.46600 | 0.26 |
| ACC3NS-0 | 3 | N-S | +1.52750 | 0.29 |

$T_0$=predominant period

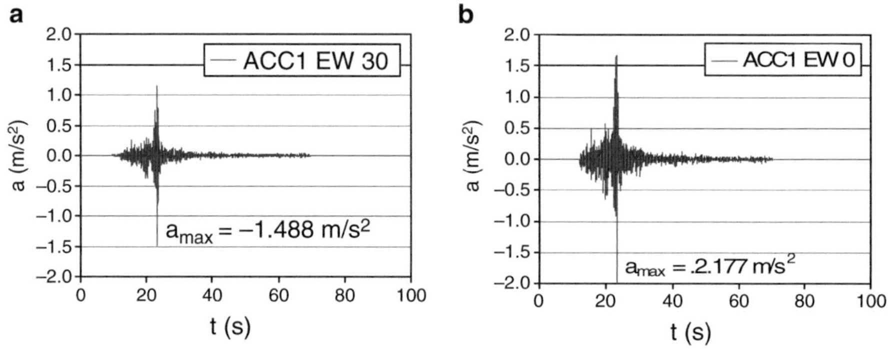


Fig. 8.7 Synthetic accelerograms given as input motion for the DSSI analyses, at 30 m below the soil surface (Fig. 8.7a) and at the soil surface (Fig. 8.7b)

## 8.4 Seismic Improvement of the Cathedral

According to the geotechnical and structural investigations, and with refer to the design spectrum, different remedial works were planned (Fig. 8.10). The structural investigations (Binda et al. 1999) showed that the 1990 earthquake caused severe cracks in the columns. These cracks increased from that time on, thus in 1996 the severe collapse, shown in Fig. 8.1 occurred. The reasons of this collapse was due to: the poor material used inside the columns; the realization of stairs inside one of the main columns supporting the dome; the substitution of the original roof with a reinforced concrete roof. In the meantime great irregularities in the foundations were found, characterized by different sizes and embedments and no connections.

The main remedial works consisted in: increasing the foundation embedment and size; using tie rods along the foundations (Fig. 8.11a) and inverted arches between the foundations (Fig. 8.11b); underpinning the external wall foundations to reinforce the foundations (Fig. 8.12); reconstructing not only the collapsed columns but also the other columns (Fig. 8.13); using tie rods to reinforce lateral arches (Fig. 8.14); using carbon fibres to reinforce the central arch.

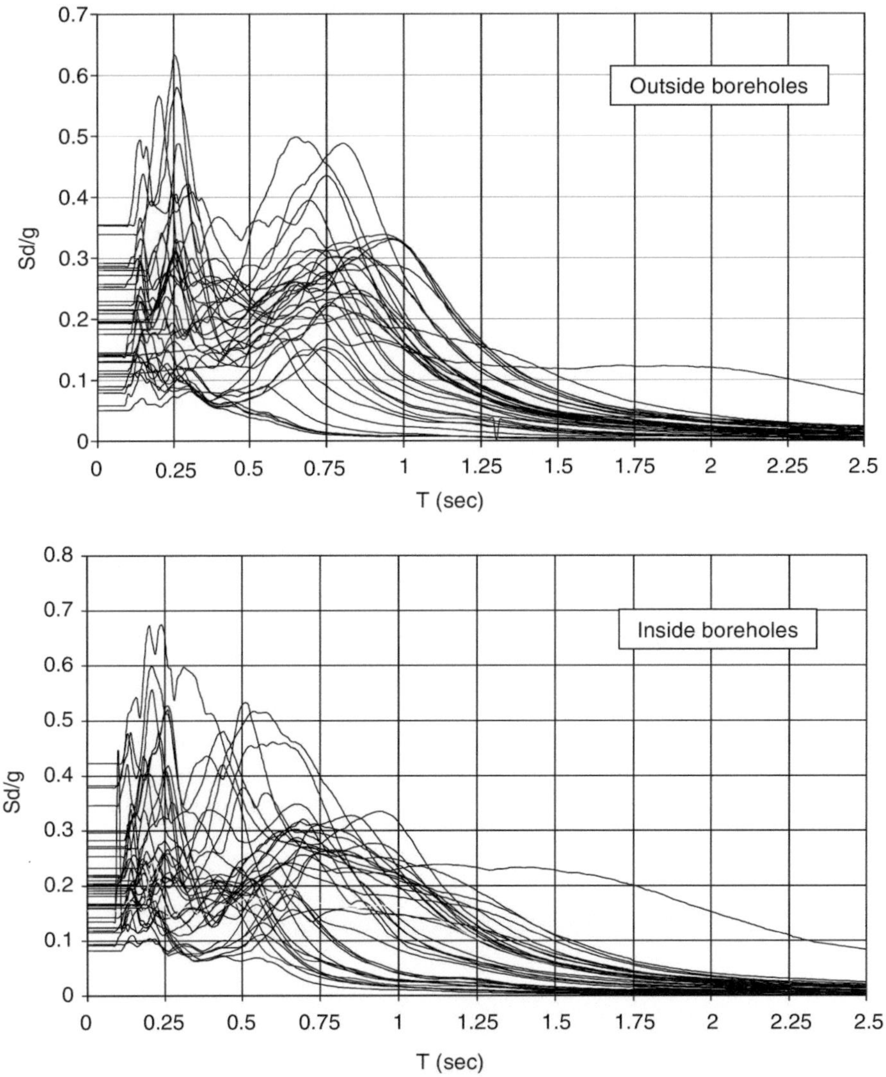


**Fig. 8.8** Inelastic design spectra for Noto Cathedral foundation soil (After Massimino and Maugeri 2003b)

## 8.5 Pseudo-Static Analysis

The results of the geotechnical characterization were utilised to develop a geotechnical model to perform a pseudo-static analysis, which allowed evaluating loads, displacements and rotations of the cathedral (Massimino and Maugeri 2003b; di Prisco et al. 2006).

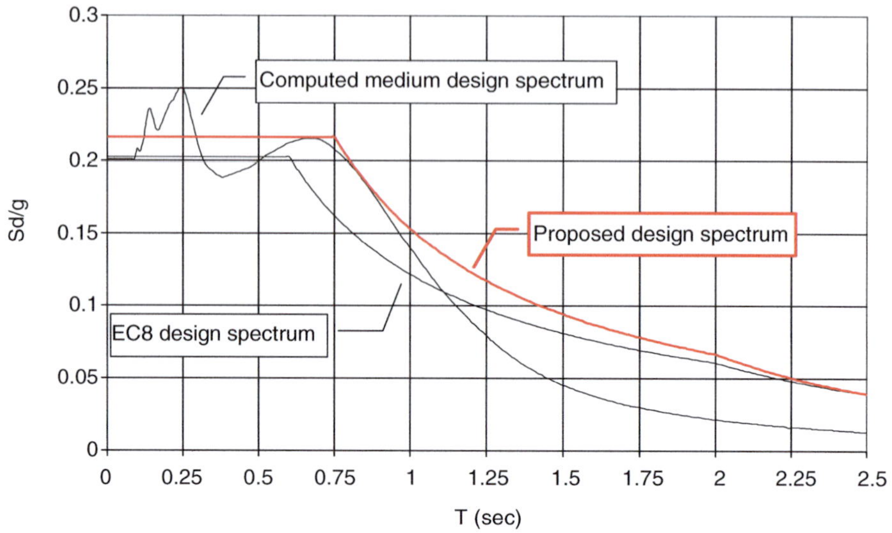


**Fig. 8.9** Design spectrum proposed for Noto Cathedral (After Massimino and Maugeri 2003a)

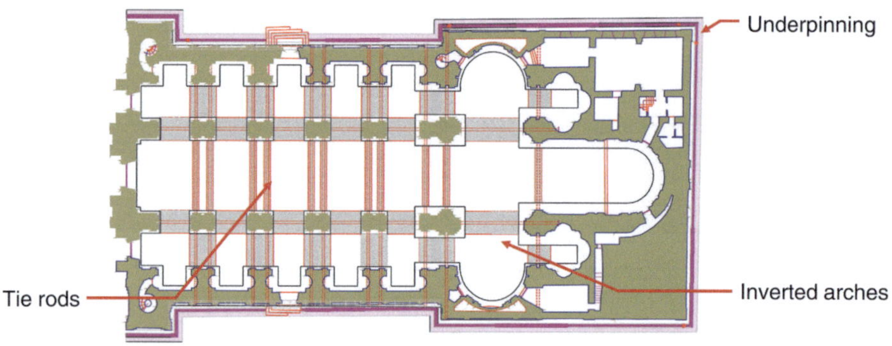


**Fig. 8.10** Remedial works on Noto Cathedral

**Fig. 8.11** Connection between the foundations: (**a**) by tie rods; (**b**) by inverted arches at the foundation level

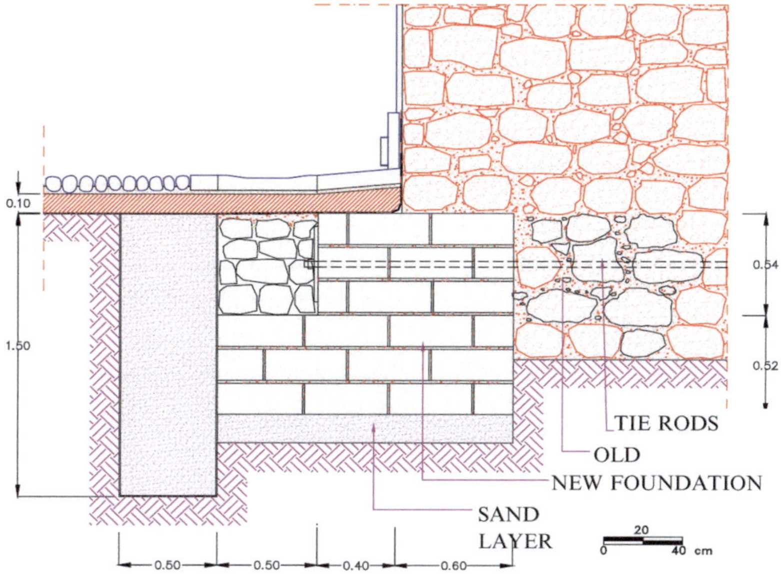


**Fig. 8.12** Underpinning of the external wall foundations

**Fig. 8.13** Different stages of the reconstruction of a column: (**a**) demolition of a damage column; (**b**) plan view of the new column; (**c**) elevation of the new column

**Fig. 8.14** Tie rods installed in the arches and in the foundations of the lateral naves

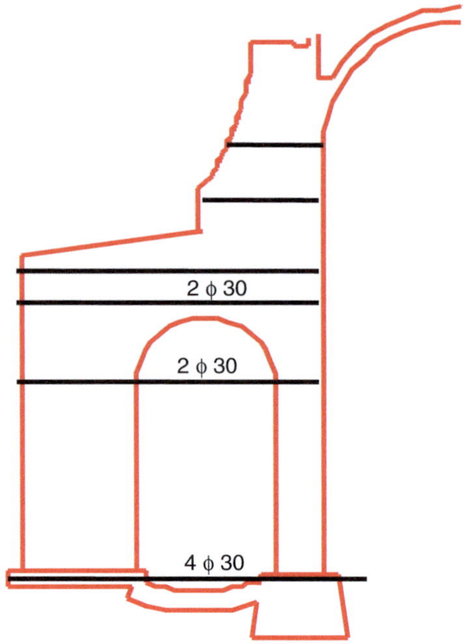


The analyzed cross-section across the naves is shown in Fig. 8.15; this section was subdivided in single elements (Fig. 8.16a) and for each element an equivalent 1-D element was considered (Fig. 8.16b). As regards the foundation soil deformability, according to the Gazetas method (Gazetas 1991), three springs per each node of the foundation were considered, in order to allow vertical displacements, horizontal displacements and rotations of the foundation. These springs are characterized by vertical, horizontal and rocking stiffness's evaluated for embedded foundations (Table. 8.3); in particular, two different typologies of foundation were considered: foundation at the external walls (nodes 1 and 43 in Fig. 8.16b) and foundation at the columns (nodes 11 and 33 in Fig. 8.16b). Moreover, in order to take into account the non linearity of the foundation soil, a secant shear modulus was used, determining the corresponding $E_{50}$.

For the analysis of the behaviour of the frame of Fig. 8.16b, a vertical load $W_i$, due to the weight of the structure (static analysis), and an horizontal load $H_i = k_h \cdot W_i$, which simulated the earthquake force (pseudo-static analysis), were considered per each node of the frame. The seismic coefficient $k_h$ was fixed equal to 0.22, with refer to the design spectrum shown in Fig. 8.9.

On the basis of the results shown in Table 8.4, the bearing capacity for the foundation of Noto Cathedral was estimated, both for static and pseudo-static conditions (Table. 8.5). The bearing capacity was evaluated both according to the Brinch-Hansen (1970) and according to the Paolucci and Pecker (1997), in order to take into account the soil inertial effect. Moreover, the bearing capacity was also evaluated with refer to the ultimate limit states according to EC7 (2004) and EC8 (2003).

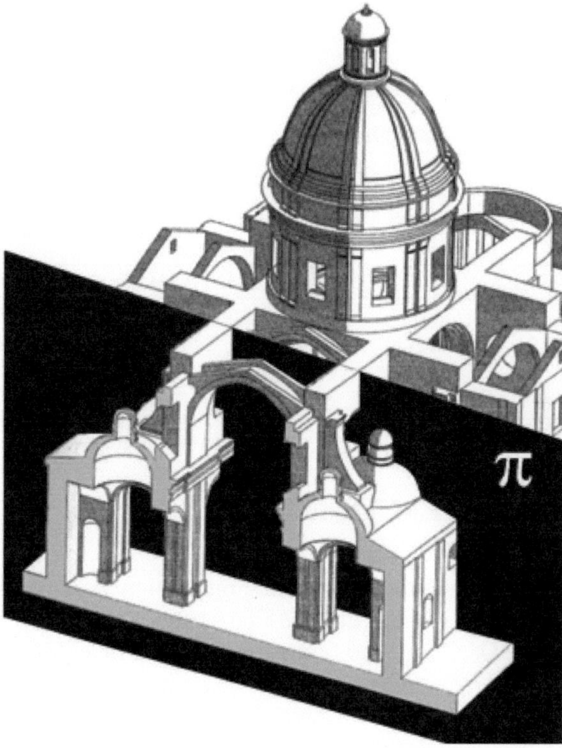


**Fig. 8.15** The analyzed cross-section across the naves

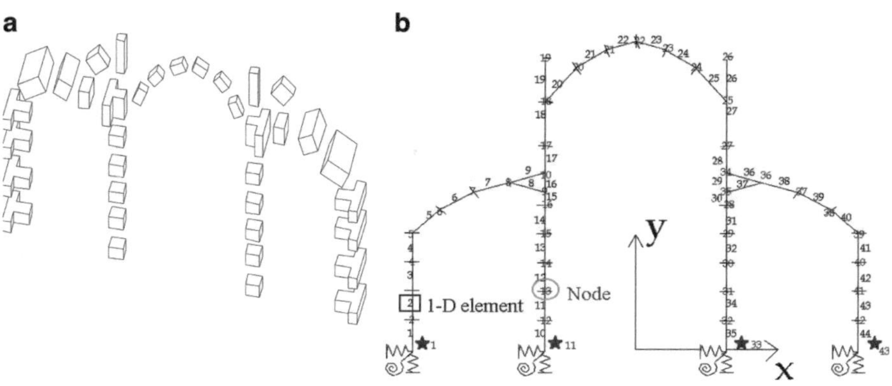


**Fig. 8.16** Schematization of the frame: (**a**) single elements; (**b**) adopted configuration (After Massimino and Maugeri 2003b)

**Table 8.3** Vertical, horizontal and rocking stiffnesses for the foundation of the frame shown in Fig. 8.15b (After Massimino and Maugeri 2003b)

| | $K_{z,\,emb}$ (kN/m) | $K_{y,\,emb}$ (kN/m) | $K_{rx,\,emb}$ (kN/m) |
|---|---|---|---|
| Foundation at the external walls | 1,192,521 | 1,249,358 | 9,281,965 |
| Foundation at the columns | 1,182,569 | 1,423,707 | 7,672,130 |

**Table 8.4** Loads on the foundation nodes (After Massimino and Maugeri 2003b)

| | Static conditions | | | Pseudo-static conditions | | |
|---|---|---|---|---|---|---|
| Nodes | N (kN) | H (kN) | M (kNm) | N (kN) | H (kN) | M (kNm) |
| 1 | 5638.00 | 271.37 | 26.54 | 4349.13 | 1283.46 | 4709.50 |
| 11 | 5117.48 | 117.91 | 252.59 | 6290.41 | 904.04 | 3393.49 |
| 33 | 5117.48 | 117.91 | 252.59 | 3939.29 | 668.35 | 2888.63 |
| 43 | 5638.00 | 271.37 | 26.54 | 6923.29 | 1825.18 | 4656.29 |

## 8.6 Dynamic Analysis by FEM Modelling

Different dynamic numerical analyses of soil-structure interaction were performed (Massimino and Maugeri 2003b; di Prisco et al. 2006; Abate et al. 2009). In this paper the recent analysis of the DSSI for Noto Cathedral by means of a FEM numerical code is presented.

A 3-D analysis would be very complex for a monument and could give not easily understandable results. So, a 2-D analysis was performed. This modelling considered the same schematic cross-section across the naves shown in Fig. 8.15.

Two different numerical models were adopted for studying the seismic response of the cathedral, by means of the ADINA finite element code (Bathe 1996): (i) a realistic FEM model (soil-structure model), involving the structure and the interacting subsoil; (ii) a simplified FEM model, considering the structure fixed at the base, utilised as a benchmark.

As regards the modelling of the structure, its geometry is simplified, considering some parts of the cathedral (the domes of the side chapels and the wood covering of the nave) only as loads applied at the nodes of the FEM model (Fig. 8.17).

As mentioned before, the mechanical behaviour of the masonry structure is assumed to be linear elastic ($E$=5,000,000 kPa; $\nu$=0.23).

In particular, a first comparison between the two configurations of the structure (Figs. 8.18 and 8.19) without and with tie rods and inverted arches, subjected to the accelerogram characterised by the highest PGA ("ACC1 EW-30"), was made. A further comparison between the two FEM models with and without interacting soil, subjected to the previously described 12 accelerograms (6 synthetic accelerograms at 30 m below the soil surface for the first model and 6 synthetic accelerograms at 0 m on the soil surface for the second model), were developed.

Figure 8.20 shows the adopted mesh and the applied boundary and loading conditions for the simplified FEM model (without the interacting soil) utilised as a benchmark; in this case the configuration of the structure after the remedial works was considered.

**Table 8.5** Foundation bearing capacity evaluation (After Massimino and Maugeri 2003b)

| Static conditions | | | | | Pseudo-static conditions | | | | |
|---|---|---|---|---|---|---|---|---|---|
| | | Brinch-Hansen 1970 | | EC7, 1994 | | Brinch-Hansen 1970 | | EC8, 1994 | |
| Nodes | $q_{es,s}$ (kPa) | $q_{lim,s}$ (kPa) | $F_{s,s}$ | $q_{lim,s}$ (kPa) | $q_{es,s}$ (kPa) | $q_{lim,e}$ (kPa) | $F_{s,e}$ | $q_{lim,e}$ [a] (kPa) | $q_{lim,e}$ [b] (kPa) |
| 1 | 133.68 | 1008.01 | 4.08 | 531.35 | 228.32 | 523.85 | 2.29 | 382.30 | 290.53 |
| 11 | 181.73 | 1384.80 | 7.62 | 761.93 | 366.30 | 1064.97 | 2.91 | 821.07 | 623.98 |
| 33 | 181.73 | 1390.18 | 7.65 | 761.93 | 264.78 | 1011.23 | 3.82 | 777.94 | 591.20 |
| 43 | 248.10 | 1011.33 | 4.08 | 531.35 | 287.61 | 589.89 | 2.05 | 427.94 | 325.22 |

[a]$q_{lim,e}$ computed ignoring soil inertial effects
[b]$q_{lim,e}$ computed considering soil inertial effects according to the Paolucci & Pecker method (1997)

Fig. 8.17 Schematization of the applied loads to the frame on the lateral and main naves

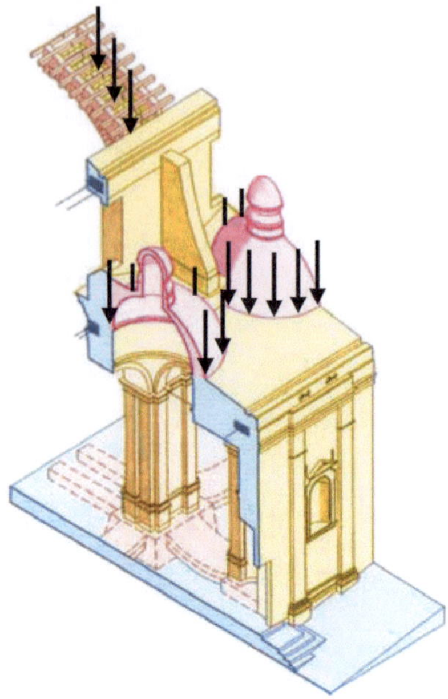

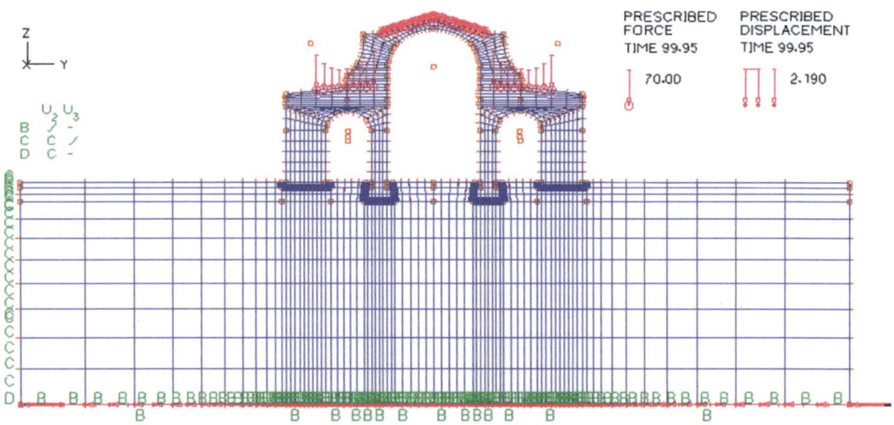


**Fig. 8.18** Numerical soil-structure model showing boundary and loading conditions, according to the configuration without tie rods and inverted arches (After Abate et al. 2009)

As far as the remedial works are concerned, both the tie rods were connecting the foundation nodes and those connecting the superstructure nodes shown in Fig. 8.14 are taken into account. In the utilised FEM code these tie rods are modelled as truss with initial strains, in order to reproduce their pre-stress. The mechanical characteristics of the designed Dywidag tie rods are reported in Table 8.6. Finally, in the configuration with the remedial works, also the inverted arches are modelled, according to the same properties of the whole structure.

As regards the modelling for the foundation soil, the horizontal bottom boundary is chosen at a depth equal to 30.00 m, according to the bedrock estimation (Cavallaro et al. 2003); the two lateral boundaries are 30.00 m from the structure, in

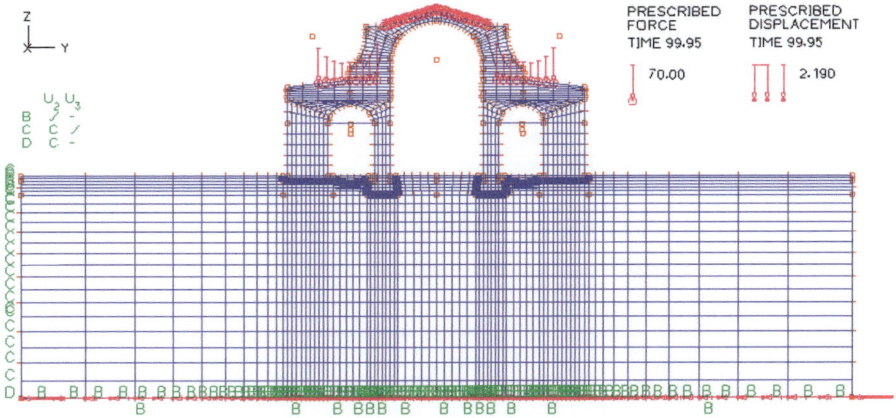


**Fig. 8.19** Numerical soil-structure model showing boundary and loading conditions, according to the configuration with tie rods and inverted arches (After Abate et al. 2009)

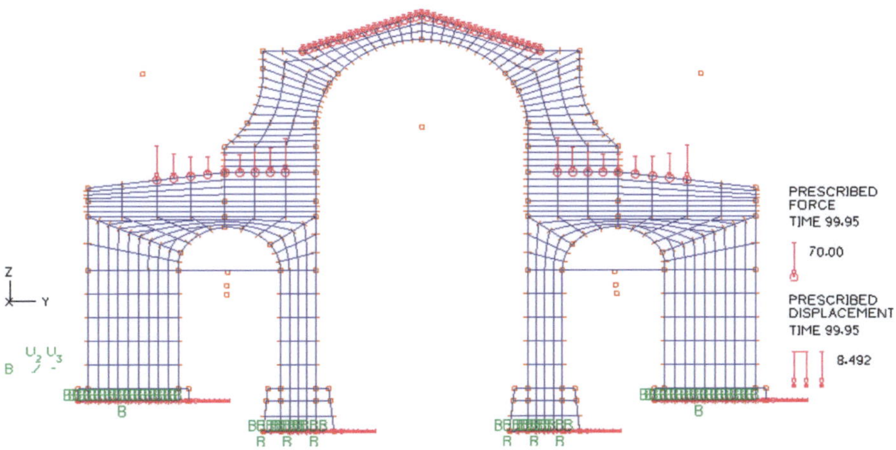


**Fig. 8.20** Numerical simplified model of the structure fixed at the base, showing boundary and loading conditions, according to the configuration with tie rods (After Abate et al. 2009)

**Table 8.6** Mechanical characteristics of the utilised tie rods (After Massimino and Maugeri 2003b)

| $\phi$ (mm) | E (MPa) | $\nu$ | $\gamma_{steel}$ (kN/m$^3$) | $\sigma$ (MPa) |
|---|---|---|---|---|
| 30 | 206,000 | 0.17 | 77 | 1,100 |

order to avoid any boundary effect. For the horizontal bottom boundary the vertical displacements are fixed equal to zero, while a horizontal displacement time-history is given. For the lateral boundaries the vertical displacements are free, while the horizontal displacements are constrained in order to have a stable model minimizing, at the mean time, the horizontal boundary effects: more precisely, each point of the left lateral boundary is constrained to have the same horizontal displacement of the point of the right lateral boundary located at the same depth.

**Table 8.7** Mechanical characteristics of the foundation soil

| $G_{0\,av}$ (kPa) | $G_{0\,50\%}$ (kPa) | D (%) | ν |
|---|---|---|---|
| 127,500 | 65,000 | 10 | 0.3 |

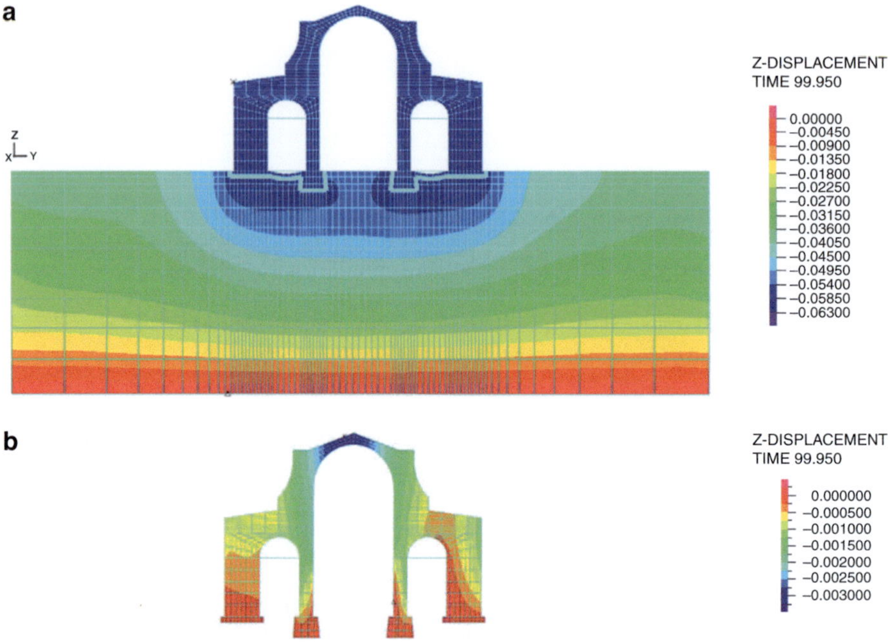


**Fig. 8.21** Vertical displacement band plot: (**a**) soil-structure model after the reconstruction; (**b**) fixed-base model after the reconstruction

The soil is modelled as linear elastic material (Table 8.7), characterised by average $G_0$ and D, estimated, as mentioned before, according to Figs. 8.3, 8.4 and 8.5.

Special contacts are fixed at the soil-foundation interface; these allow us to simulate frictional sliding at the soil-foundation interface and uplifting of the foundations from the soil.

It is necessary to fix the Rayleigh damping factors α and β, which allows the user to define the damping matrix [C] as a linear combination of the mass matrix [M] and the stiffness matrix [K], as follows:

$$[C] = \alpha\,[M] + \beta\,[K] \tag{8.1}$$

where $\alpha = D_{min} \cdot \omega$ and $\beta = D_{min} / \omega$; the angular frequency of the input ω was obtained investigating the frequency content of the utilised input accelerograms (see Tables 8.1 and 8.2).

Figures 8.21, 8.22 and 8.23 shows the two models with and without the soil, in the configuration after the reconstruction, with refers to the accelerogram characterised

**Fig. 8.22** Horizontal displacement band plot: (**a**) soil-structure model after the reconstruction; (**b**) fixed-base model after the reconstruction

**Fig. 8.23** Vertical stresses band plot: (**a**) soil-structure model after the reconstruction; (**b**) fixed-base model after the reconstruction

**Table 8.8** Displacements, stresses and strains at the central arch for the configurations shown in Figs. 8.18 and 8.19

| | $u_{seismic}$ (cm) | $w_{seismic}$ (cm) | $\sigma_{y, max}$ (kPa) | $\sigma_{z, max}$ (kPa) | $\varepsilon_{y, max}$ (%) | $\varepsilon_{z, max}$ (%) |
|---|---|---|---|---|---|---|
| "before" | 0.76 | 0.06 | 330.27 | 66.38 | 0.0064 | 0.0006 |
| "after" | 0.63 | 0.06 | −125.43 | 39.20 | 0.0026 | 0.0011 |

**Table 8.9** Final horizontal and vertical displacements of the central arch for the soil-structure model and the simplified model

| | | $a_{max}$ (m/s$^2$) | $u_{seismic}$ (cm) | $w_{seismic}$ (cm) |
|---|---|---|---|---|
| Soil-structure model | ACC1EW-30 | 1.488 | 0.63 | 0.06 |
| | ACC1NS-30 | 0.399 | 0.87 | 0.12 |
| | ACC2EW-30 | 0.987 | 1.72 | 0.12 |
| | ACC2NS-30 | 0.407 | 0.63 | 0.06 |
| | ACC3EW-30 | 1.284 | 1.07 | 0.06 |
| | ACC3NS-30 | 1.077 | 0.36 | 0.06 |
| Fixed-base model | ACC1EW-0 | 2.177 | 0.19 | 0 |
| | ACC1NS-0 | 0.431 | 0.14 | 0 |
| | ACC2EW-0 | 1.146 | 0.15 | 0 |
| | ACC2NS-0 | 0.540 | 0.08 | 0 |
| | ACC3EW-0 | 1.466 | 0.16 | 0 |
| | ACC3NS-0 | 1.527 | 0.06 | 0 |

by the highest PGA ("ACC1 EW-30"). In particular, Figs. 8.21 and 8.22 show the vertical and horizontal displacement band plots, respectively; Figs. 8.23 and 8.24 show the vertical and horizontal stress band plots, respectively.

As regards the comparisons between the described FEM models, they were performed with refer to the behaviour of the central arch, because this can be considered the most critical part of the cathedral.

The comparison between the configurations shown in Figs. 8.18 and 8.19, subjected to the accelerogram characterised by the highest PGA ("ACC1 EW-30", $a_{max} = 1.488$ m/s$^2$), was presented both in terms of final displacements and in terms of maxima values of stresses and strains (Table 8.8). It is possible to see that thanks to the remedial works all the values of displacements, stresses and strains suffer a benefit reduction.

The comparison between the two models with and without interacting soil (Figs. 8.19 and 8.20) was also presented in terms of maxima displacements and of maxima stresses and strains (Tables 8.9 and 8.10, respectively).

From Table 8.9 it is possible to observe that considering the structure without the soil the values of the displacements of the arch are underestimated, and this can be dangerous for the evaluation of the dynamic behaviour of the structure. From Table 8.10 it is possible to observe that considering the structure resting on the soil the values of stresses are lower because of the interaction effects, while for the same reason the values of the horizontal strain are greater; there isn't any significant difference for the vertical strains for the two types of the FEM models.

**Table 8.10** Maxima values of stresses and strains of the central arch for the soil-structure model and the simplified model (After Abate et al. 2009)

| | | $\sigma_{y,\,max}$ (kPa) | $\sigma_{z,\,max}$ (kPa) | $\varepsilon_{y,\,max}$ (%) | $\varepsilon_{z,\,max}$ (%) |
|---|---|---|---|---|---|
| Soil-structure model | ACC1EW-30 | −125.43 | 39.20 | −0.0026 | 0.0011 |
| | ACC1NS-30 | −311.73 | −0.12 | −0.0062 | 0.0007 |
| | ACC2EW-30 | −254.22 | 11.93 | −0.0051 | 0.0008 |
| | ACC2NS-30 | −273.32 | 7.99 | −0.0054 | 0.0008 |
| | ACC3EW-30 | −180.75 | 27.76 | −0.0036 | 0.0009 |
| | ACC3NS-30 | −176.82 | 27.54 | −0.0036 | 0.0009 |
| Fixed-base model | ACC1EW-0 | 404.73 | 105.52 | 0.0078 | 0.0012 |
| | ACC1NS-0 | −35.12 | 20.14 | −0.0007 | 0.0005 |
| | ACC2EW-0 | 112.28 | 48.20 | 0.0021 | 0.0007 |
| | ACC2NS-0 | −31.79 | 20.62 | −0.0007 | 0.0005 |
| | ACC3EW-0 | 219.46 | 70.11 | 0.0042 | 0.0009 |
| | ACC3NS-0 | 66.76 | 208.72 | 0.0009 | 0.0040 |

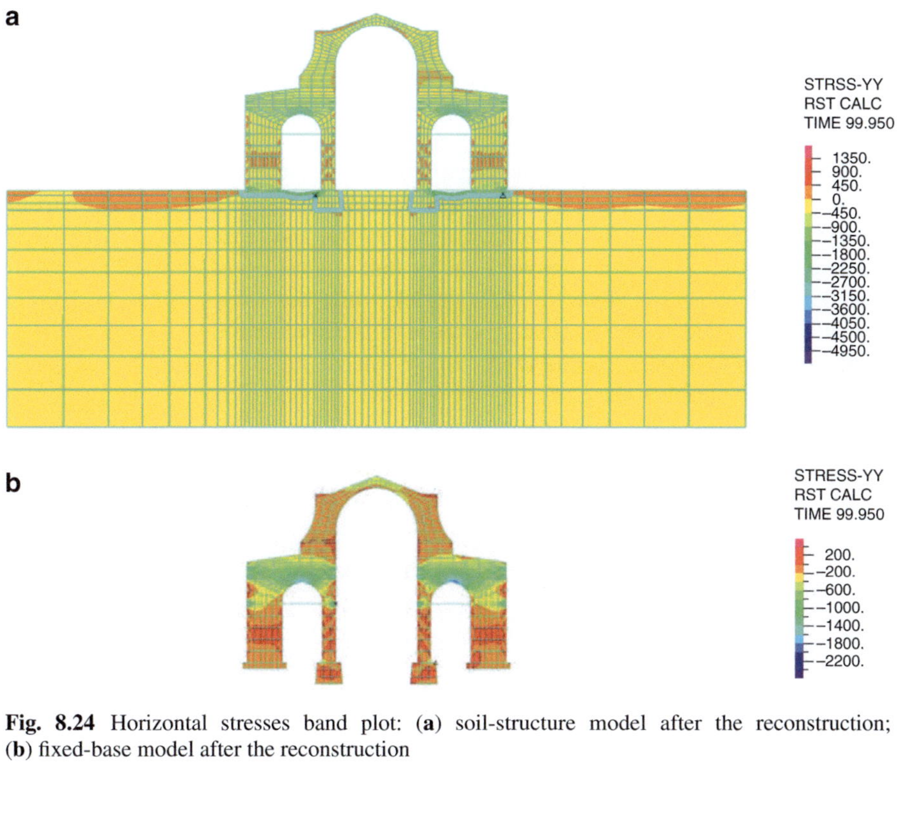


**Fig. 8.24** Horizontal stresses band plot: (**a**) soil-structure model after the reconstruction; (**b**) fixed-base model after the reconstruction

## 8.7 Conclusions

When dealing with a monument, a deterministic modelling based on the maximum credible scenario earthquake is preferable to the probabilistic approach. When the return period of the earthquake is very long, as in the case of the 1,693 Val di Noto earthquakes, using the probabilistic approach the seismic action could be significantly underestimated. When the return period is very short, as, for instance, in the case of 1997 Umbria-Marche earthquake, the opposite could occur.

As regards the Noto Cathedral, the retrofitting was designed to resist not only to the repetition of the 1990 earthquake but to resist to the 1,693 maximum credible scenario earthquake; moreover, the retrofitting work was carried out not only for the superstructure, as usually, but also for the foundations.

The present paper deals with pseudo-static and 2-D FEM analyses of soil-structure interaction. These analyses were carried out for the 1,693 scenario earthquake and they were performed by FEM modelling.

By the pseudo-static analysis, the bearing capacity was investigated taking into account the soil inertial effects. As regards the effects of the foundation soil deformability, three springs per each node of the foundation were considered, in order to evaluate vertical displacements, horizontal displacements and rotations of the foundation.

The dynamic SSI analysis allowed us to investigate the real structural response to the seismic action, while the fixed-base model, usually adopted by the practitioners, is not able to capture the effects of soil deformability on stresses and strains of the superstructure.

The analyses show that the most critical part of the cathedral was the arch of the central nave if the SSI is considered, while ignoring SSI the most critical part of the cathedral are the base of the arches of the central and lateral naves. Moreover, the soil-structure interaction analysis shows that the values of stresses are lower than those obtained for the base-fixed model, while the values of the horizontal strains and displacements are greater because of the interaction.

Finally, thanks to the remedial works all the values of displacements, stresses and strains suffer a benefit reduction.

## References

Abate G, Massimino MR, Maugeri M (2009) Effects of DSSI on Noto Cathedral behaviour: evaluation by means of FEM analyses. In: Proceedings of the “1st international conference on protection of historical buildings”, Rome, 21–24 June 2009

Bathe KJ (1996) Finite element procedures. Prentice Hall, Englewood Cliffs, 1996

Binda L, Baronio G, Gavarini C, De Benedictis R, Trincali S (1999) Investigation on materials and structures for the reconstruction of the partially collapsed Cathedral of Noto. In: Proceedings of the 6th international conference structural studies, repairs and maintenance of historical buildings, STREMAH99. Dresden, Germany, pp 323–332

Binda L, Saisi A, Tiraboschi C, Valle S, Colla C, Forde M (2001) Indagini soniche e radar sui pilastri e sulle murature della Cattedrale di Noto. In: Proceedings of the "Symposium on Ricostruendo la Cattedrale di Noto e la Frauenkirche di Dresda", 10–12 Feb 2000 (In Italian)

Bottari A, Saraò A, Teramo A, Termini D, Careni P (2003) The January 11, 1693 South-Eastern Sicily earthquake: macro seismic analysis and strong motion modelling at Noto. In: Maugeri M, Nova R (eds) Geotechnical analysis of seismic vulnerability of monuments and historical sites. Pàtron Editore, Bologna

Brinch-Hansen J (1970) A revised and extended formula for bearing capacity. Danish Geotechnical Institute Bulletin XXVIII, pp 5–11

Campoccia I, Massimino MR (2003) Grade-3 microzonation at Noto. In: Maugeri M, Nova R (eds) Geotechnical analysis of seismic vulnerability of monuments and historical sites. Pàtron Editore, Bologna

Cavallaro A, Massimino MR, Maugeri M (2003) Noto Cathedral: soil and foundation investigation. Constr Build Mater J 17:533–541

di Prisco C, Massimino MR, Maugeri M, Nicolosi M, Nova R (2006) Cyclic numerical analyses of Noto Cathedral: soil-structure interaction modelling. Ital Geotech J XL:49–64

EC7-Part 1 (2004) Geotechnical design, general rules. ENV 1997, European Committee for Standardization, Brussels

EC8-Part 1 (2003) Design of structures for earthquake resistance – Part 5: General rules, seismic actions and rules for buildings. European Prestandard, ENV 1998, European Committee for Standardization, Brussels

Gazetas G (1991) Foundation vibrations. In: Fang HY (ed) Foundation engineering handbook. Van Nostrand Reinhold, New York, pp 553–593

Giuffrè A, Giannini R (1982) La risposta non lineare delle strutture in cemento armato. Progettazioni e particolari costruttivi in zone sismiche. ANCE-AIDIS

Massimino MR, Maugeri M (2003a) Design spectrum by synthetic accelerograms for Noto Cathedral (Siracusa, Italy). In: Maugeri M, Nova R (eds) Geotechnical analysis of seismic vulnerability of monuments and historical sites. Pàtron Editore, Bologna, pp 287–308

Massimino MR, Maugeri M (2003b) Retrofitting of the foundation of Noto Cathedral (Italy) according to Eurocodes. In: Maugeri M, Nova R (eds) Geotechnical analysis of seismic vulnerability of monuments and historical sites. Pàtron Editore, Bologna, pp 309–330

Maugeri M, Castelli F, Massimino MR (2006) Analysis, modelling and seismic improvement of foundation of existing buildings. Ital Geotech J 4/2006:52–122. Bologna: Pàtron Editore (in Italian with English Summary)

Paolucci R, Pecker A (1997) Seismic bearing capacity of shallow strip foundation on dry soils. Soils Found 37(3):95–105

Priolo E (1999) 2-D Spectral element simulation of destructive ground shaking in Catania. J Seismol 3(3):289–309

Priolo E (2000) 2-D Spectral element simulation of the ground motion for a catastrophic earthquake in the Catania project: earthquake damage scenarios for a high risk area in the Mediterranean editors: Faccioli and Pessina. CNR-Gruppo Nazionale per la difesa dai terremoti-Roma 2000, pp 67–75

Rovelli A, Boschi E, Coco M, Di Bona M, Berardi R, Longhi G (1991) Il terremoto del 13 Dicembre 1990 nella Sicilia Orientale: analisi dei dati accelerometrici. In: Contributi allo studio del terremoto della Sicilia Orientale del 13 Dicembre 1990 a cura di Boschi e Basili. Istituto Nazionale di Geofisica, pubbl. n. 537, pp 85–101

# Chapter 9
# Study on the Method for the Seismic Design of Expressway Embankments

**Susumu Yasuda and Kazuyori Fujioka**

**Abstract** The behaviour of road embankments during past earthquakes in Japan are reviewed and methods for the seismic design of expressway embankments are discussed. There are several grades of roads in Japan. Though basic design methods for the embankments are similar, detailed methods are different. For example, the standard inclinations of slopes differ. The embankments for national expressways, which are of the highest grade, have been designed and constructed with special care. Nevertheless, the 2004 Niigataken-chuetsu earthquake damaged the embankment of the Kan-etsu Expressway due to very strong shaking. Three types of failure occurred on the Kan-etsu Expressway. Type 1 failure was serious slides of the embankment on slope. Type 2 failure was settlement of the embankment on level ground without deformation of the ground. Type 3 failure was settlement of the embankment on level ground with deformation of the ground. If road embankments are designed to withstand very strong shaking motion, their probable deformation must be considered in their design. And it is necessary to introduce performance based design by considering the deformation of the embankments. Newmark's method of estimating deformation was applied to Type 1 failures and residual deformation analysis was applied to Type 2 and 3 failures. Base on these studies, relevant method for the seismic design of expressway embankments are discussed.

S. Yasuda (✉)
Tokyo Denki University, Saitama, Japan
e-mail: yasuda@g.dendai.ac.jp

K. Fujioka
Nippon Expressway Research Institute, Co., Ltd, Machida, Japan
e-mail: k.fujioka.aa@c-nexco.co.jp

M.A. Sakr and A. Ansal (eds.), *Special Topics in Earthquake Geotechnical Engineering*, Geotechnical, Geological and Earthquake Engineering 16,
DOI 10.1007/978-94-007-2060-2_9, 

## 9.1 Introduction

In the design of road embankments, the inclination of slopes, and the height and density of embankments are usually decided by empirical methods without stability analysis. Even though stability analysis is conducted, seismic force is rarely considered because embankments are not difficult to restore if they settle or slide during earthquakes. However, recently, it has become necessary to consider seismic force in the design of embankments for high-grade roads in Japan, such as national expressways. Moreover, Level 1 and Level 2 shaking motion must be considered in designing embankments of important roads, such as emergency transport roads. Current method of analysing slope stability cannot be applied to Level 2 shaking motion. Therefore, in this paper, new design methods based on the deformation of embankments are discussed.

## 9.2 Classification of Roads in Japan and Methods for Their Design

### *9.2.1 Classification of Roads in Japan*

In Japan, about 1.2 million km of roads cover an area of 377,800 km$^2$. The roads are classified into 6 groups as shown in Table 9.1. National expressways are constructed and administrated by the Japan Highway Public Corporation (JH). In 2005, the JH was privatized and renamed Nexco. Nexco composes three companies, West Nippon Expressway Co., Ltd., Central Expressway Co., Ltd. and East Nippon Expressway Co., Ltd. The first national expressway, Meishin Expressway, was opened for traffic in 1963. Since then many other national expressways have been constructed. Figure 9.1 shows the present national expressway network, 8,476 km of expressway are in operation, 1,098 km under construction, and about 4,400 km are being planned. The Kan-etsu Expressway and Hokuriku Expressway, which were damaged during two recent earthquakes, were opened in 1973 and 1972, respectively.

**Table 9.1** Classification of roads in Japan

| Classification of roads | Grade |
|---|---|
| (1) High-Standard Arterial Roads | High |
| (a) National Expressways | ⇕ |
| (b) Other Expressways | |
| (2) Urban Expressways | |
| (3) Ordinary Roads | |
| (a) National Highways | |
| (b) Prefectural Roads | |
| (c) Municipal Roads | Low |

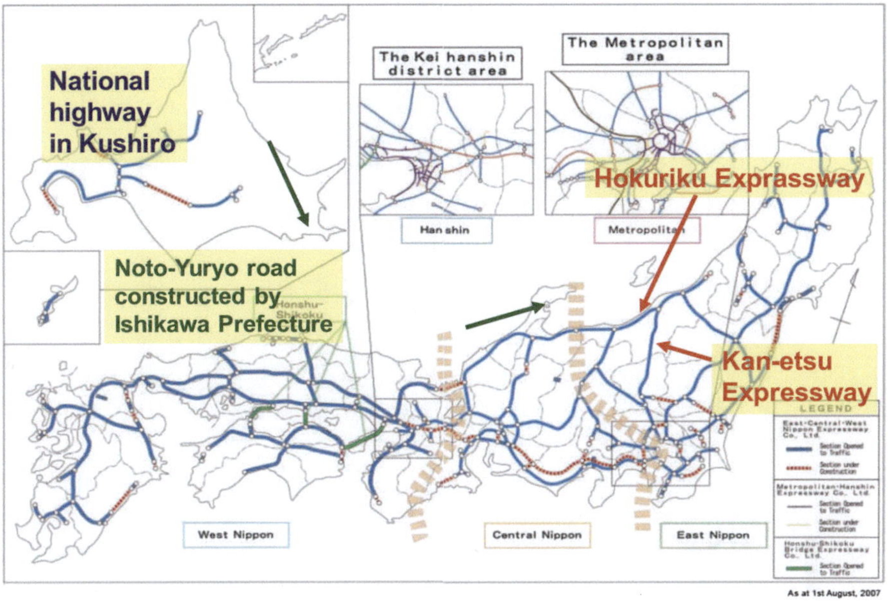


**Fig. 9.1** National expressway network by Nexco

The Noto-yuryo road which was seriously damaged in 2007, was constructed and administrated by a public corporation belonging to Ishikawa Prefecture and is classified as "other expressways".

## *9.2.2 Methods of Designing Road Embankments in Japan*

In the design of a road embankment, the selection of appropriate banking material is important because bad material is inefficient to handle and leads to embankment settlement and instability. The inclination of slopes and the height of an embankment must be designed to prevent slope failure. Proper drainage must be designed to ensure during the construction of an embankment and to maintain embankment stability. In addition, banking materials must be appropriately placed, spread and compacted. General recommendations and requirements are summarized briefly in Table 9.2. However, design details differ by road type. In general, strict design is required for national roads, especially for national expressways. For example, following specifications apply to national expressways constructed by Nexco and to the Noto-yuryo road constructed by Ishikawa Prefecture:

(a) National expressways constructed by Nexco
Standard inclination of a slope: 1:1.8

**Table 9.2** General recommendations for the design of road embankments in Japan

| Item | Recommendations | Restrictions |
|---|---|---|
| Banking material | In general, materials with the following properties are appropriate for embankment<br>(1) Easy to place, spread and compaction.<br>(2) Shear strength and bearing capacity are high, and compressibility is low.<br>(3) Water absorption swelling is low.<br>(4) Stable against erosion and shear strength does not decrease due to saturation.<br>Well-graded gravelly or sandy soils satisfy these conditions and are recommended to use as banking materials. | Bentonite, solfataric soil, acidic clay and highly organic soil can not be used to construct the embankment. The following soils should not be used:<br>(a) Volcanic cohesive soil: Trafficability of the soil is low. The soil is unstable against slope failure, and causes long-term settlement of embankment.<br>(b) Some sedimentary soft rocks, such as mudstone, shale and tuff: These rocks cause slaking after filling. |
| Density of embankments | Bunking materials must be appropriately placed, spread, and compacted. In general, compaction control standard is decided before banking. Several compaction control methods are available as follows:<br>(i) Soil density measurement.<br>(ii) Degree of saturation.<br>(iii) Compaction method. | Compaction cannot be controlled by density measurement if the natural moisture content of banking material is greater than the optimum moisture content. Compaction cannot be controlled by strength if the strength of embankment material decreases due to saturation by rainfall or water seepage after compaction. |
| Gradient and height of slope | In general, inclinations of slopes are designed empirically without conducting slope stability analyses, because embankments are very long. In the empirical approach, appropriate inclination of a slope is decided based on the height and the material of the embankment. If the embankment is high and/or material is soft, low inclination is selected. | Slope stability analysis is necessary if conditions of the foundation ground and/or the embankment are bad as follows:<br>(a) Foundation ground is soft.<br>(b) Embankment is constructed in landslide area.<br>(c) Embankment is too high.<br>(d) Banking material is bad such as high moisture clay or volcanic ash.<br>(e) Houses exist near the embankment and the houses may be damaged if the embankment is deformed. |

| | | |
|---|---|---|
| Drainage facilities | During the construction of embankments, seepage water and surface water must not be contained in the construction site temporarily. In the design of permanent drainage, two kinds of liners must be considered. The first ones are liners against surface water. Subsurface drainage, slope surface drainage are these facilities. The second ones are facilities against seepage water into embankments. Horizontal drains, horizontal blankets, drainage pipes, gabions are these lines. | Special care for drainage is necessary in the following locations:<br>(1) Small river crossings or boundaries between cuts and embankments where flows of surface water concentrate.<br>(2) Valleys, boundaries between cuts and embankments where much spring water spews out. |

In-situ tests to check the density of soil during compaction work: 15 points per day

(b) Noto-yuryo road constructed by Ishikawa Prefecture
Standard inclination of a slope: 1:1.5 (1:1.8 for the lower part of slope if the embankment is high)
In-situ tests to check the density of soil during compaction work: 1 point per day

Therefore, the embankments of national expressways are well designed and are not easily damaged during earthquakes.

## 9.3 Damage to the Embankments of General Roads During Past Earthquakes

In Japan, the embankments of general roads are frequently damaged during earthquakes and heavy rains. Based on past experience, the damage to road embankments is classified into the three levels shown in Table 9.3. Recent earthquakes caused seriously damage to several road embankments in Japan. For example, the 1993 Kushio-oki earthquake caused severe damage to the embankments of national highways in and around Kushiro City. Damage classified as "Serious on sloping ground" in Table 9.3 occurred at four sites. Photo 9.1 shows the damage at Toridoushi in Kushiro Town where the embankment built on a small valley failed. The banking material was volcanic soil. The lower part of the embankment must have been saturated because the water table was only 3.5 m below the road surface.

**Table 9.3** Classification method of the level of damage to road embankments in Japan

| Level of damage | Schematic diagram | Definition of damage |
|---|---|---|
| Minor | 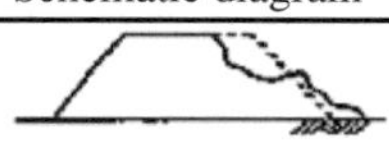 | Surface slide of embankment at the top of slope only |
| |  | Minor cracks on the surface of a road |
| Medium | 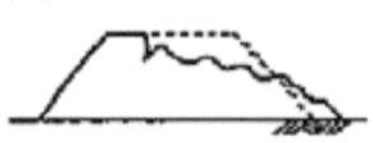 | Deep slide of embankment or slump involving traffic lines |
| |  | Medium cracks on the surface of a road and/or settlement of embankment |
| Serious | 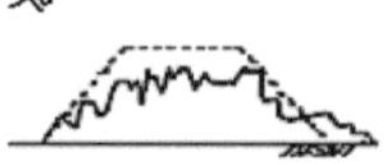 | Serious slump of embankment |
| | 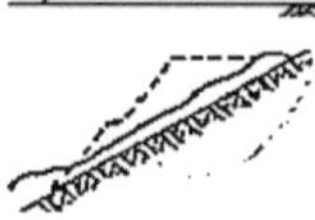 | Serious slide of embankment |

**Photo 9.1** Damage to the embankment of national highway at Toridoushi in Kushiro Town during the1993 Kushiro-oki earthquake

The embankments of the Noto-yuryo road were very seriously damaged during the 2007 Noto-hanto earthquake. Huge slides occurred at 11 sites shown in Fig. 9.2. Photos 9.2 and 9.3 show the slides at sites No.9 and No.32, respectively. A soil cross section and plan at site No.32, estimated after the earthquake, are shown in Figs. 9.3 and 9.4. The height, thickness and length of the collapsed soil were about 30, 12 and 80 m, respectively. The embankment soil was weathered tuff. Sasaki et al. (2007) studied soil condition at damaged and not damaged sites and found that the groundwater level at the damaged sites tended to be high near the slope toe of the embankment. And the degree of compaction of the embankment varied between 80% and 90%, but no significant difference was observed between damaged and undamaged sites.

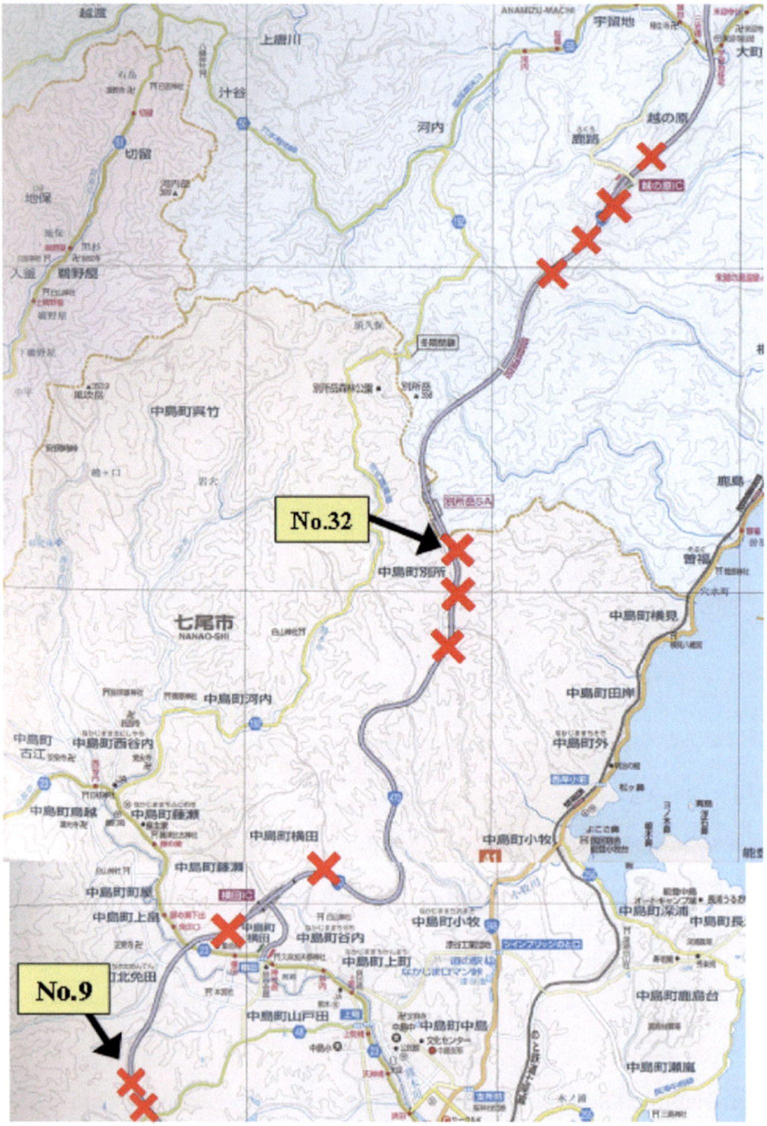


**Fig. 9.2** Map of sites along Noto-yuryo road seriously damaged during the 2007 Noto-hanto earthquake

**Photo 9.2** Slide of the embankment of Noto-yuryo road at site No.9

**Photo 9.3** Slide of the embankment of Noto-yuryo road at site No.32

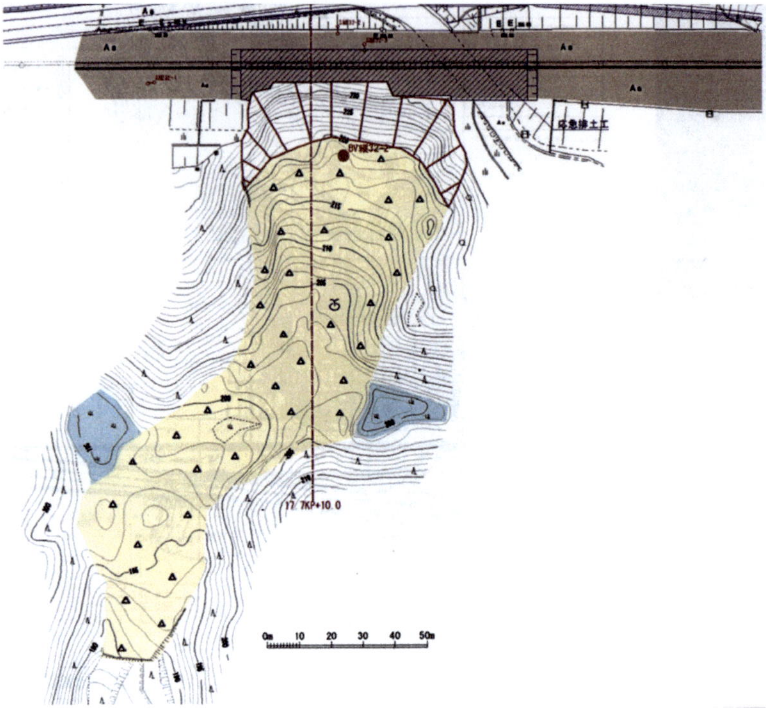


Fig. 9.3 Plan at site No.32 (Ishikawa Prefecture)

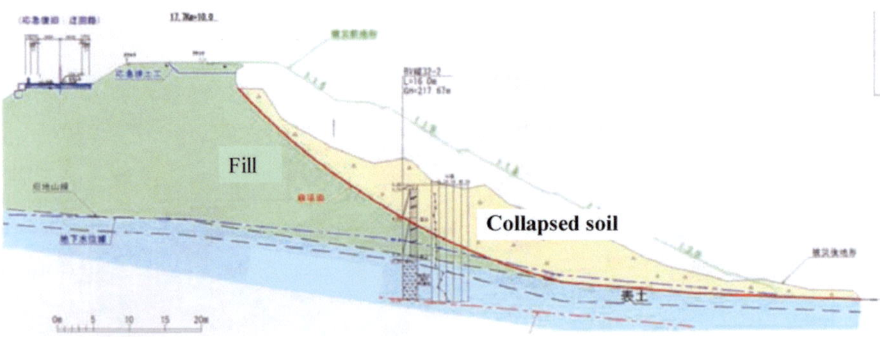


Fig. 9.4 Cross section at site No.32 (Ishikawa Prefecture)

## 9.4 Behaviour of the Embankments of National Expressways During Two Earthquakes

### *9.4.1 The 2004 Niigataken-chuetsu Earthquake*

On October 23, in 2004, the Niigataken-chuetsu earthquake, of magnitude 6.8, occurred and caused serious damage to many structures and slopes in Japan. 580 km along six national expressways were closed due to the earthquake. Damaged expressway embankments were filled with banking materials and all expressways were able to open for emergency vehicles about 19 h after the earthquake because expressway bridges and tunnels were not seriously damaged. About 13 days after the earthquake all expressways were opened to normal traffic.

The most serious damage occurred in the following sections:

(a) between Horinouchi IC and Echigokawaguchi IC (8.8 km), and between Yamamotoyama Tunnel and Yamaya PA (5.5 km) of Kan-etsu Expressway as shown in Fig. 9.5, and
(b) between Ohzumi PA and Nagaoka JCT (6.0 km) of Hokuriku Expressway

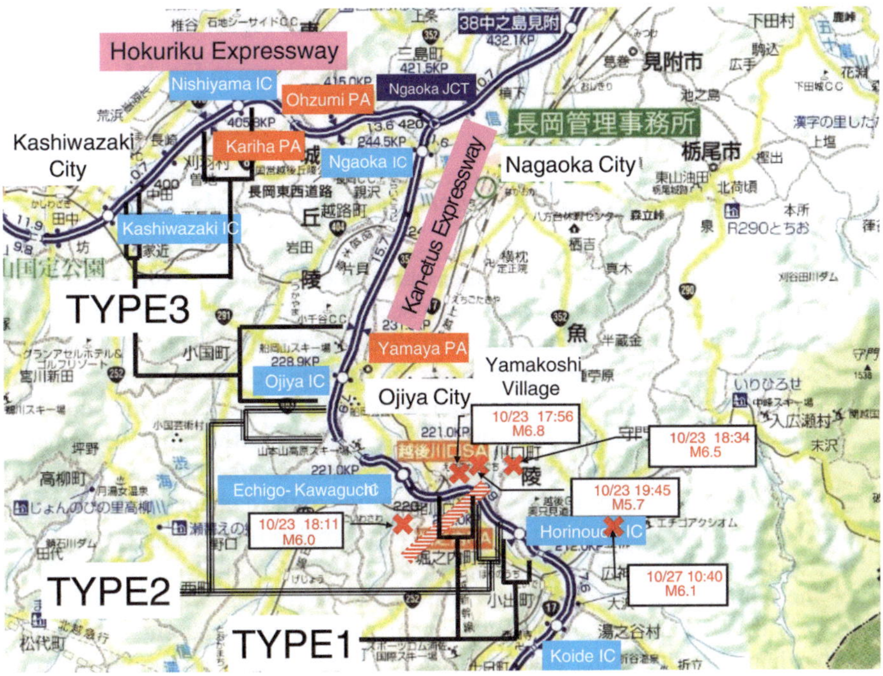


**Fig. 9.5** Route map of Kan-etsu and Hokuriku Expressways and zones damaged during the Niigataken-chuestu and Niigataken-chuetsu-oki earthquakes

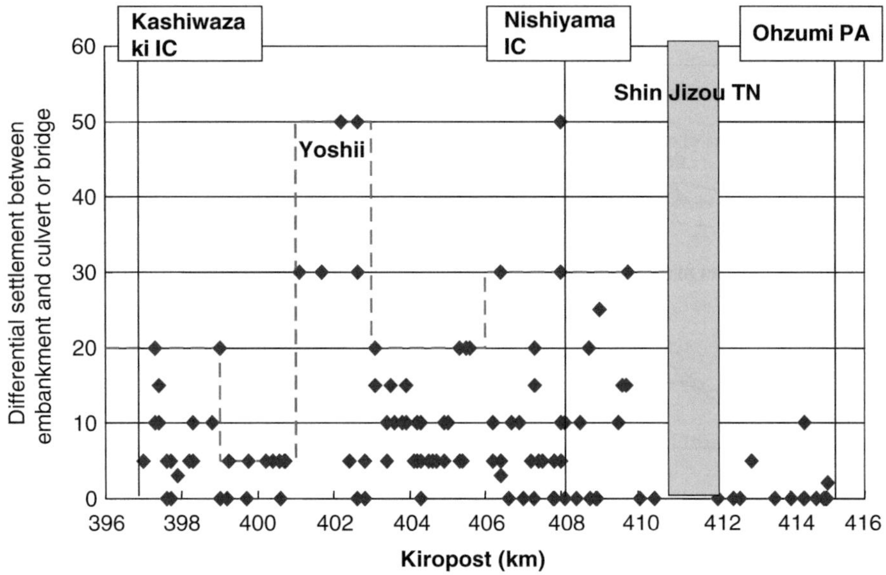


**Fig. 9.8** Distribution of the differential settlements measured after the earthquake

## 9.5 Detailed Study on the Damaged Embankments of Kan-Etsu Expressway

### *9.5.1 Study on Soil Conditions and Failure Mechanisms*

#### 9.5.1.1 Type 1: Serious Slide of the Embankment on Sloping Ground

Serious slides of embankments occurred between Koide IC and Kawaguchi IC, where the Kan-etsu Expressway had been constructed mainly by the cut and fill method. As the expressway crosses several small valleys, embankments were constructed by filling soils on the valleys. Severe slides occurred as shown in Photo 9.4. A detailed soil investigation was conducted after the earthquake. The estimated soil cross section at 215.9 KP is shown in Fig. 9.9. An embankment was constructed on a gentle slope of about 7°. The height and inclination of the embankment were 12.0 m and 1:1.8, respectively. The subsurface soil of the sloping ground was a thin sandy layer with SPT *N*-values of about 10. Dense gravelly soil underlaid the subsurface soil. Filled material was silty sand with gravel. The SPT *N*-values and fines contents of the fill were 4–5% and 20–30%, respectively. According to the measurement of the ground water level conducted after the earthquake, the water level was slightly above the original ground surface. However, it is estimated that the water level had been higher than the measured level because almost a half section of the embankment had disappeared at the time of measurement, due to the slide.

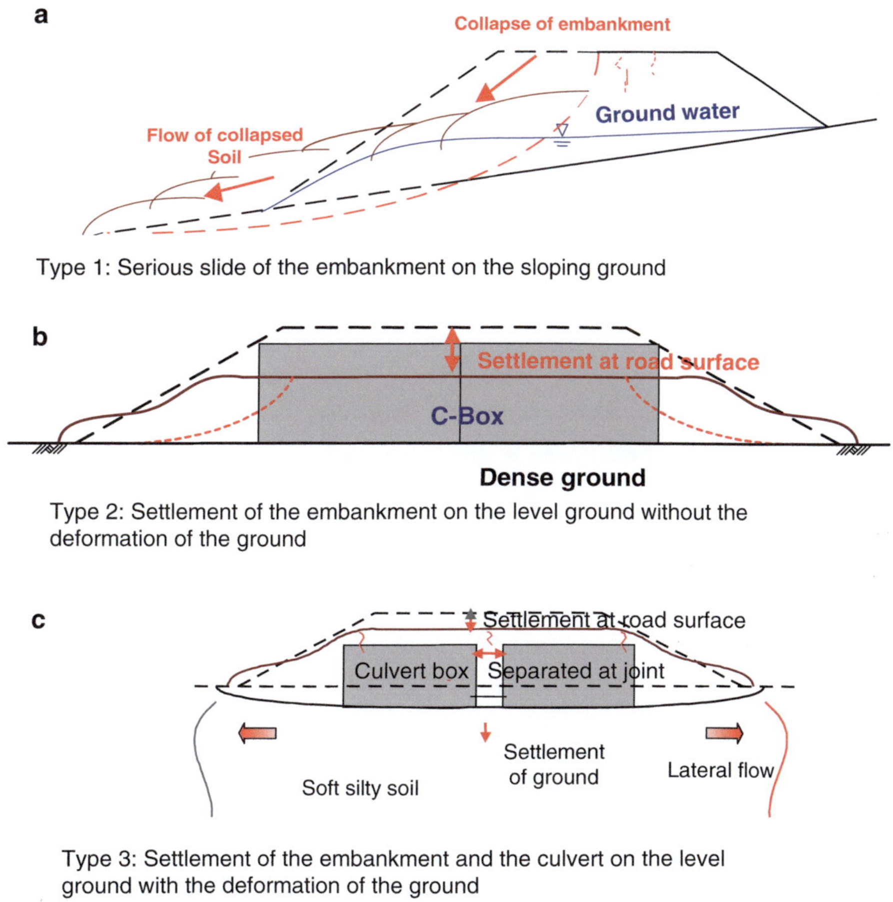


**Fig. 9.6** Classification of the damage to the embankment of Kan-etsu Expressway according to the mechanism of failure (**a**) Type 1: Serious slide of the embankment on the sloping ground, (**b**) Type 2: Settlement of the embankment on the level ground without the deformation of the ground, (**c**) Type 3: Settlement of the embankment and the culvert on the level ground with the deformation of the ground

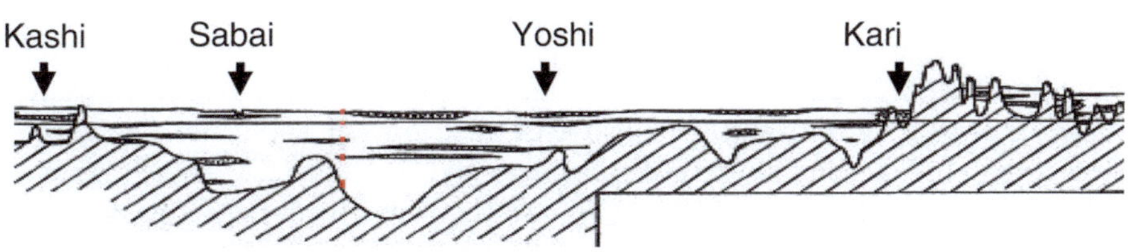


**Fig. 9.7** Brief soil cross section of damaged area along the Hokuriku Expressway

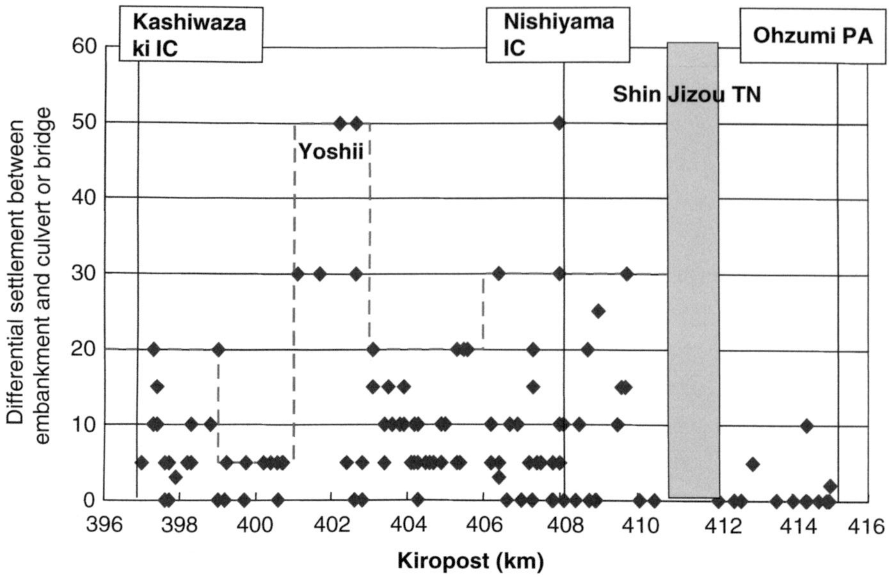


**Fig. 9.8** Distribution of the differential settlements measured after the earthquake

# 9.5 Detailed Study on the Damaged Embankments of Kan-Etsu Expressway

## *9.5.1 Study on Soil Conditions and Failure Mechanisms*

### 9.5.1.1 Type 1: Serious Slide of the Embankment on Sloping Ground

Serious slides of embankments occurred between Koide IC and Kawaguchi IC, where the Kan-etsu Expressway had been constructed mainly by the cut and fill method. As the expressway crosses several small valleys, embankments were constructed by filling soils on the valleys. Severe slides occurred as shown in Photo 9.4. A detailed soil investigation was conducted after the earthquake. The estimated soil cross section at 215.9 KP is shown in Fig. 9.9. An embankment was constructed on a gentle slope of about 7°. The height and inclination of the embankment were 12.0 m and 1:1.8, respectively. The subsurface soil of the sloping ground was a thin sandy layer with SPT *N*-values of about 10. Dense gravelly soil underlaid the subsurface soil. Filled material was silty sand with gravel. The SPT *N*-values and fines contents of the fill were 4–5% and 20–30%, respectively. According to the measurement of the ground water level conducted after the earthquake, the water level was slightly above the original ground surface. However, it is estimated that the water level had been higher than the measured level because almost a half section of the embankment had disappeared at the time of measurement, due to the slide.

**Photo 9.4** Slide of embankment at 215.9 KP of Kanetsu Expressway

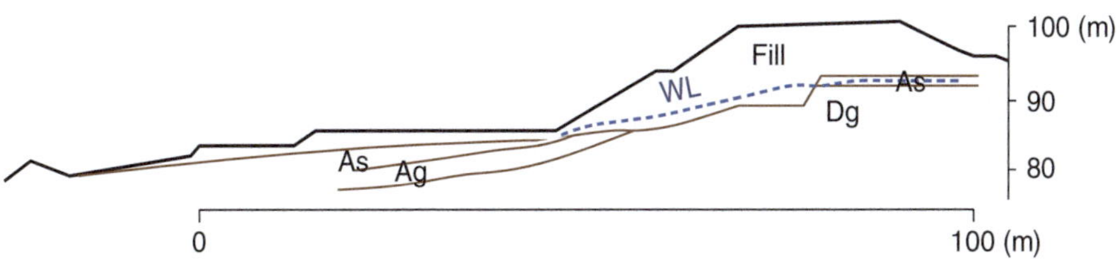


**Fig. 9.9** Cross section at 215.9 KP of Kan-etsu Expressway

#### 9.5.1.2 Type 2: Settlement of the Embankment on Level Ground Without Obvious Deformation of the Ground

Figure 9.10 shows the soil cross section from Yamamotoyama Tunnel to Yamaya PA. The surface soil is gravel near Yamamotoyama Tunnel, then, gradually changes

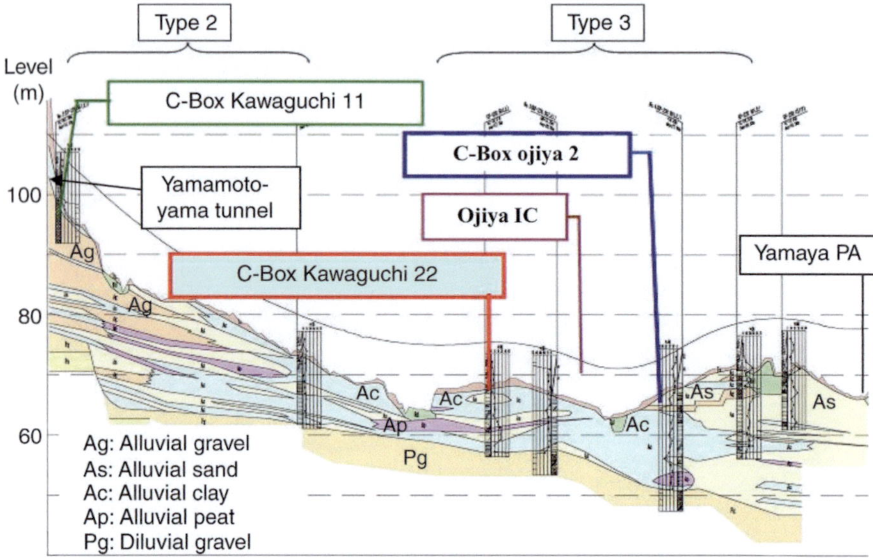


**Fig. 9.10** Soil cross section between Yamamotoyama Tunnel and Yamaya PA

**Photo 9.5** Differential settlement between embankment and culvert at C-Box Kawaguchi 11

to soft clayey soil or loose sandy soil towards Yamaya PA. Type 2 and Type 3 failures occurred near Yamamotoyama Tunnel and around Ojiya IC, respectively.

Large differential settlement of 70 cm occurred between the embankment and the culvert box at C-Box Kawaguchi 11, as shown in Photo 9.5. The culvert box was

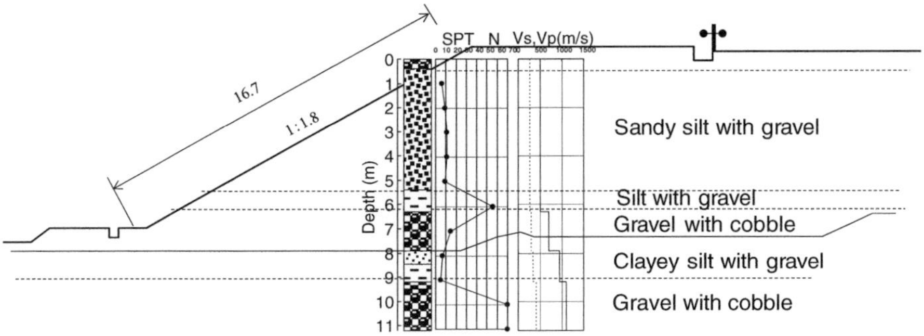


**Fig. 9.11** Soil cross section adjacent to the culvert box at C-Box Kawaguchi 11

constructed of two concrete boxes. The culvert itself settled only 10 cm and the joint of the two concrete boxes opened 10 cm, as schematically shown in Fig. 9.6b.

Figure 9.11 shows boring data and the estimated soil cross section of the embankment adjacent to C-box Kawaguchi 11. The subsurface soil of the foundation ground is dense gravel with the an SPT *N*-value of more than 50. The height of the embankment is about 10 m. Fill materials are sandy silt with gravel, gravel with cobbles and clayey silt with gravel. The fines content of these soils was 50–60%. The measured water level was about 3 m higher than the bottom of the embankment. However, it is not clear whether the water was perched.

The differential settlements between embankments and culverts, and the settlements of culverts themselves between Yamamotoyama Tunnel and Yamaya PA, were measured and are plotted on Fig. 9.12. The differential settlement was about 50–70 cm near Yamamotoyama Tunnel and the settlements of culverts were about 10–20 cm. On the contrary, culverts near Ojiya IC settled much more than this.

#### 9.5.1.3 Type 3: Settlement of the Embankment and Culvert on Level Ground with Deformation of the Ground

Around Ojiya IC, differential settlement of several tens cms occurred between embankments and culverts as shown in Photo 9.6. Moreover, culverts settled several tens cms and stretched as shown in Photos 9.7 and 9.8. Embankment soil fell through the opened joints. The deformation of embankments, culvert boxes and ground are schematically shown in Fig. 9.6c. Both toes of the embankments spread in a lateral direction and caused the lateral displacement of adjacent ground as shown in Photo 9.9.

Detailed soil investigations were conducted at C-Box Kawaguchi 22 and C-Box Ojiya 2 to demonstrate the mechanism of the Type 3 failure (Inagaki et al. 2005). The subsurface soil conditions at C-Box Kawaguchi 22 and C-Box Ojiya 2, investigated after the earthquake, are shown in Figs. 9.13 and 9.14, respectively. The embankment

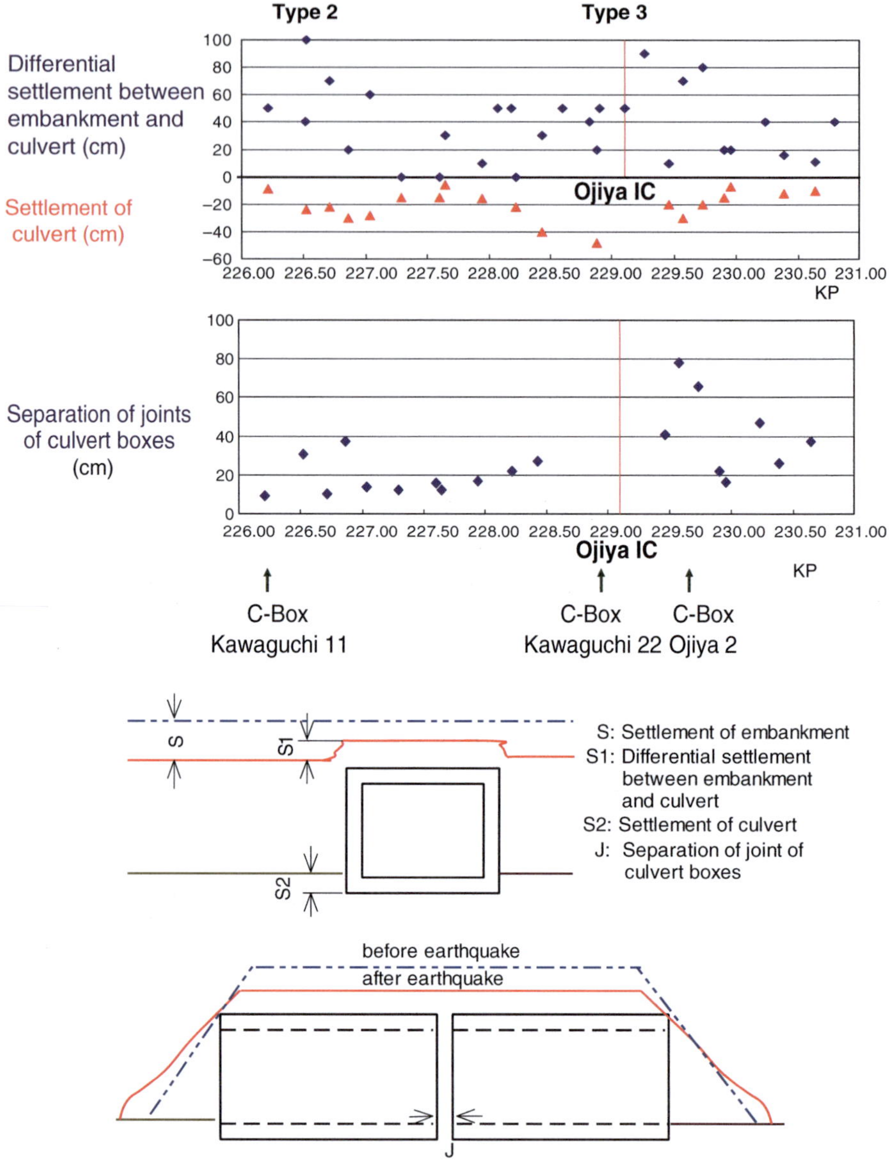


**Fig. 9.12** Distribution of differential settlements between embankments and culverts, settlements of culverts and separation of joints of culverts in level ground

soils at the two sites are clayey soils with 70–80% of fines. The SPT *N*-values of the embankment soils are 5–10. The heights of the embankments at the two sites were 5.6–6.8 and 5.3–5.6 m, respectively. The water levels at the sites were about 3 m higher than the bottom of the embankments, though the embankments were constructed on level ground.

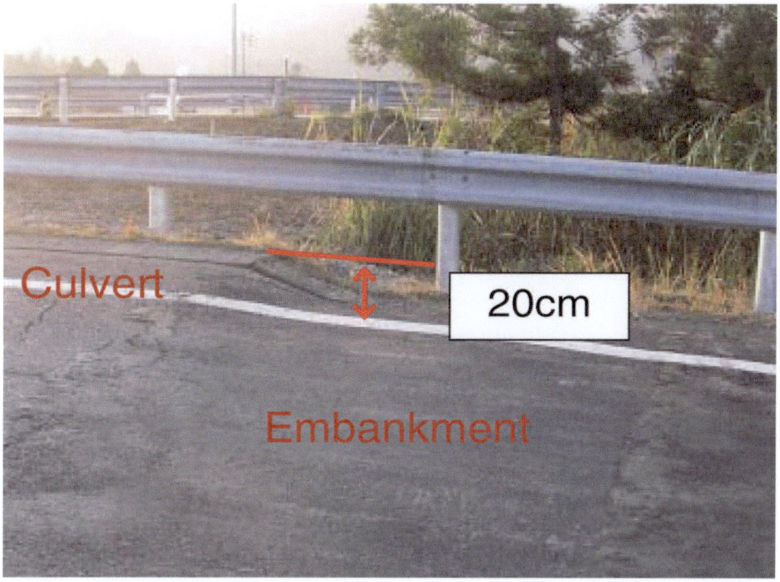


**Photo 9.6** Differential settlement at C-Box Kawaguchi 22

**Photo 9.7** Settlement of culvert at C-Box Kawaguchi 22

At C-Box Kawaguchi 22, thick soft silty layers, with the SPT *N*-values of about 5 are deposited from the ground surface to a depth of 16 m. A thin soft, silt, layer 2 m thick is deposited under the embankment at C-Box Ojiya 2. Then, silty sand, silt, sandy silt and silt layers, with SPT *N*-values of 10–20, are deposited to a depth of 24 m. Figure 9.15 shows the estimated cross section at C-Box Kawaguchi 22.

**Photo 9.8** Separation of culvert joint at C-Box Ojiya 2

**Photo 9.9** Moved ground adjacent to C-Box Ojiya 2

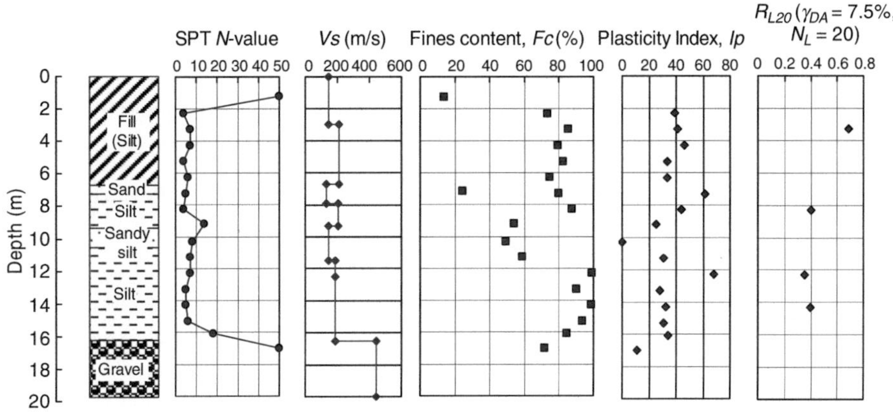


**Fig. 9.13** Soil profile and test results at C-Box Kawaguchi 22

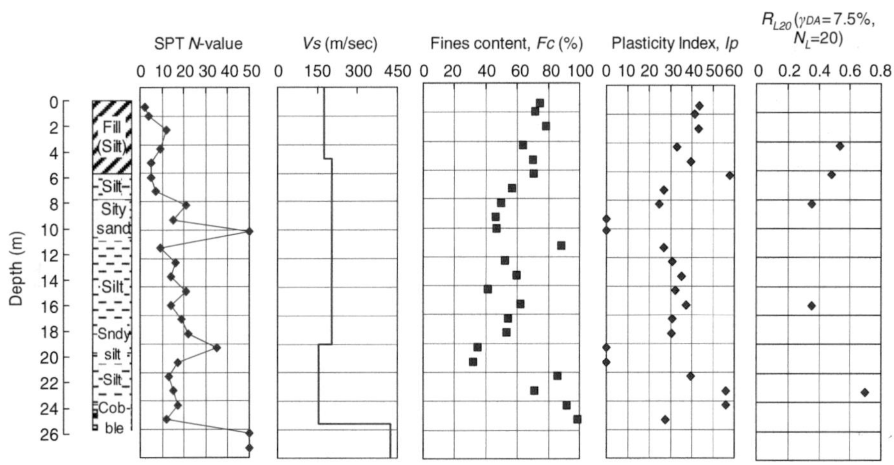


**Fig. 9.14** Soil profile and test results at C-Box Ojiya 2

As mentioned before, large settlements of embankments and culverts occurred at these sites. Differential settlements between embankments and culverts were 20 and 70 cm at C-Box Kawaguchi 22 and C-Box Ojiya 2, respectively. As the settlements of culverts were 48 and 30 cm, the total settlements of embankments were 68 and 100 cm, respectively. The joints at C-Box Kawaguchi 22 and C-Box Ojiya 2 opened by 119.5 and 78 cm, respectively.

## 9.5.2 Study on the Effect of Seismic Force on the Damage

As mentioned above, differential settlements between embankments and culverts, and the settlements of culverts themselves were measured after the earthquake.

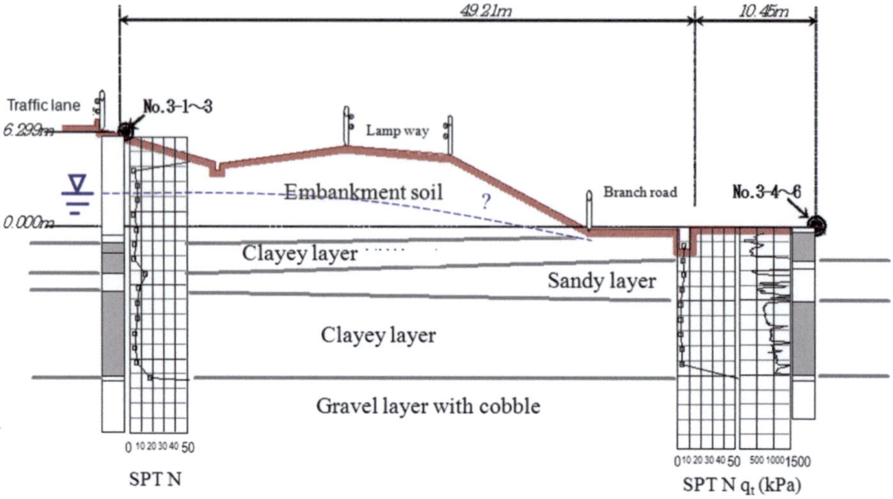


**Fig. 9.15** Cross section of the embankment adjacent to C-Box Kawaguchi 22

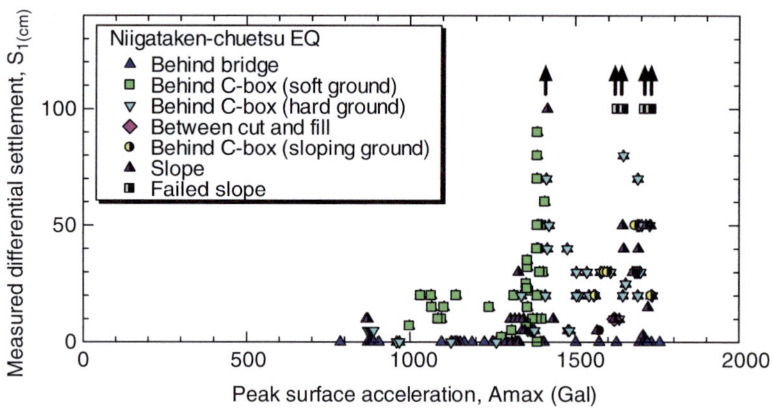


**Fig. 9.16** Relationship between differential settlement and maximum surface acceleration

Moreover, differential settlements between embankments and bridges, and differential settlements between fill and cut zones were measured. The relationships between these differential settlements and peak surface accelerations are plotted in Fig. 9.16. The peak surface acceleration was estimated based on the relationship between the epicentral distance and the peak surface accelerations recorded by K-net (National Research Institute for Earth Science and Disaster Prevention) during the Niigataken-chuetsu earthquake as shown in Fig. 9.17. Differential settlements increased with the peak acceleration as shown in Fig. 9.16. If the peak surface acceleration was less

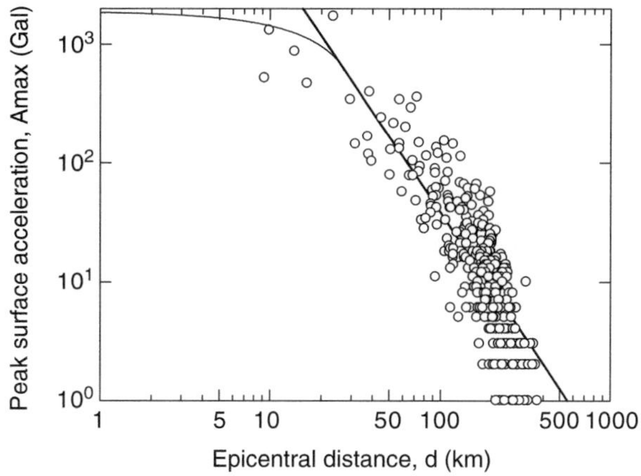


**Fig. 9.17** Relationship between epicentral distance and maximum surface acceleration

than about 700–800 gals, no obvious differential settlement occurred. Therefore, it must be said that the embankments of national expressways appear stable against moderate shaking, such as Level 1 earthquake motion even though the embankments are not designed to withstand earthquakes. However, against very strong shaking, such as Level 2 earthquake motion, the embankments settle and slide, and the severity of the settlement and sliding increases with acceleration.

## 9.6 Analyses on the Deformation of the Damaged Embankments

### *9.6.1 Necessity of Introduction of Deformation Analysis*

In the current seismic design of embankments, the stability of the embankments is evaluated based on a safety factor against sliding $F_s$, as mentioned through slope stability analyses. However, since the 1995 Kobe earthquake, it has become necessary to evaluate not only the safety against sliding but also against deformation in Japan. One reason is that the calculated $F_s$ becomes less than 1.0 under Level 2 earthquake motion such, as a seismic coefficient $k_h = 0.3$ or 0.4, even though the embankments and underlying ground are not loose. However, in road embankments, though large settlement interrupts traffic, minor settlement is not serious. Vehicles can pass with low speed over an embankment where minor settlement has occurred. And a damaged embankment can be restored quickly by filling soils. Then, new performance-based design methods based on the settlement of the embankments must be introduced in the seismic design. In this paper, Newmark's method and a residual deformation method were applied to Type 1 failures and to Type 2 and Type 3 failures, respectively. The adaptability of these methods is discussed.

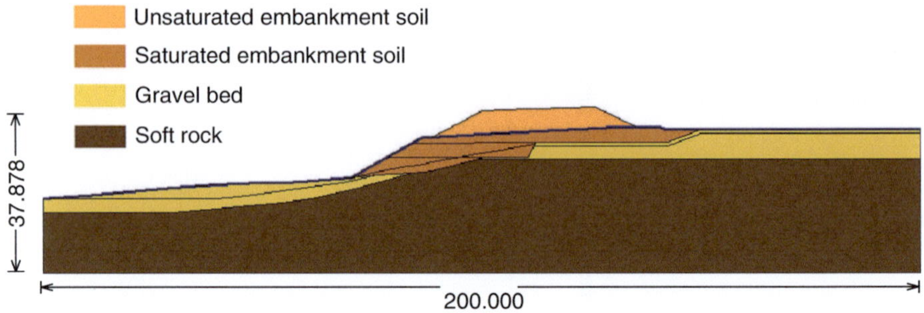


**Fig. 9.18** Model of the cross section at 215.9 KP of Kanetsu Expressway

## 9.6.2 *Analyses for Type 1 Failure*

The soil cross section at 215.9 KP of Kan-etsu Expressway was selected to evaluate the deformation of Type 1 failure. Analyses were conducted in the following four steps:

(i) Step 1: Static stress distribution in the cross section was analysed by the static finite element method.
(ii) Step 2: Dynamic stress distribution in the cross section was analysed by seismic response analysis.
(iii) Step 3: Distribution of safety factor in the section was calculated based on the static and dynamic stresses and soil strength.
(iv) Step 4: A slip surface was assumed and the sliding deformation along the slip surface was estimated by Newmark's method.

In the first step, the actual cross section was modelled for the analyses as shown in Fig. 9.18. Three water tables were assumed. First, it was assumed that water table was lower than the bottom of the filled soil. The second assumed level was that measured after the earthquake. However, as mentioned before, it was judged that the actual water table during the earthquake was higher than the measured table. The water table shown in Fig. 9.18 was assumed as the third level. Soil densities, deformation modulus and other values necessary for a static FEM were determined based on the soil investigations and tests conducted after the earthquake.

In the second step, the "FLUSH" software program for seismic response analysis was used. An acceleration wave was recorded on the ground surface at Kawaguchi IC during the earthquake. Based on the recorded wave, the acceleration wave at the depth of base rock was estimated, then the estimated wave was used in the analyses. The peak acceleration of the wave was 667 Gals, as shown in Fig. 9.19. The shear modulus and other values necessary for FLUSH were determined based on the soil investigations and tests. Figure 9.20 shows the distribution the peak acceleration

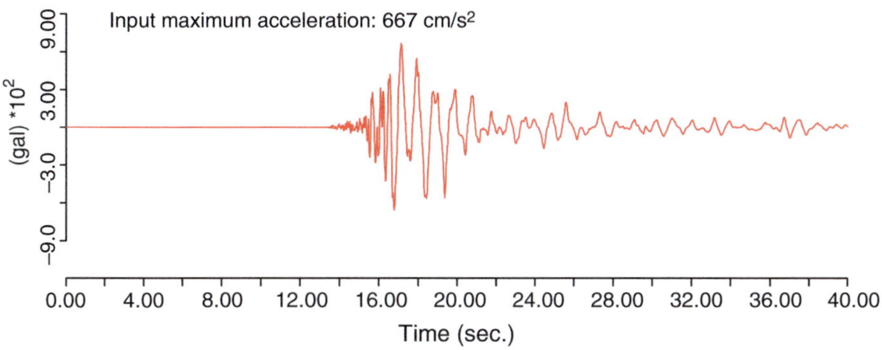


**Fig. 9.19** Input acceleration wave

**Fig. 9.20** Distribution of peak horizontal acceleration

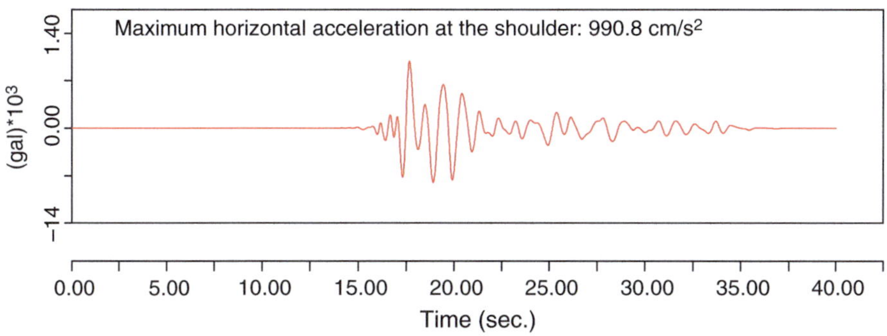


**Fig. 9.21** Analyzed time history of acceleration at the shoulder of the embankment

under the third assumed water table. The peak acceleration increased gradually from the bottom to the top of the embankment. Figure 9.21 shows the time history of analysed acceleration at the shoulder of the embankment. The peak acceleration at the shoulder was 990.8 cm/s$^2$.

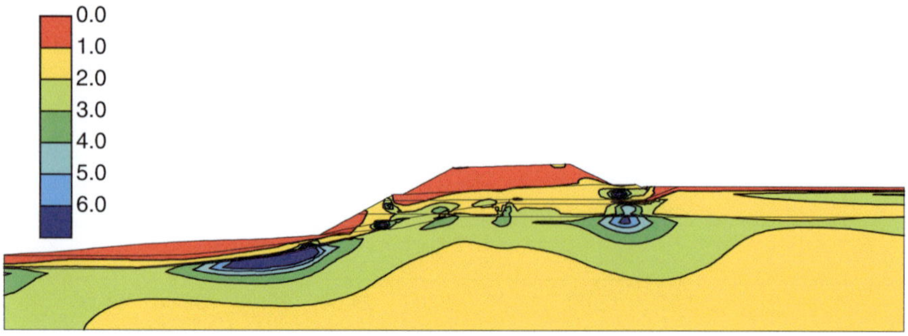


**Fig. 9.22** Distribution of safety factor against sliding

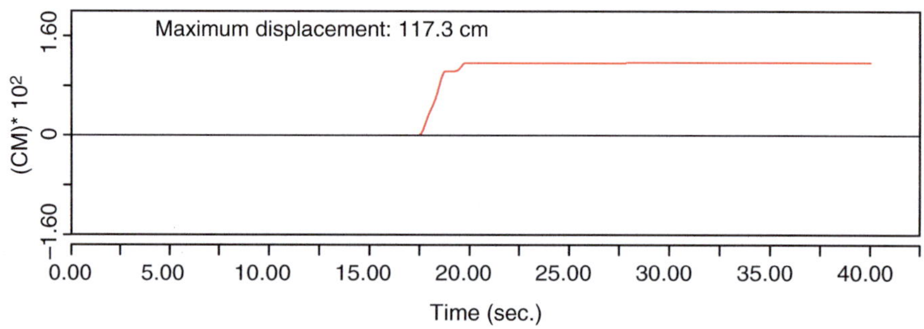


**Fig. 9.23** Analyzed time history of sliding displacement

The safety factor against sliding $F_s$, at each finite element mesh is calculated by the following equation in the third step:

$$Fs = \frac{shear\ strength}{static\ shear\ stress + dynamic\ shear\ stress}$$

Figure 9.22 shows the distribution of the $F_s$ thus calculated under the third assumed water table. $F_s$ is less than 1.0 in some zone of the embankment.

An appropriate slip surface is assumed in the fourth step. In the design of embankments, several slip surfaces must be assumed and the maximum displacement along the slip surfaces is selected. In this study, the slip surface during the earthquake was clear as shown in Fig. 9.9. The displacement along the surface was calculated by Newmark's method. Figure 9.23 shows the time histories of sliding displacement under the third water table.

The final displacement under the three assumed water tables were 1.7, 2.6 and 117 m, in that order. The actual displacement is unknown. However the displacement might be more than 10 m because the slide was long. Therefore, it can be said that sliding displacement such as Type 1 failure can be evaluated by these procedures.

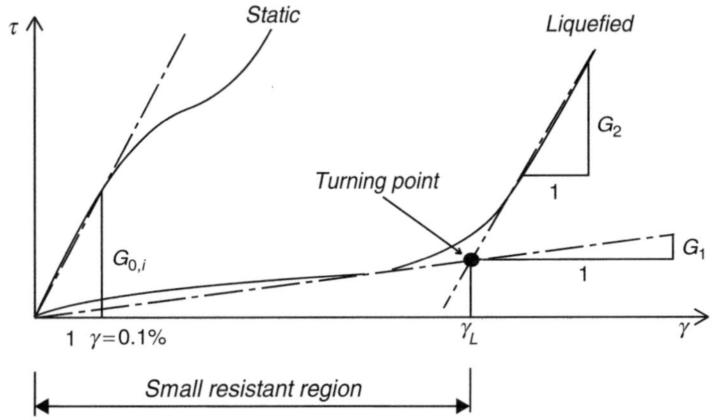


**Fig. 9.24** Schematic diagram of stress-strain curve of liquefied soil

### *9.6.3 Analyses for Type 2 and 3 Failures*

Type 2 and 3 failures were not slide failure but slump failures. As slope stability analyses cannot be applied to slump failures, an analytical computer program to estimate residual deformation named "ALID (Yasuda et al. 1999, 2003)" was applied. In this method, static finite element method is applied in the following steps:

1. Step 1: Static stress distribution and deformation of the embankment and underlying ground before the earthquake are evaluated by static FEM based on the stress-strain relationships of intact soils.
2. Step 2: Distribution of undrained cyclic shear strength $\tau_l$ in the embankment and the ground is determined by laboratory tests and/or in-situ SPT.
3. Step 3: Distribution of cyclic shear stress $\tau_d$ induced by the earthquake is estimated by seismic response analyses or simple empirical methods.
4. Step 5: Distribution of safety factor against liquefaction or softening $F_l$, which is the ratio of $\tau_l$ to $\tau_d$, is calculated.
5. Step 6: Stress-strain relationships of liquefied or softened soils with different $F_l$ were obtained from laboratory tests or empirical relationships.
6. Step 7: Deformation of the ground due to liquefaction or softening is evaluated by static FEM by using the stress-strain relationships of liquefied or softened soils.
7. Step 8: Deformation of the embankment and the ground due to the dissipation of excess pore pressure is evaluated by the relationship between volume strain and $F_l$.

Figure 9.24 shows a schematic diagram of stress-strain curves of intact soil and liquefied or softened soil. The shear strain increases with very low shear stress up to

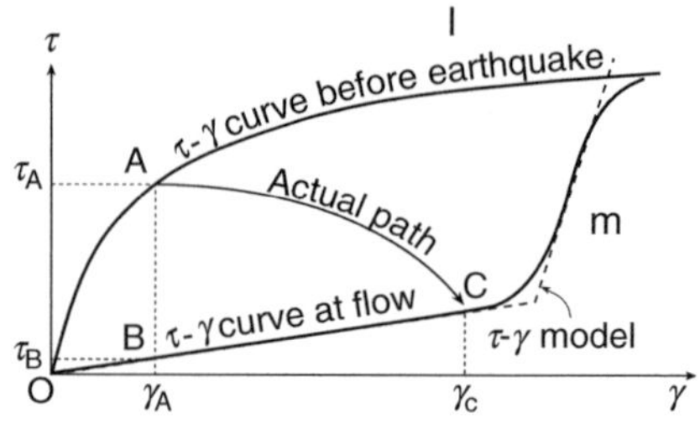


**Fig. 9.25** Concept of stress-strain curves used for analysis

very large strain. Then, after a resistance transformation point, the shear stress increases comparatively rapid with shear strain, following the decrease of pore water pressure. Figure 9.25 shows the concept of the stress-strain curves which are used in ALID. Line *l* denotes a backbone curve at the beginning of the earthquake. Point A is supposed to be the initial state of a soil element in the ground. Suppose that the backbone curve moves from *l* to *m* due to liquefaction or softening, then strain should increase in order to hold the driving stress. Since driving stress, however, decreases according to the change of geometry, the actual strain increment from state *l* to *m* is from A to C. Namely, the strain increment caused by the change of material property is $\gamma_C - \gamma_A$. Ground deformation stops when new material property comes into balance with a new driving stress, which is shown as point C in the figure. Many cyclic torsional tests were conducted to obtain the empirical stress-strain curves of liquefied sands and softened clays.

Sites C-Box Kawaguchi.22 and C-Box Ojiya 2 on the Kan-etsu Expressway, were analyzed. Cyclic torsional shear tests were carried out to obtain cyclic shear strength and shear modulus after cyclic loading. Then the “ALID” was applied to evaluate the deformation of embankments and grounds. The deformations of embankments and grounds before and after the earthquake are shown in Fig. 9.26a, b. Ground under the embankments spread laterally, and embankments stretched and settled the same as the actual deformation shown in the Photo 9.9. The analyzed settlement at the center of the surface of roads at Ojiya 2 and Kawaguchi 22 were 110.8 and 68.6 cm, respectively. Figure 9.27a–c compare analyzed settlements and horizontal displacements at both toes, with the measured ones, respectively. Analyzed values coincide fairly well with the actual values, implying that this method can be applied for Type 2 or 3 failures.

### *9.6.4 Effect of Several Factors on the Settlement of Embankments*

Using ALID, the effects of several factors on the settlement of embankments can be estimated. Then analyses were carried out under several conditions shown in Table 9.4:

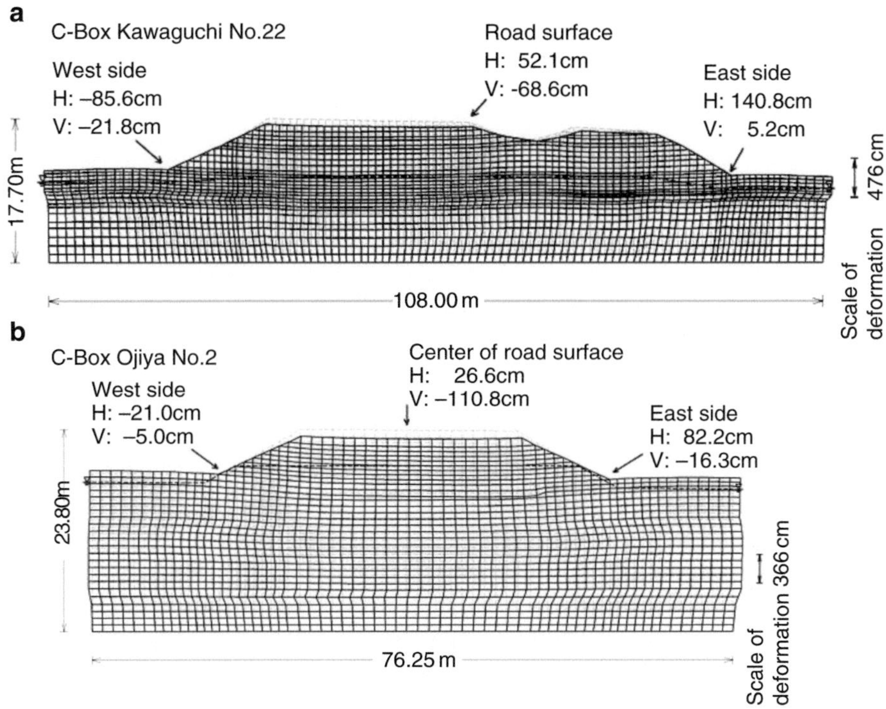


**Fig. 9.26** Analyzed deformation of embankments and grounds (**a**) C-Box Kawaguchi No.22, (**b**) C-Box Ojiya No. 2

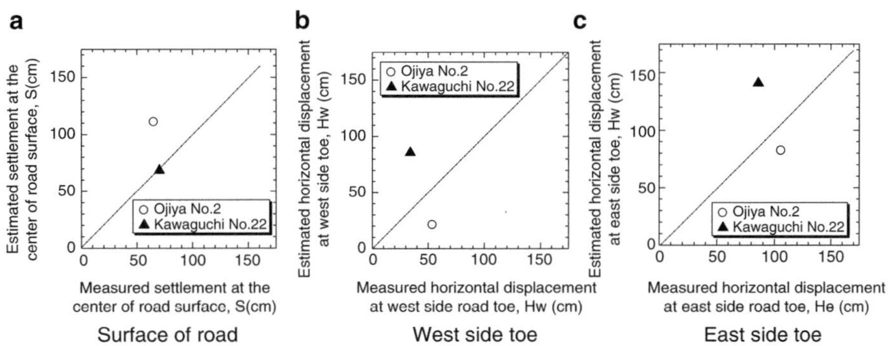


**Fig. 9.27** Comparison of analyzed and measured displacements (**a**) Surface of road, (**b**) West side toe, (**c**) East side toe

**Table 9.4** Condition of analyses by ALID

| Factor | Basic condition | Alternative conditions |
|---|---|---|
| Height of embankment | 7 m | (i) 10 m, (ii)14 m |
| Level of water table | 1.75 m higher than the bottom of embankment | (i) 1 m lower than the bottom, (ii) 3.5 m higher than the bottom, (iii) 7 m higher than the bottom |
| Embankment soil | Clayey soil at Atsugi ($I_p$=20) | (i) Sandy soil at C-Box Ojiya 2 ($I_p$=7), (ii) Clayey soil at Noto ($I_p$=22.4), (iii) Clayey soil at Anamizu ($I_p$=24.6) |
| Ground soil | Clayey ground at C-box Ojiya 2 | (i) Gravelly ground at C-Box Kawaguchi 22, (ii) Loose sand ground (SPT $N_1$=6) |
| Peak acceleration at ground surface | 600 Gals | (i) 250 Gals, (ii) 500 Gals, (iii) 750 Gals, (iv) 1,000 Gals, (v) 1,250 Gals, (vi) 1,500 Gals |

Figure 9.28a–e shows relationships among the factors and analyzed settlements. The settlement increased with height and the water table, as shown in Fig.9.28a, b. The settlement decreased with $I_p$ of the embankment soil as shown in Fig. 9.28c. This means that the shear modulus of sandy soil is easily decreased by earthquakes, so sandy soil is not appropriate for embankments. As shown in Fig. 9.28d, settlement is fairly affected by the soil of the ground. If the soil is loose and liquefiable, the settlement becomes very large. Peak acceleration surely affects the settlement as shown in Fig. 9.28e.

## 9.7 Conclusions

Appropriate seismic design methods for road embankments were discussed based on a review of the damage during past earthquakes, soil investigations and analyses. The following conclusions were derived through these studies:

1. There are several kinds of roads in Japan. The basic design of road embankments is similar. However, detailed design and resistance against earthquakes differ. National expressways are well designed and are of the highest grade in Japan. Nevertheless, the embankments of national expressways were damaged by the 2004 Niigataken-chuetsu and 2007 Niigateken-chuetsu-oki earthquakes because shaking was very strong.
2. Three types of failure occurred in the Kan-estu Expressway during the 2004 Niigateken-chuetsu earthquake. Type 1 failure was serious slides of the embankment on the sloping ground. Type 2 failure was settlement of the embankment on level ground without the deformation of the ground. Type 3 failure was settlement of the embankment on level ground with deformation of the ground. In Type 1, a high water table might have caused the slide. In Types 2 and 3, settlement

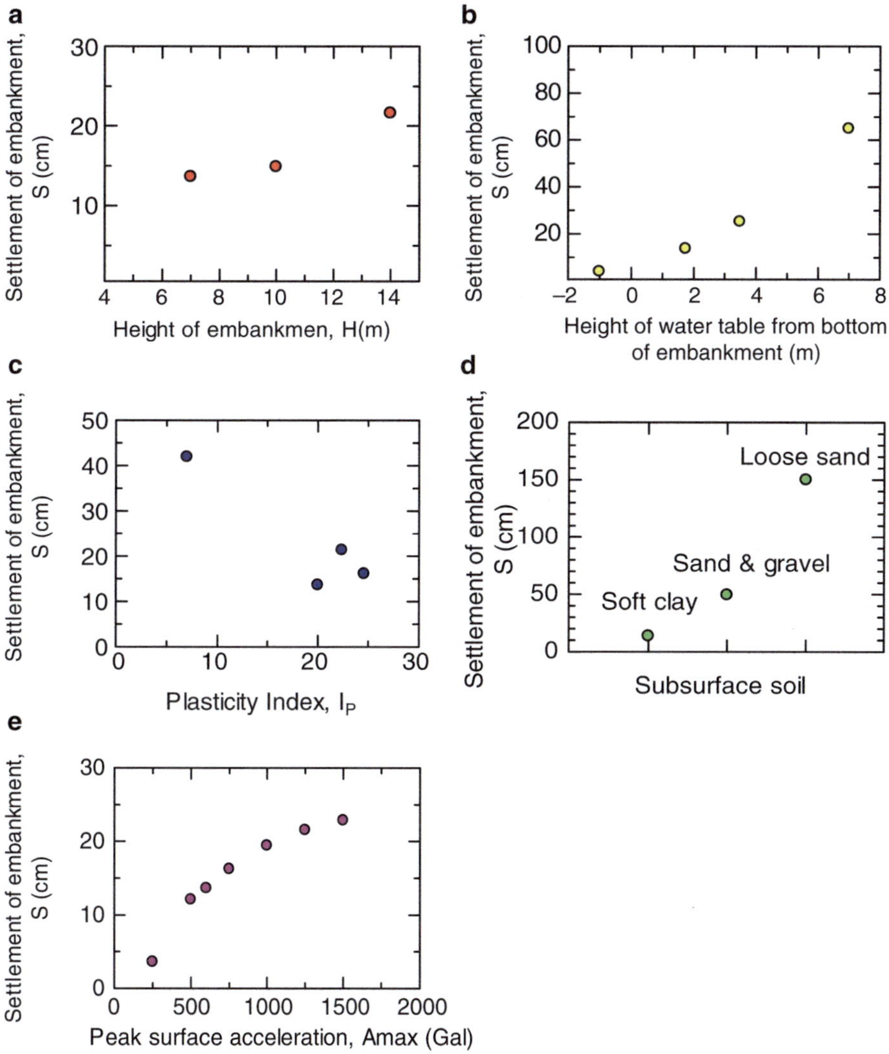


**Fig. 9.28** Effect of five factors on the settlement of embankments on level grounds analyzed by ALID

of the embankment seemed to have been induced due to the reduction of shear modulus of the fill materials. A reduction of the shear modulus of soils in the foundation ground also seemed to influence Type 3 failure.

3. Deformation of the damaged embankments could be evaluated by Newmark's method and ALID for Type 1 failure and Type 2 and 3 failures, respectively. However more studies are necessary to determine an effective seismic design for road embankments.

## References

Inagaki M, Itakiyo K, Kasuda K, Yamada S, Yasuda S (2005) Deformation of embankments on clayey grounds during the 2004 Niigataken-chuetsu earthquake (Part 2). In: Proceeding of the 4th annual conference on Japan Association for Earthquake Engineering (in Japanese)

National Research Institute for Earth Science and Disaster Prevention (NIED). K-NET WWW service (http://www.k-net.bosai.go.jp/)

Sasaki T, Matsuo O, Sugita H, Matsuda Y (2007) Damage to road fill by the 2007 Noto-hanto earthquake. In: Proceeding of the 14th world conference on earthquake engineering (WCEE), Paper No. S26-20

Yasuda S, Yoshida N, Adachi K, Kiku H, Gose S (1999) A simplified analysis of liquefaction-induced residual deformation. In: Proceedings of the 2nd international conference on earthquake geotechnical engineering, pp 555–560

Yasuda S, Ideno T, Sakurai Y, Yosida N, Kiku H (2003) Analyses of liquefaction-induced settlement of river levees by ALID. In: Proceedings of the 12th Asian Regional conference on SMGE, pp 347–350

# Chapter 10
# Performances of Rockfill Dams and Deep-Seated Landslides During Earthquakes

**Kenji Ishihara**

**Abstract** There has been increasingly high intensity of acceleration ever recorded during earthquakes in recent times. In consistence with this trend several characteristics have been unearthed of performances of rock fill dams and of large-scale landslides. With due considerations to these, damage features of high rock fill dams during recent earthquakes in Japan and China are briefly introduced herein, together with those previously reported. Large-scale landslides are mostly associated with development of deep-seated slip planes which pass through layers of cohesive soil deposits existing at great depths. Two cases of massive landslides that occurred in Japan during recent earthquakes are introduced and some interpretation given on the basis of the laboratory test results on volcanic pumice soils which are considered to have induced the landslides. It is warned that borings as deep as 100–150 m need to be made not to miss presence of any cohesive soils underlying stiff soils or rocks to identify hidden danger of land sliding in the case where important facilities are constructed in the backland areas near hills or mountains.

## 10.1 Introduction

In recent years, there have been remarkable progresses worldwide in sophistication of instruments and strengthening of observation network for recording strong motions during earthquakes. Thus, the coverage of areas was widened and precision improved greatly to monitor motions during any scale of earthquakes.

As a consequence, the magnitude of recorded accelerations has increased remarkably year after year. At the time of the Tokachi-Oki earthquake in 1968 in Japan the peak horizontal ground acceleration recorded was 225 gal in the port of Hachinohe.

K. Ishihara (✉)
Chou University, Nittetsu ND-Tower 12FL., Kameido, Koto-ku, Tokyo 136-8577, Japan
e-mail: kenji-ishihara@e-mail.jp

M.A. Sakr and A. Ansal (eds.), *Special Topics in Earthquake Geotechnical Engineering*, Geotechnical, Geological and Earthquake Engineering 16,
DOI 10.1007/978-94-007-2060-2_10, 

It jumped up to a value of 891 gal at the time of the 1995 Kobe earthquake, and at the time of the most recent earthquake in 2008, the peak recorded acceleration reached a value as high as 4,022 gal in combined 3D absolute peak acceleration. Since earthquakes are natural phenomena, it seems unlikely that the intensity of motion itself has in fact increased so dramatically, but it is not permissible to ignore such strong motions in design consideration once they are recorded.

Thus, it would be worth while looking over characteristic features of distress incurred to soils and structures during the recent large earthquakes. In the year of 2008, two large-scale earthquakes occurred in China and Japan, both in inland mountainous regions. The major damage was a number of landslides, rockfalls, and consequent debris flow. There were several rockfill dams in the epicentral areas which suffered some damage. There were several instruments installed on or near the dam body, and these did record motions which may be utilized to develop better understanding of performances of high rockfill dams in general when subjected to strong shaking during earthquakes.

In the following pages of this paper, features of damage to high rockfill dams will be introduced and their performances discussed in the light of the motions recorded during the earthquakes.

Issues associated with landslides generally do pose difficulty for providing proper understandings in terms of soil and rock mechanics. However, large-scale landslides have been known to generate disastrous consequences such as a large number of casualties and loss of houses and infrastructures of social importance. Thus, efforts need to be expended to understand causative mechanisms and their consequences particularly for deep-seated landslides which generally involve large movements of massive soils or rocks. The deep-seated landslides are known generally to occur when cohesive materials such as old-aged clays or pumice deposits of volcanic origin are existent at great depths overlaid and camouflaged by stiff competent soils or rocks.

The presence of such weak soil deposits can not be generally identified by the current practice of in-situ soil investigation because of its being conducted only to depths of 20–50 m. In this context, it is considered of engineering significance to address warning for potential risk of landslide by introducing outcomes of some studies on the cause of large landslides associated with deep-seated slips of pumice deposits at depths 100–150 m. It is hoped that some incentives will be instigated in future for further studies on mechanism of similar type of landslides.

## 10.2 Performances of Rock-Fill Dams

### *10.2.1 Ishibuchi Dam in Japan*

#### 10.2.1.1 General

An earthquake with a magnitude of M = 7.2 rocked the northern part of Japan Main Island on June 14, 2008 and caused severe damage to bridges and dams. A number of landslides and mud flows were triggered during this event which is

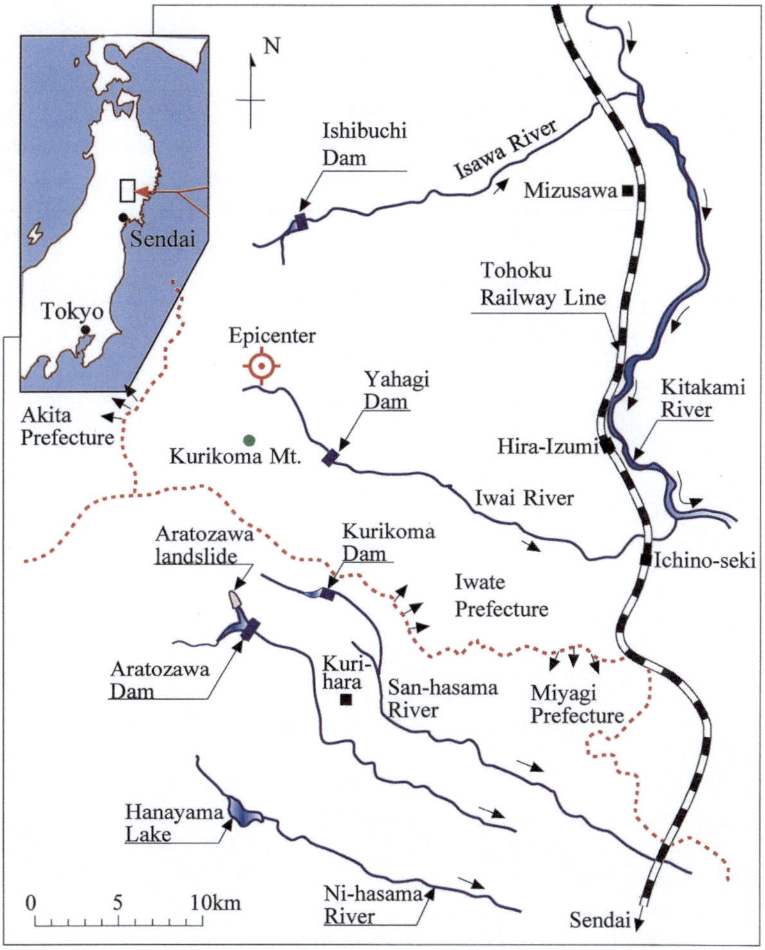


**Fig. 10.1** Location of the 2008 Iwate-Miyagi Inland earthquake

termed Iwate-Miyagi Inland earthquake. The location of its epicenter is shown in Fig. 10.1. Among various kinds of destructions, worthy of note were settlements of a concrete-faced old dam called Ishibuchi Dam whose location is shown in Fig. 10.1. The dam was constructed by impounding water in Isawa River as shown in Fig. 10.2 and its water storage capacity is 1.6 million cubic meters. A bird-eye view of the dam is shown in Fig. 10.3 and its plan and back view in Fig. 10.4. As shown in the cross section of Fig. 10.5, the dam 53 m high was constructed in 1953 with the concrete facing on the upstream side at an average slope of 1:1.3. The slope in the downstream side is 1:1.5 with two berms. The concrete facing 50 cm thick was laid on the well-compacted gravel sub-grade on the upstream face.

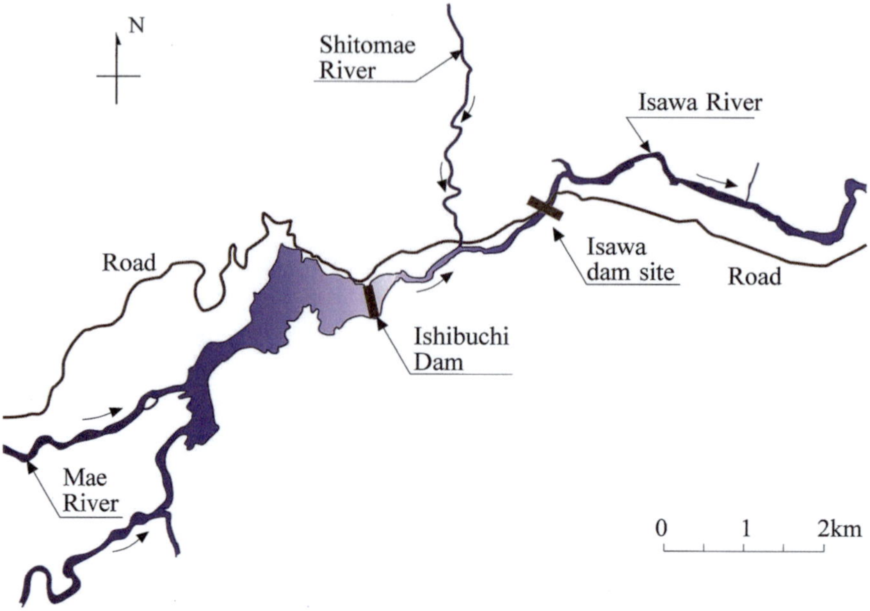


**Fig. 10.2** Site of Ishibuchi dam

**Fig. 10.3** Bird-eye view of Ishibuchi rockfill dam (From the office of Ishibuchi dam)

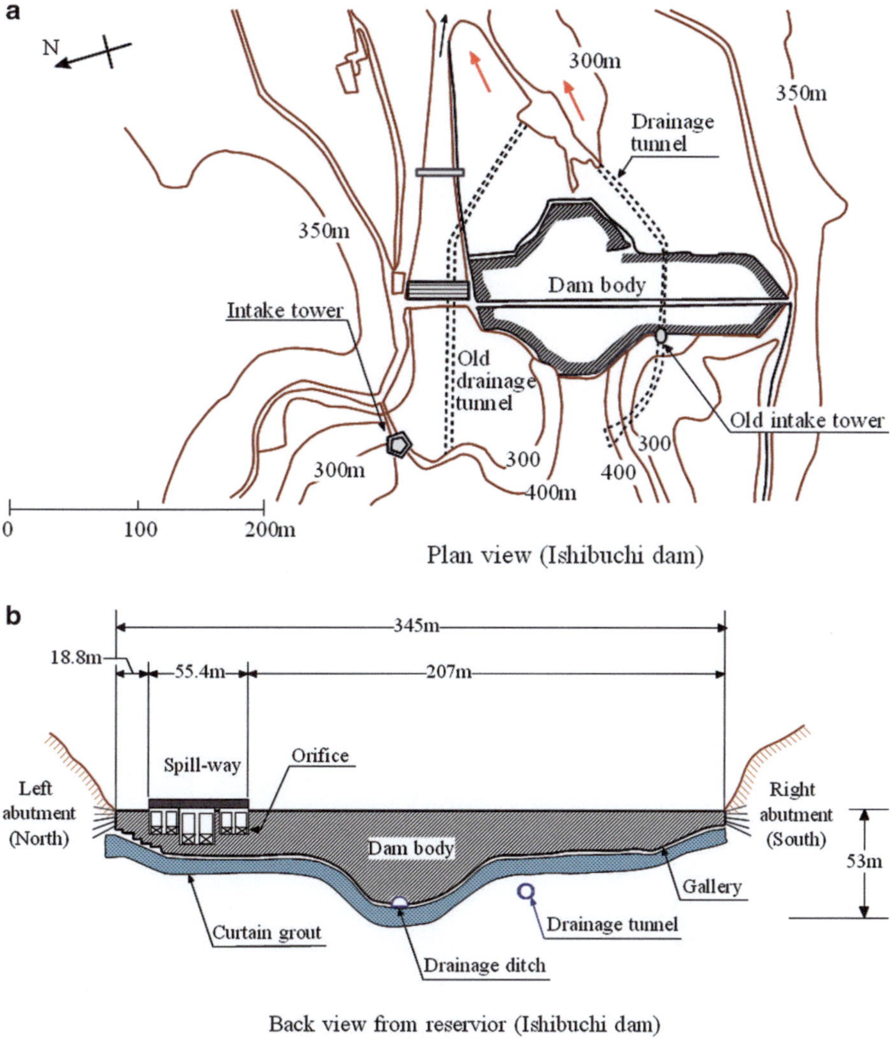


**Fig. 10.4** Plan view and back view of Ishibuchi dam

In constructing the dam, nine pillars of reinforced concrete were first erected and bridge girders were placed on the top to provide a track road for rock-carrying wagon, as displayed in Fig. 10.6. The rocks were carried from quarries by the wagon and dumped from the bridge. A photo from the left abutment in Fig. 10.7 shows the cranes placing large blocks of rocks on the downstream face. The pillars for the bridge girders were buried within the rockfills when the dam construction was completed.

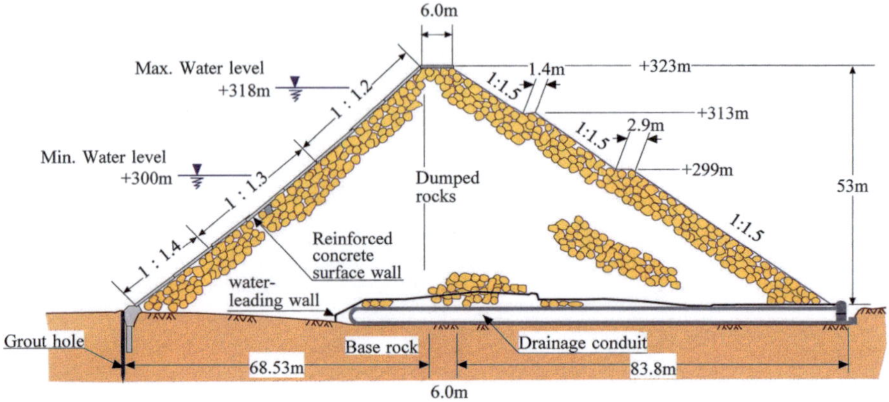


**Fig. 10.5** Cross section of Ishibuchi dam

**Fig. 10.6** A bridge with pillars and girders used for carrying and dumping rocks and gravels for Ishibuchi dam, looking over the right abutment (From the office of Ishibuchi dam)

### 10.2.1.2 Performances During the Earthquake

As the Ishibuchi Dam is located at a distance of about 10 km north from the epicenter, it was shaken very strongly. It is to be noted that the up-and down components was as high as 2.11 g at the crest and 1.78 g at the nearby terrace. The horizontal components in the direction perpendicular to the dam axis were also intense with the

**Fig. 10.7** Cranes placing surface rocks near completion of Ishibuchi dam (From the office of Ishibuchi dam)

peak acceleration of 0.95 g at the crest and 2.14 g at the terrace. At the time of the quake the elevation of impounded water was about 300 m, a condition near the lowest water level.

There was no fatal damage resulting in degrading in the integrity of the dam as a whole. The water-sealing function of the concrete facing remained practically intact. Several stones over the downstream face near the crest plunged out or fell down, but there were no cracks or no offset between the concrete blocks on the upstream face. Most conspicuous feature of the damage was overall settlement of the crest of the order 0.8 m. Because of the presence of the concrete pillars buried in the alignment of the dam axis, the settlements occurred at a certain interval in the portion where there was no pillars as shown in a photo of Fig. 10.8. Figure 10.9 shows the bulging of stone facing near the crest. The vertical strain in the fill at the center is roughly estimated as being as much as 0.8/53 = 1.5%.

## *10.2.2 Aratozawa Dam in Japan*

### 10.2.2.1 General

At the time of the Iwate-Miyagi Inland earthquake on June 14, 2008, a huge-scale landslide took place over the gentle mountain slope about 300 m upstream of the reservoir which is impounded by the Aratozawa rockfill dam. The site, 15 km south

**Fig. 10.8** Settlement features of the dam crest indicating presence of hidden pillars, Ishibuchi dam (Looking over the *right* abutment)

**Fig. 10.9** Stone face bulging out near the downstream crest, Ishibuchi dam (Looking towards the *left* abutment)

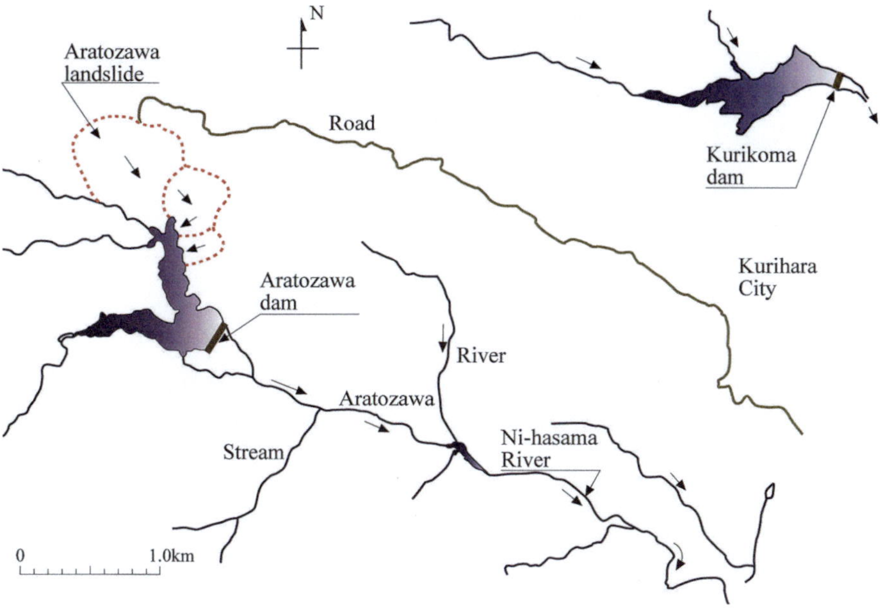


**Fig. 10.10** Area of Aratozawa dam

of the epicenter is shown in Figs. 10.1 and 10.10. The bird-eye view of the dam including the landslide is shown in Fig. 10.11. The Aratozawa dam with a center core was constructed in 1985 and its cross section is displayed in Fig. 10.12. The height is 74 m and its crest length is 250 m.

### 10.2.2.2 Performances During the Earthquake

There were sets of strong-motion recorders installed at the crest and at the basement of the dam. The peak acceleration was 950 gal at the crest in the direction perpendicular to the dam axis. At the rock outcrop near the base, the peak recorded acceleration in the same direction was 900 gal, indicating that the body of the dam had been subjected to an intense level of shaking. Fortunately, there was practically no injury to the dam body, although a few seiches (waves) generated by the landslide hit the dam and flowed over the spillway. According to a witness account, the height of the wave is purported to have been about a few meters. Although there was no clearly visible injury incurred to the dam, its body did deform slightly as a whole causing about 30 cm settlement at the crest as visualized in a photo shown in Fig. 10.13. The overall vertical strain of the dam beneath the crest is estimated roughly to have been of the order of $0.3/74 = 0.41\%$.

**Fig. 10.11** A view of the Aratozawa dam and the landslide in the north

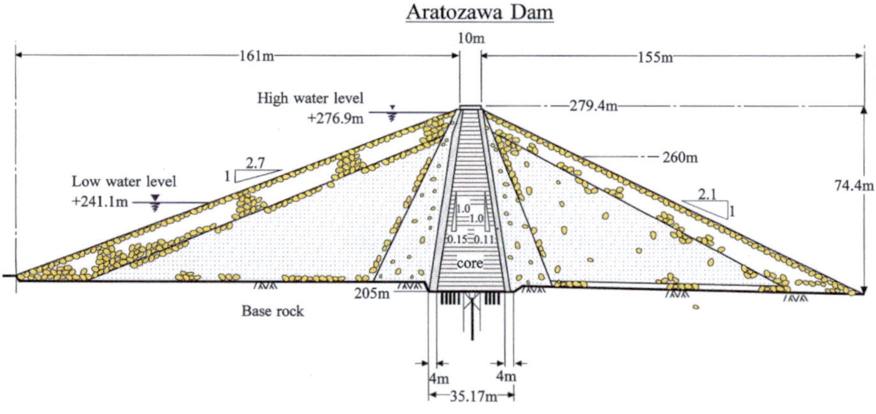


**Fig. 10.12** Cross section of Aratozawa rockfill dam

There were two small rivers upstream the dam supplying water to the reservoir. A huge landslide occurred on the north upstream side of the dam involving 70 million cubic meter of soils and rocks. This landslide is said to have caused seiches in the reservoir to a height of a few meters which propagated not only towards the dam, but also upstream back to the southern tributary, thereby

**Fig. 10.13** Crest of the Aratozawa dam after the earthquake

destroying two bridges and surrounding forest. The increased water level by waves discharged impounded water through the spillway on the right abutment. The amount of discharged water was not so much, but probably on the order of several hundred cubic meters.

## *10.2.3 Zipingpu Dam in China*

### 10.2.3.1 General

The Wengchuan earthquake of May 12, 2008 was amongst the most devastating event that rocked the inner part of the mainland China in recent years. The magnitude of the earthquake was M = 7.9 and its epicentral region is shown in Fig. 10.14. There was a rockfilled dam 156 m high with concrete facing within the area of highest intensity of shaking, about 15 km southeast of the epicenter as indicated in Fig. 10.15. The dam was constructed in 2001–2005 in the middle reaches of Min River, about 7 km upstream of the famous world heritage, Dujiang Yan as shown in Fig. 10.15. The dam serves for multiple purposes such as irrigation, flood control

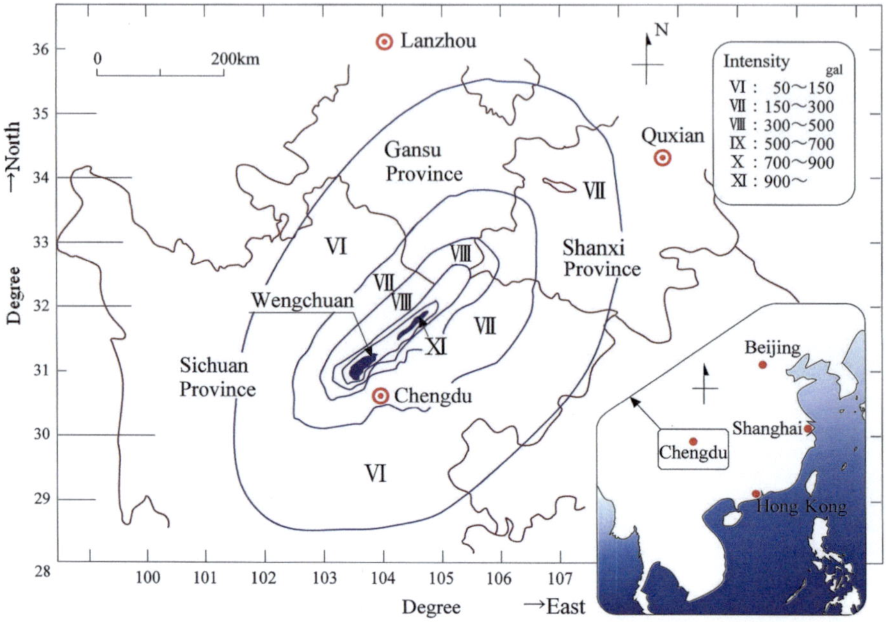


**Fig. 10.14** Location of Wengchuan earthquake of May 12, 2008 in China

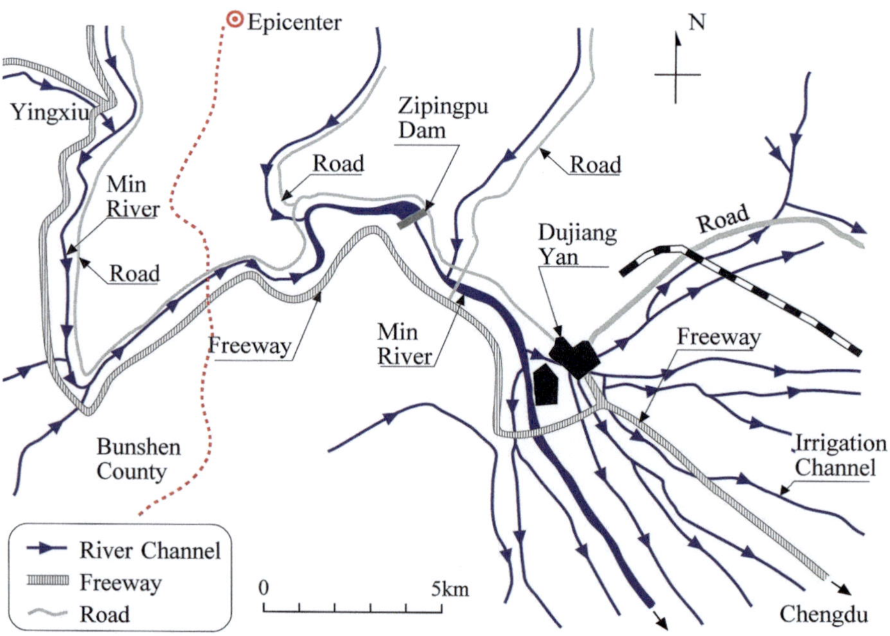


**Fig. 10.15** Location of Zipingpu rockfill dam

**Fig. 10.16** Bird-eye view of the Zipingpu Dam (From the office of Zipingpu dam)

and electric power supply with its capacity of water storage as much as one billion cubic meters. The capacity of electric power generation is 750,000 kW.

The bird-eye view from the reservoir side after the quake is shown in Fig. 10.16. The length at the crest level is about 600 m. The plan view and backside view of the Zipingpu Dam are shown in Fig. 10.17. The cross-section of the Zipingpu Dam is displayed in Fig. 10.18. The materials were taken from the limestone quarries on the mountain flank about a few kilometers upstream of the dam. The rocks with a maximum size of 0.8–1.0 m were compacted and water-proof reinforced concrete slabs 30–80 cm thick were placed on top of the 3 m-thick filter zone on the upstream face of the dam.

As shown in the cross section in Fig. 10.18, the upstream slope is 1:1.4 and the slope on the downstream side is 1:1.5 near the crest. It is noted that there is a curtain wall installed to a depth of about 50 m beneath the upstream toe. There is also a small dumped fills of fine sands or silts near the toe of the upstream side which are expected to act as self-healing agent by clogging fissures or cracks in case they happen to develop on the concrete face particularly near the toe. As indicated in Fig. 10.18, the elevation of the crest is 884 m and the maximum elevation for water storage is set at an elevation 877 m. At the time of the Weng-Chuan Earthquake, the water storage was 830 m in elevation. From the backside view of the dam shown in Fig. 10.17b, it is noted that on the left (eastern) abutment, the rockfill was placed on a steep slope of the bedrock which is generally deemed unfavorable to maintain stability not only from seismic but also from static point of view. Thus, instruments had been installed to monitor the displacement. On the west side, the rock surface was cut almost vertically and concrete wall was constructed to ensure the integrity of the dam.

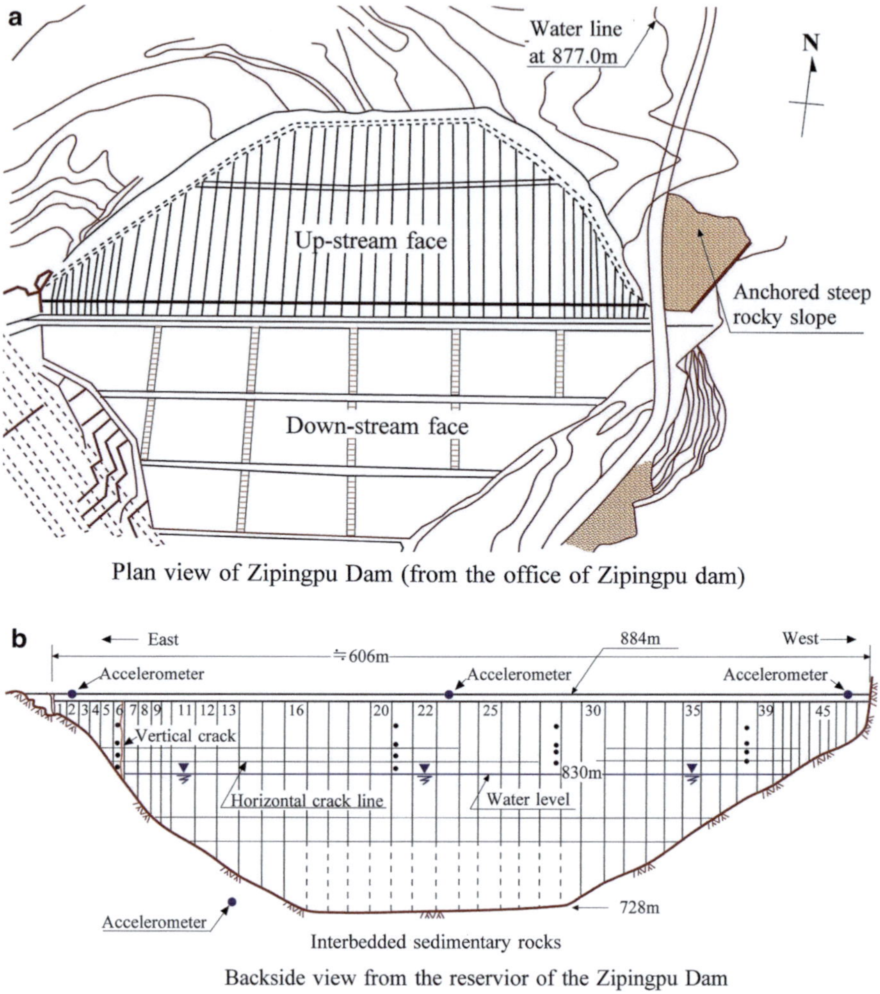


**Fig. 10.17** Plan view and backside view of the Zipingpu Dam (From the office of Zipingpu dam)

#### 10.2.3.2 Performances During the Earthquake

There were 4 strong motion recorders installed on different portions of the dam, namely, at the east, west and central parts at the crest level and on the rock at the basement as indicated in Fig. 10.17b. Except the last one on the rock, three components of the acceleration were obtained with the peak values as shown in Table 10.1. It is noted that the peak crest accelerations in the vertical and horizontal directions across the dam body were as high as 2.0 g. Although actual records of time history are not available, it is said that these peaks occurred as a

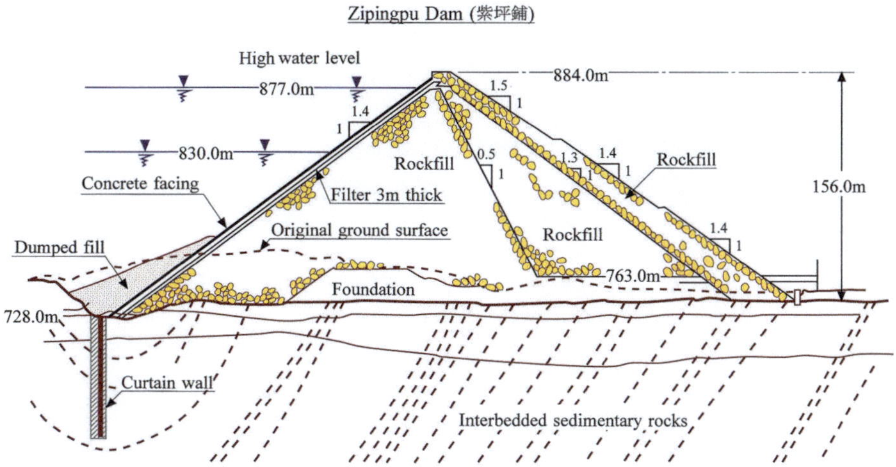


**Fig. 10.18** Cross section of the Zipingpu Dam (From the office of Zipingpu dam)

**Table 10.1** Recorded peak accelerations at Zipingpu dam (times gravity acceleration)

| | Perpendicular to the dam axis | Parallel to the dam axis | Vertical up and down |
|---|---|---|---|
| Center of the dam | 2.061 g | 1.635 g | 2.065 g |
| Left abutment | 0.811 | 0.591 | 0.715 |
| Right abutment | 1.371 | 1.177 | 0.654 |

single spike or impulse with a high frequency and the duration of entire shaking was of the order of 15 s. It is also noted that the horizontal component of acceleration in the direction of the dam axis was very high as well, which is indicative of the fact that high compressive force might have acted in the direction of the dam axis.

The Zipingpu Dam safely withstood the strong shaking by the earthquake as a whole except for some minor injuries as described below. The maximum settlement of 74 cm was incurred at the center of the crest with a downstream horizontal displacement of 30 cm there. The average vertical strain through the height of the dam is estimated to have been as $0.74/156 = 0.47\%$. At the time of the earthquakes, water level in the reservoir was 830 m as indicated in Fig. 10.17b. A long horizontal offset across the dam occurred about 5 m above the water line over the length of 600 m on the upstream concrete facing in which the topmost part of the concrete slab was displaced by 25 cm toward the reservoir. The location of the offset is shown in Fig. 10.17b. It is conceived likely that, the lower part of the concrete face was supported by counterweight of impounded water, whereas the upper part was more susceptible to the horizontal movement. In addition, there were injuries involving two vertical cracks observed in the

**Fig. 10.19** Damage to parapet wall

concrete facing near the left (east) abutment as indicated in Fig. 10.17b, where it is to be noticed, as mentioned above, that the rock abutment beneath the fill has a steep slope.

The vertical cracks can readily develop near the steep abutment because of uneven settlement likely to occur potentially in the rockfills placed on the steep slope. It is to be remembered that the rockfill near the abutment has been generally recognized as the weakest point in the overall stability of filled dams particularly when rockfills are to be placed on steeply dipping rock abutment. The parapet wall along the crest on the downstream side was damaged severely, as shown in Fig. 10.19, by the strong shaking. Also, the rocks on the downstream armored surface were destroyed and slipped down the slope. In view of the intense shaking near the epicenter, the Zipingpu dam is deemed, by and large, to have performed nicely without any serious distress.

## *10.2.4 Crest Settlements of High Dams*

For the high embankment dams constructed of roller-compacted rockfills, effects of strong shaking have been known to manifest primarily in the form of surface cracking or local slides near the crest of the dam. In many instances, these are associated with overall deformation of the dam body as represented by the settlements of the crest. Thus, it seems reasonable to adopt the crest settlement as an index quantity indicative of the degree of damage to the dam during earthquakes.

During major earthquakes in the recent past, several sets of acceleration records have been obtained by means of the instruments installed on high embankment dams. Thus, it would be of engineering significance to collect data on seismic records and crest settlements during major earthquakes. The outcome of such data compilation was presented by Ishihara et al. (1989). The observed data from the three dams mentioned above are shown in Table 10.2, together with similar information compiled before.

The data in Table 10.2 are displayed in Fig. 10.20 where it may be seen that the crest settlement tends to increase roughly with increasing intensity of shaking at the crest level. Among the ten dams investigated, Ishibuchi dam is the oldest which was constructed in 1953 when the compaction technique by heavy machines was still in an infant stage. Excluding this dam, the crest settlement during earthquakes may be considered to lie in the range of shaded area in Fig. 10.20, if an embankment dam is designed by the state-of-the-art criteria and accordingly constructed by means of heavy compaction machines.

## 10.3 Large Scale Landslides

### *10.3.1 Aratozawa Landslide in the 2008 Earthquake*

At the time of the Iwate-Miyagi Inland Earthquake of June 14, 2008, a large scale landslide occurred at a place about 15 km south of the epicenter. The location is shown in Figs. 10.1 and 10.10. The landslide was triggered in the mountain areas having a gentle slope. The forest area was dreadfully devastated producing some major separate ridges with spreading of broken blocks of soil and rock masses and uprooted trees. The length of the slide was about 1.5 km and width was 0.8 km as shown in plan view of Fig. 10.21. The bird eye photo is shown in Fig. 10.22. The sliding mass of soils and rocks traveled about 300 m towards east-south, as illustrated by the displaced roadway shown in Fig. 10.23. More exact plan view is shown in Fig. 10.24. The landslide is conjectured to have been triggered by the intense shaking as well as by the offset due to fault ruptures which appeared at the headwall portion as well as another hidden fault line crossing at the center of the slide area. A view forwards northeast from the bottom of the headwall is displayed in Fig. 10.25.

After the landslide, detailed investigations were performed by traversing over the devastated area, and also by using air-photos. There are five lines drawn each in the longitudinal and transverse directions as shown in Fig. 10.24 which indicate

**Table 10.2** Earthquake-induced crest settlements of rock-fill dams

| Name of dams | Year built | Earthquake | Date of Eqs. | Magnitude, M | Epicenter distance | Peak acc. amax(gal) | Height H(m) | Settelment S(cm) | S/H(%) |
|---|---|---|---|---|---|---|---|---|---|
| La Villita (Mexico) | 1966 | – | 1985/9/19 | 8.1 | 10 | 696 | 60 | 10–33 | 0.16–0.55 |
| El Infier-nillo (Mexico) | 1964 | – | 1985/9/19 | 8.1 | 10 | ≒450 | 146 | 10–12 | 0.07–0.08 |
| Matahina (New Zealand) | 1964 | Edge-cumbe | 1987/3/2 | 6.3 | 23 | 420 | 86 | 3.4–80 | 0.04–0.93 |
| Namioka (Japan) | 1983 | Nihonkai-Chubu | 1983/5/26 | 7.7 | 146 | 223 | 52 | 5.7 | 0.11 |
| Makio (Japan) | 1961 | Naganoken-seibu | 1984/9/14 | 6.8 | 5 | ≒700 | 107 | ≒50 | ≒0.47 |
| Ishibuchi (Japan) | 1953 | Iwate-Miyagi | 2008/6/14 | 7.2 | 12 | ≒1000 | 53 | ≒80 | ≒1.5 |
| Aratozawa (Japan) | 1998 | Iwate-Miyagi | 2008/6/14 | 7.2 | 13 | ≒1000 | 74 | ≒30 | ≒0.41 |
| Zipingpu (China) | 2008 | Weng-Chuan | 2008/5/12 | 7.9 | 10 | ≒2000 | 156 | ≒74 | 0.47 |

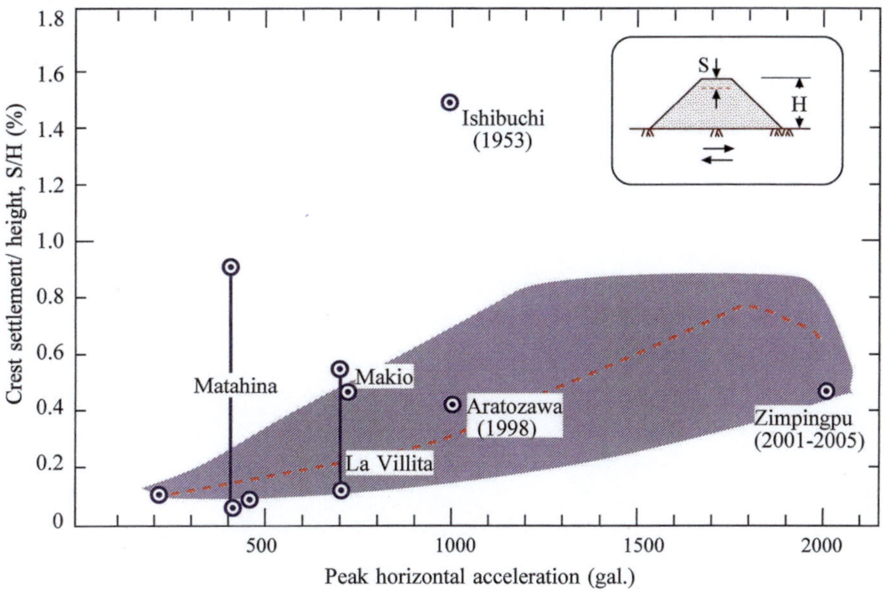


**Fig. 10.20** Crest settlement of rock-fill dams

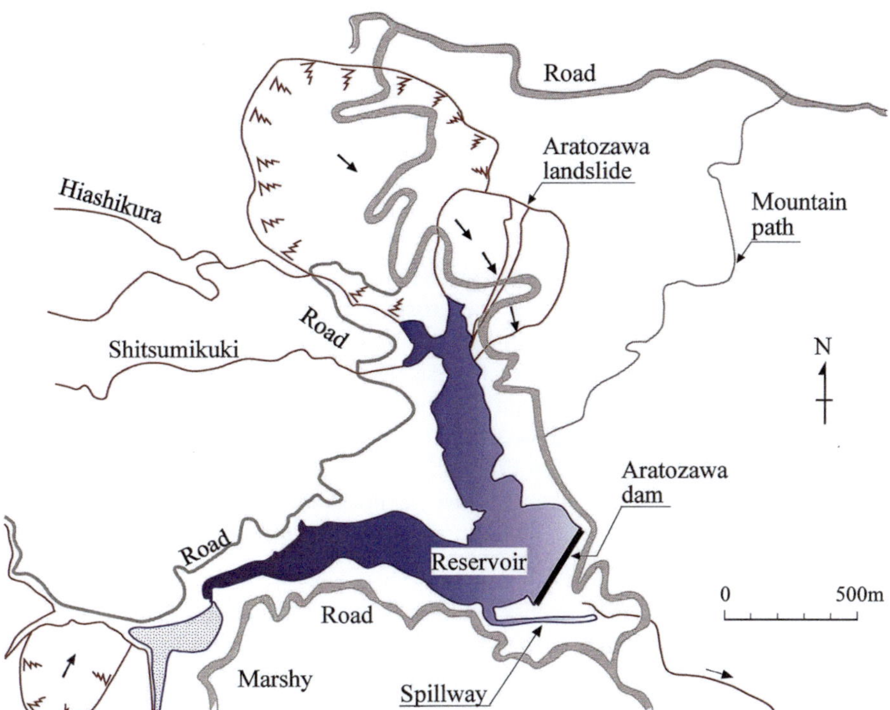


**Fig. 10.21** Aratozawa area

**Fig. 10.22** Aerial view of Aratozawa landslide

reference alignments for site investigation. Mainly at the cross points of these lines, boring was performed as designated by BV1 through BV23. To identify precise points, the GP observation was made at 9 points. Surveillance cameras were also installed at two intact places, one near the edge of the headwall cliff and the other near the downhill toe of the slide area. To monitor earth movements, the wire type sensors were installed near the top and also at the toe of the slide zone encroaching the reservoir (Fig. 10.26).

On the basis of the boring data, soil profiles were established along the cross section A-A′, through E-E′. Three of the cross sections established by Momikura (2009), involving the main stream of the slide, that is, B-B′, C-C′ and D-D′ sections are demonstrated in Fig.10.27. The cross sections A-A′, B-B′, C-C′ and D-D′ are not exactly coincident with the main direction of the slide. As such, another cross section F-F′ was chosen as indicated in Fig. 10.27 and its feature established by Momikura (2009) as displayed in Fig. 10.27. In the cross sections D-D′ and F-F′, the locations of fault rupture likely to have occurred are also shown. Looking over the cross sections together with the plan view in Fig. 10.24, one can visualize the main features of the slide as described below.

1. While the extraordinarily intense shaking is a main triggering factor, two fault rupture lines running across the slide area transversely seem to be associated with the occurrence of such a huge scale mass movement. The locations of these fault offset were inferred from the fault lines manifested on the ground surface in the intact forest area close to the slide. These are indicated in Fig. 10.27.

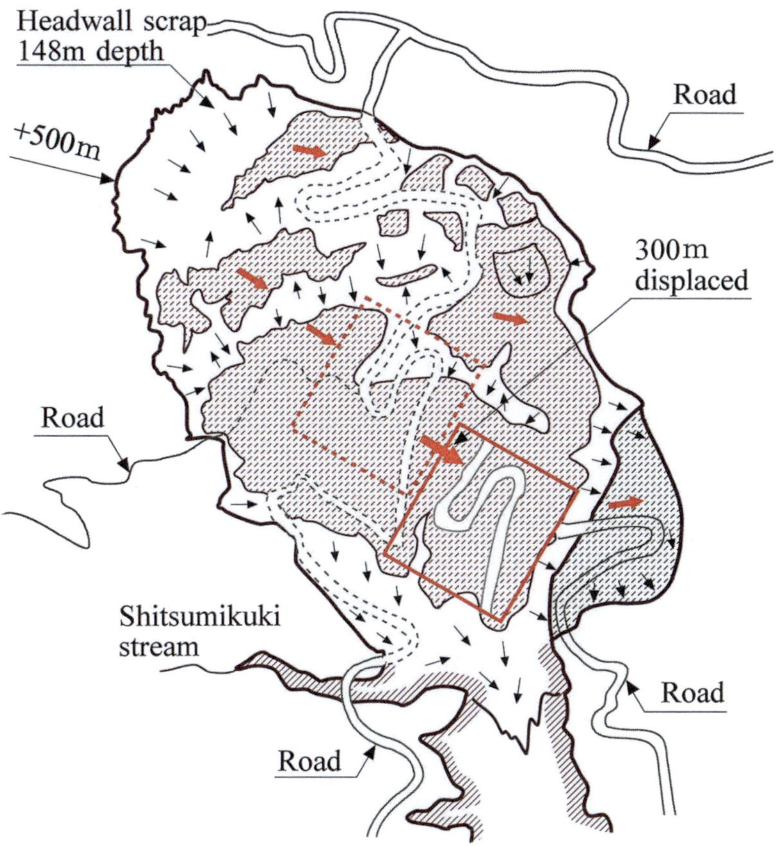


**Fig. 10.23** Movement of massive soils and rocks in Aratozawa landslide

2. There was a large-scale fracture of the rock formation exposed on the cliff of the headwall scarp to a depth of about 150 m, as displayed in Fig. 10.25. The upper part is comprised of welded hard tuff underlain by the tuff of pumice origin. Although not seen on the surface, softer pumice seemed to be buried below the debris and this pumice is considered as the cause of the landslide.
3. The sliding mass was carried downhill through a distance of about 300 m. The original forest area is seen in Fig. 10.22 perching on the sliding blocks of soil mass.
4. There seems to be another slide which might have been driven by the second fault rupture as indicated by point A to A′ in Fig. 10.27. Because of possible separation of the two main soil blocks, one involved in the uttermost uphill slide and the other one downstream, a depression was created as shown in Fig. 10.27.
5. The second slide moved downstream, but due probably to slight buttress actions by the underlying stiff rocks, the sliding mass in the north portion appears to be slightly pushed up.

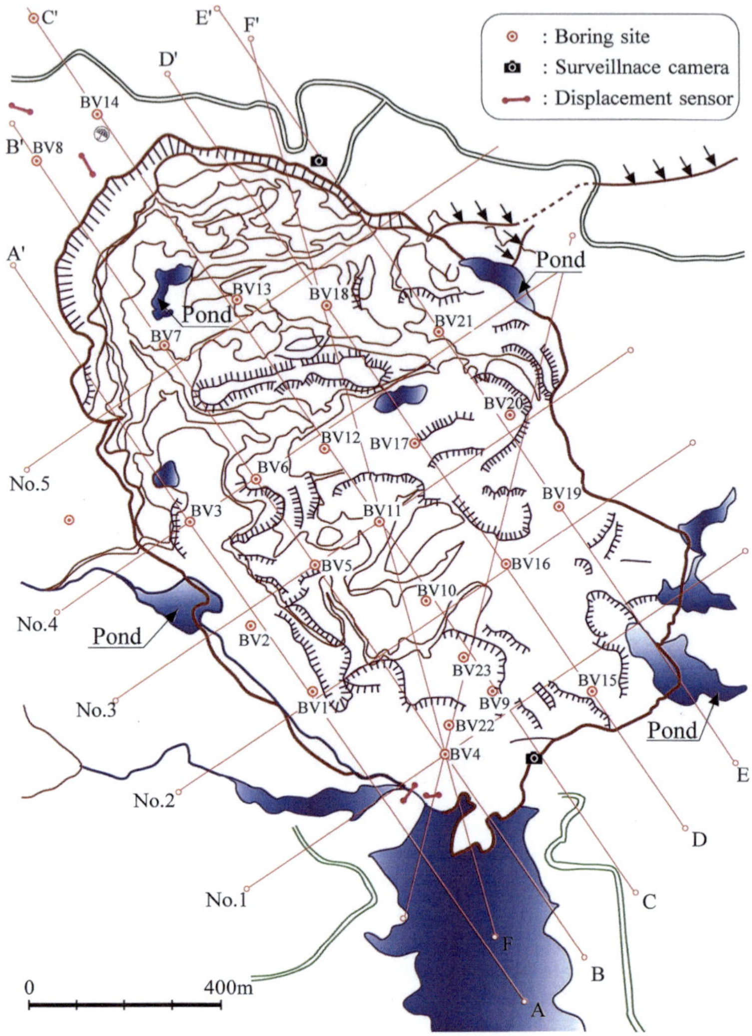


Fig. 10.24 Plan view of Aratozawa landslide

6. The portion 300 m long near the lake of the dam slid and spread over into the lake bottom. However, the lateral spreading was not extensive as shown in Figs. 10.23 and 10.27.

It is envisaged that the soft rocks and stiff soils to a depth of about 50 m from the ground surface are composed of layers of rocks and conglomerate of volcanic origin. Underneath these layers, there exist layers of silts of lacustrine origin. It is conjectured that volcanic ashes of pumice fell over the old ancient lake and

**Fig. 10.25** Headwall scarp of Aratozawa landslide looking toward the east

**Fig. 10.26** *Upper part* of the Aratozawa landslide viewed from northeast corner

sedimented in geologically long era, and later on volcanic lava flowed out and deposited over such the pumice-containing lacustrine sediments. There are a lot of large broken blocks of sedimented soft rocks spreading over the devastated area. The soft rocks are composed of cemented pumice of various shape about 10–30 cm in diameter entrapped in the silty or sandy rocks. There are sometimes broken pieces of such soft rocks consisting predominately of pumice.

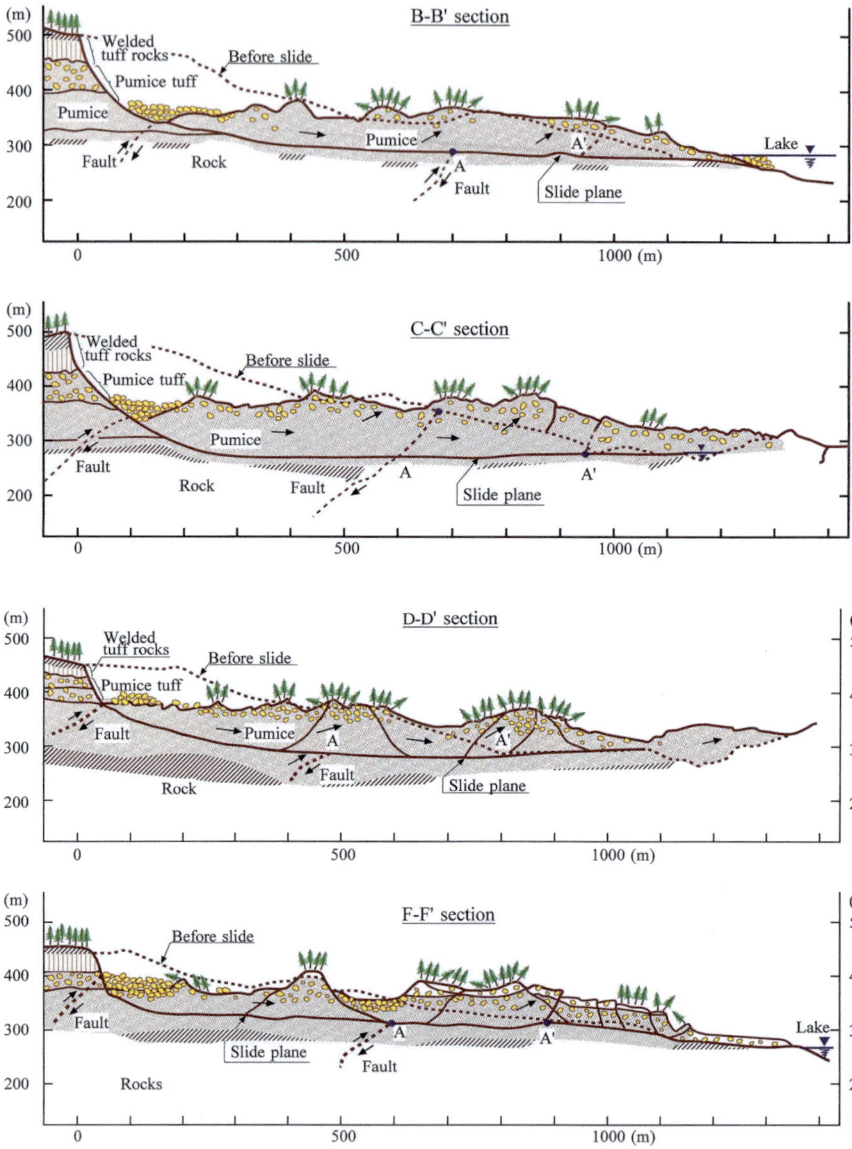


**Fig. 10.27** Cross sections of Aratozawa landslide (After Momikura)

Over the devastated ground surface, fractured pieces of pumice or powdered pumice were seen spreading here and there, indicating the possibility for the pumice deposits to have been responsible for triggering and inducing large displacement at depths in the landslide area.

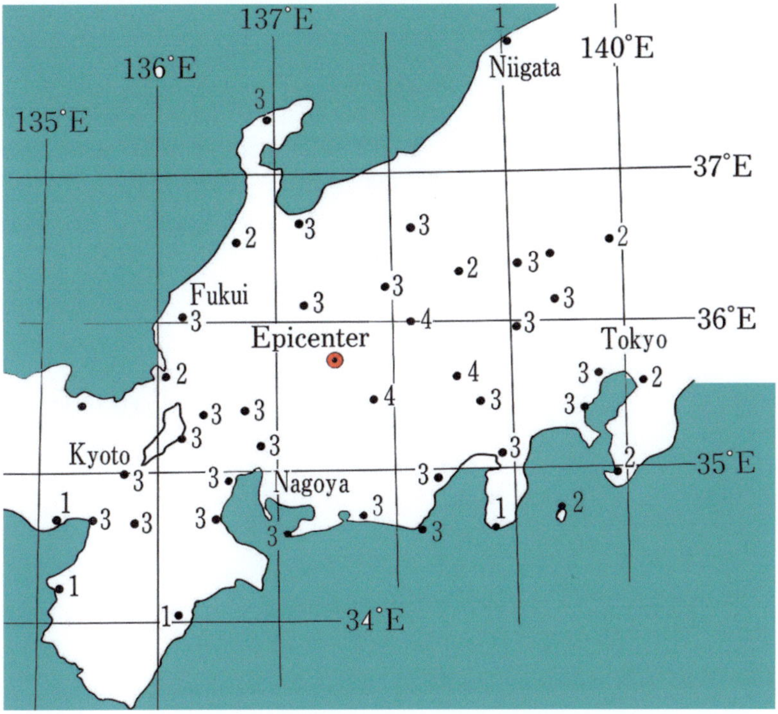


**Fig. 10.28** Location of the epicenter of Naganoken-seibu earthquake of 1984

## *10.3.2 Mount Ontake Landslide in the 1984 Naganoken-seibu Earthquake*

A strong earthquake shook the central part of Japan mainland at 8:48 a.m. on September 14, 1984 with its epicenter located at the southern foot of Mt. Ontake about 100 km northeast of Nagoya, as shown in Fig. 10.28. Its magnitude 6.8 placed this event among the greatest to have occurred in this locality in the historic time (Ishihara et al. 1986) (Fig. 10.29).

A large-scale landslide leading to the debris avalanche was triggered on the south flank of the Mt. Ontake at a site about 2.5 km southeast of the peak as shown in Fig. 10.30. A cataclysmic debris flow was originated and traveled down the valley through a distance of 12 km as shown in Fig. 10.30. The topmost part of the head-wall scarp was located at an altitude of 2,550 m and the lowest elevation of the slide scarp was at an altitude of 1,850 m. The slide covered an area about 300 m wide and 1,300 m long.

The geological conditions over the basal failure surface and in its neighborhood were provided by Sakai (2008) as shown in Fig. 10.31. The cross section in the longitudinal direction through A-A′ is shown in Fig. 10.32. As can be seen, the

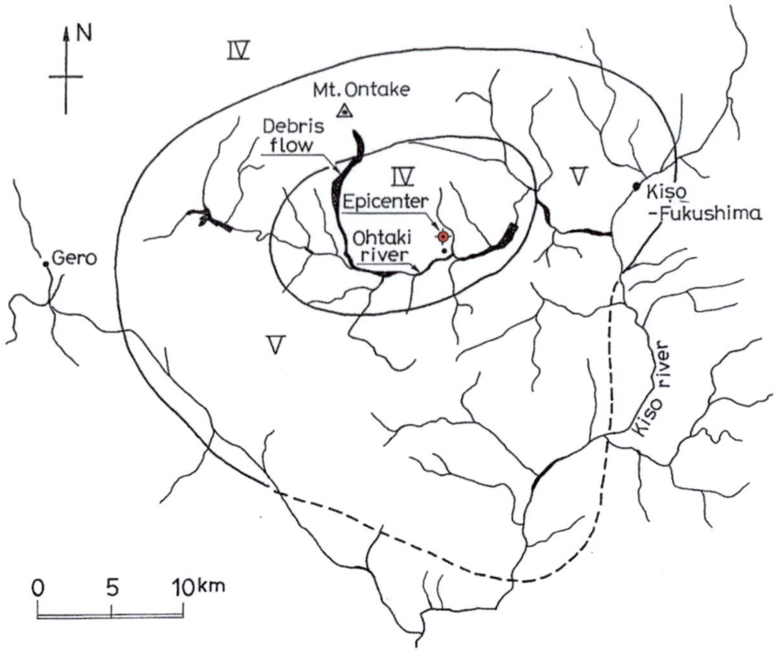


Fig. 10.29 Distribution of intensity of shaking in Nagano-seibu earthquake of 1984

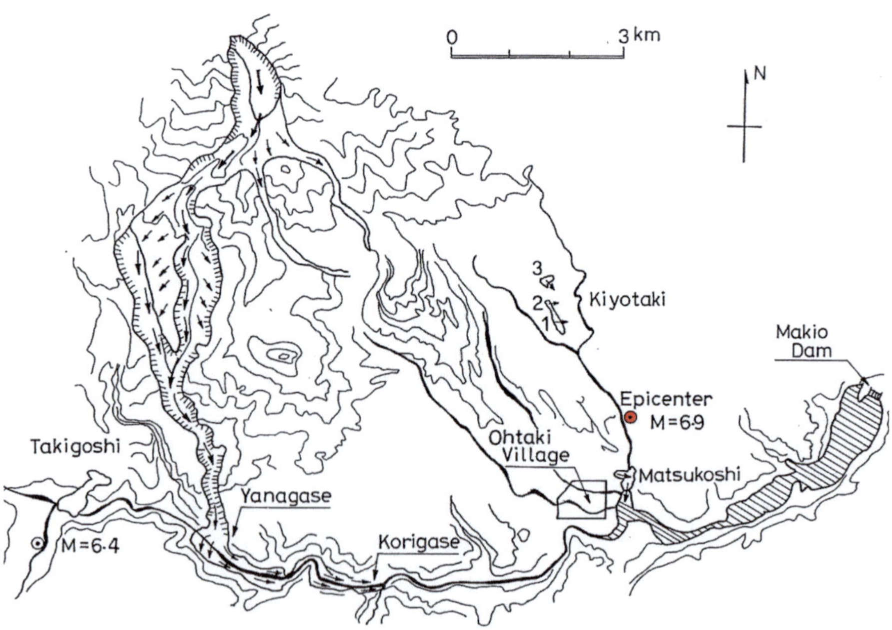


Fig. 10.30 Runout route of the Ontake debris avalanche

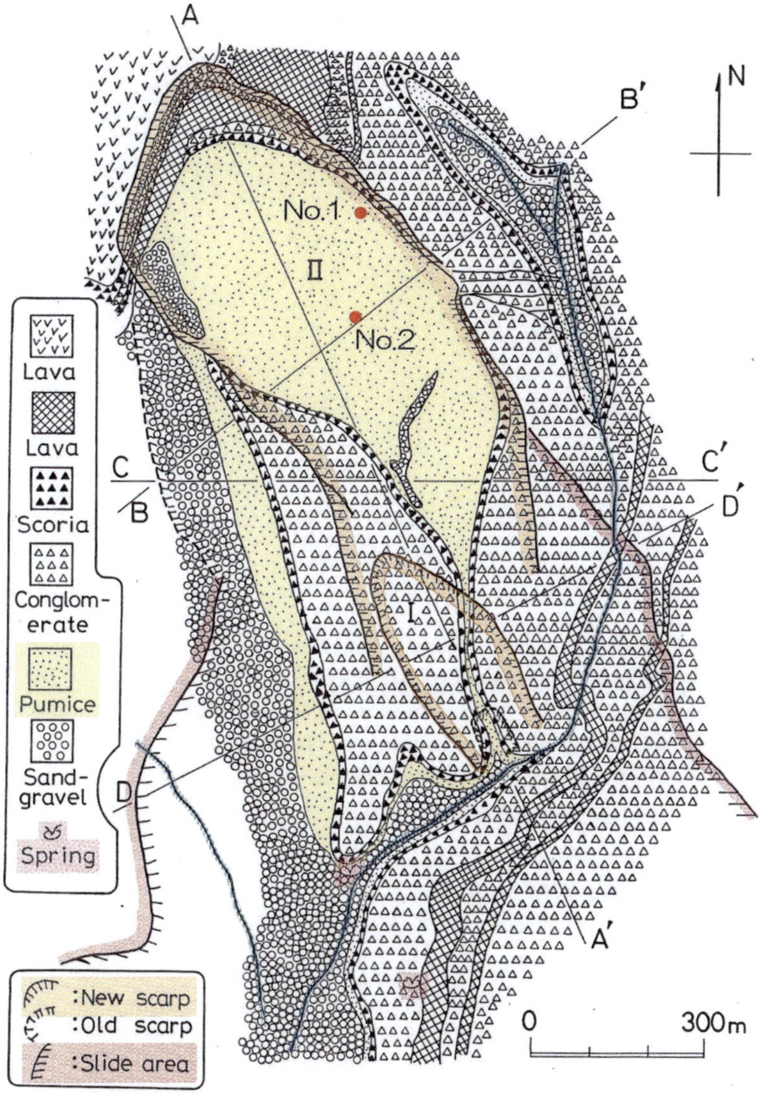


**Fig. 10.31** Geological regime in the headwall in Mt. Ontake landslide (Sakai 2008)

headwall bluff about 130 m high is formed by steeply dipping fractured plane which cuts through several layers of materials from volcanic origin. Deposition of lava flow with different compositions exists near the surface, underlaid by thin layers of scoria and pumice with yellowish color. Over the wide area below the headwall bluff yellowish pumice deposits were seen being exposed on the surface as indicated in Fig. 10.32.

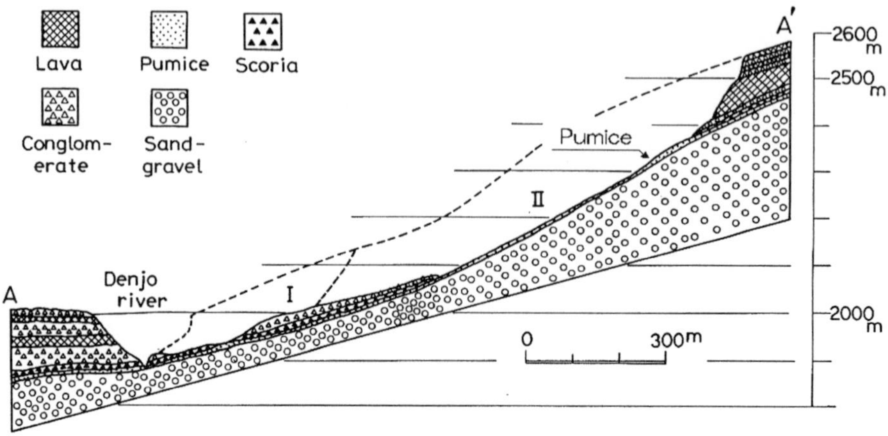


**Fig. 10.32** Geological regime of Mt. Ontake slide zone in a longitudinal cross section A-A′

The pumice was overly consolidated, but because of its weakness in strength, it is deemed most likely that the slide was triggered by the failure of the pumice deposit. Underlying the pumice tephra there exists a breccias formation consisting of cemented sand and gravel. The sliding surface downslope cuts through the conglomerate and scoria deposit as indicated in Fig. 10.32.

## *10.3.3 Sampling and Lab. Tests*

### 10.3.3.1 Aratozawa Pumice

It is apparent that the landslide at Aratozawa took place along a deep-seated sliding plane through an old deposit of pumice which had been laid probably in Pleistocene era as a result of volcanic eruption of Kurikoma Mountain. (see Fig. 10.1) The pumice deposit about 5–10 m thick exists at depths of 100–150 m overlaid by layers of rocks of volcanic origin.

Because of abundance of underground water, the pumice deposits are deemed to have been almost water-saturated. It is also with reasons to believe that the pumice had been normally consolidated to a high overburden pressure as great as $2.0 \times (100 \sim 150) = 200 \sim 300\text{t}/\text{m}^2 = 2 \sim 3\text{MPa}$.

Since there was no trace of visible sliding plane exposed on the surface, it was not possible directly to recover undisturbed samples. However, there were many spots over the devastated hill area where crushed blocks of pumices had been pushed over or howled up on the ground surface. Thus, disturbed samples were recovered

**Fig. 10.33** Sampling site at a lower part of Aratozawa landslide

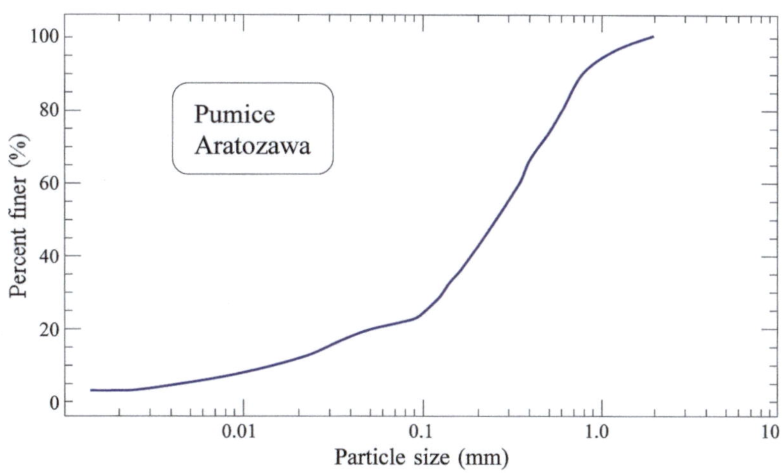


**Fig. 10.34** Grain size distribution curve of the pumice from Aratozawa

from broken pieces of a typical deposit of the pumice near the lower part of the landslide. The views of sampling site are shown in Fig. 10.33. The grain size distribution of the pumice sample tested is shown in Fig. 10.34, where it can be seen that the pumice contains about 25% fines. The specific gravity is Gs = 2.44 and plasticity is non-plastic as indicated in Table 10.3.

**Table 10.3** Physical properties of tuffacious pumice at two landslide sites in Japan

| Site | Specific gravity Gs | Unit Weight γt (kN/m³) | Water Content w (%) | Plasticity index Ip | Fines content Fc (%) |
|---|---|---|---|---|---|
| Aratozawa pumice | 2.44 | – | – | $N_p$ | ≑25 |
| Mt. Ontake Headwall Scarp (No. 2) | 2.56 | 12.8–13.2 | 100–110 | $N_p$ | ≑65 |

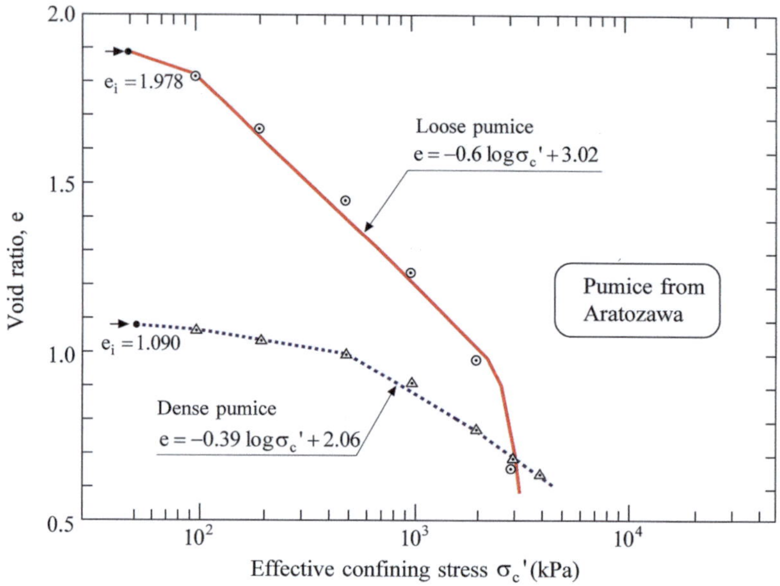


**Fig. 10.35** Consolidation curves for pumice samples form Aratozawa

The samples were saturated and consolidated to high confining stresses in the lab triaxial apparatus, and tested under undrained conditions. The results of tests on loose samples prepared with a dry density of $\gamma_d = 1.18\text{t}/\text{m}^3$ or a void ratio of $e \cong 1.98$ are presented in Figs. 10.35, 10.36 and 10.37. The outcome of the tests on dense samples with an initial void ratio of $e_i = 1.090$ is presented in Figs. 10.38 and 10.39.

Figure 10.35 shows consolidation curves where the void ratio, e, is plotted versus the confining stress, $\sigma_c'$, identified as $\sigma_c' = (\sigma_{1c}' + \sigma_{3c}')/2$, for two sets of pumice samples prepared at different densities. The loosely compacted samples with an initial void ratio of $e_i = 1.978$ indicate a high value of compressibility index of $C_c = 0.60$ when subjected to high confining stresses in the range of 0.5–2 MPa. At higher confining stresses, the value of $C_c$ is seen increasing further up as shown in Fig. 10.35. Even in the dense samples, the value of $C_c$ is shown to be as much as 0.39 in the range of high confining stresses. It is to be noticed thus that the pumice soil from Aratozawa is highly compressible such as clays, in spite of its being

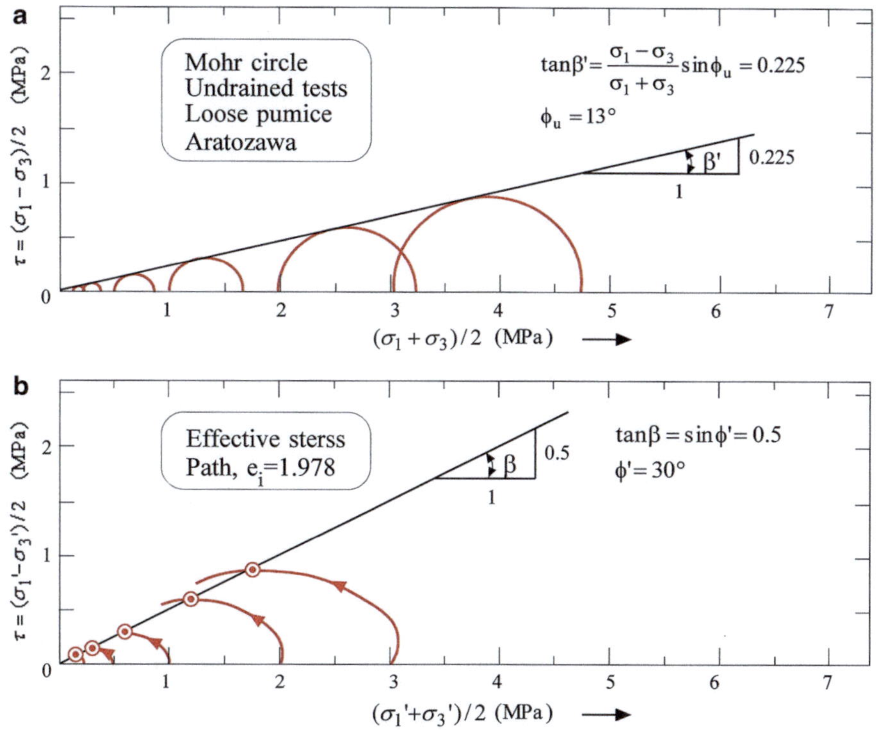


**Fig. 10.36** Mohr circles and effective stress paths for loose samples of pumice from Aratozawa

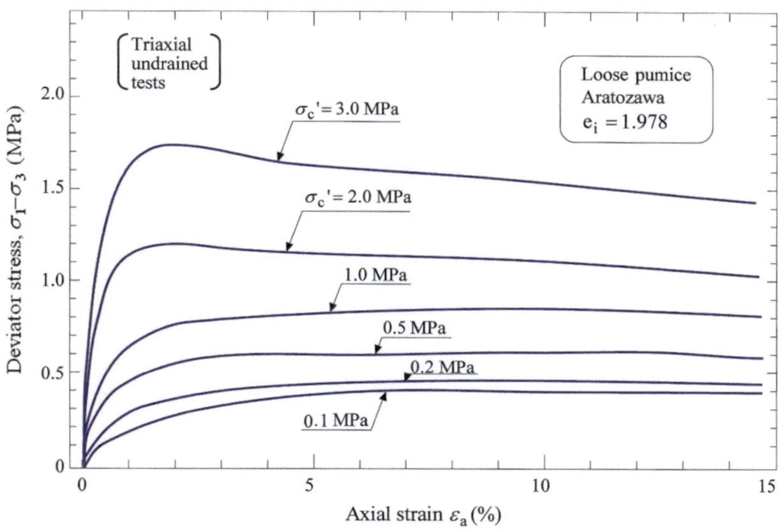


**Fig. 10.37** Stress -strain curves for the loose samples of pumice from Aratozawa

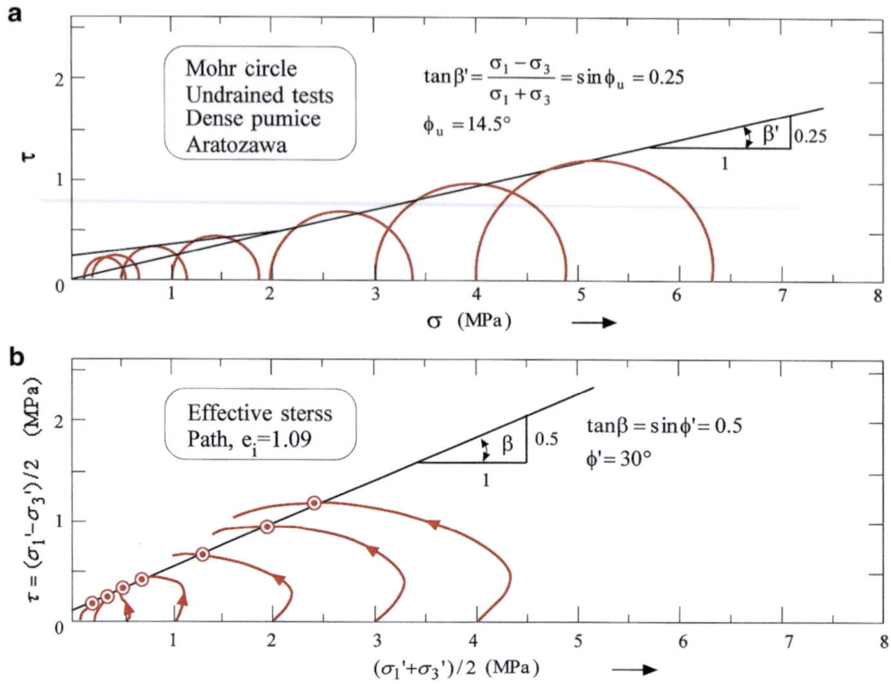


**Fig. 10.38** Mohr circles and effective stress paths for dense samples of pumice from Aratozawa

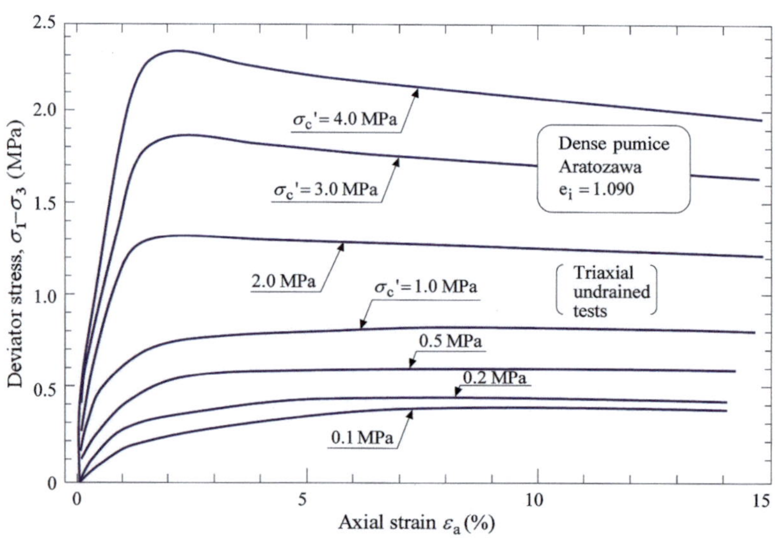


**Fig. 10.39** Stress–strain curves for the dense sample of pumice from Aratozawa

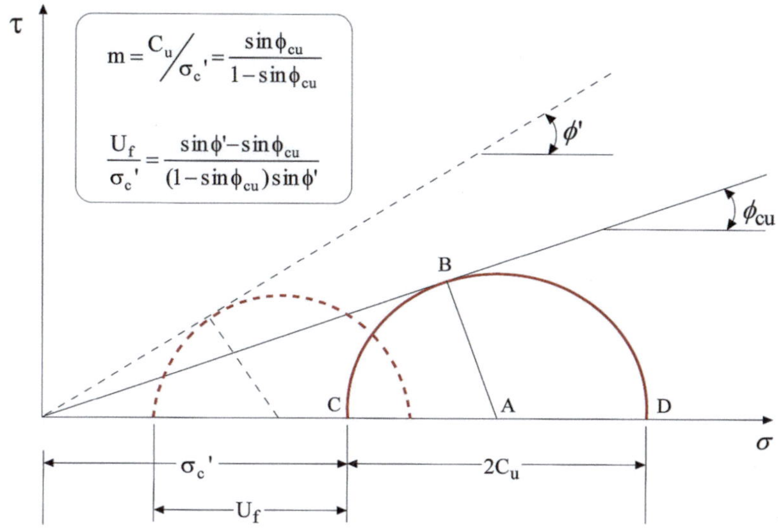


**Fig. 10.40** Mohr circles for total stress and effective stress and the relation between them

categorized as silty sand in terms of grain size distribution. This characteristic behaviour seems to stem from the fact that the material is composed of particles containing macro or micro pores. It is envisioned, therefore, that the crushing or collapse had taken place within the porous particles themselves under high confining stresses during consolidation.

Thus, it appears reasonable to interpret the deformation characteristics of the pumice undergoing high confining stresses in terms of the conventionally developed concepts for cohesive soils. The effective stress paths for two sets of tests are depicted in Figs. 10.36b and 10.38b, where it can be seen that positive pore water pressures developed during undrained loading in both loose and dense samples in the tests with high consolidation pressures. However, for the dense samples tested with low consolidation pressures, the samples are shown to induce negative pore water pressure.

The behaviour at this stage is similar to that of over consolidated clays mobilizing some cohesion at low consolidation pressures. The stress–strain relations during undrained shear stress application in the tests of loosely prepared pumice samples are displayed in Fig. 10.37, where it can be seen that large axial strains, $\varepsilon_a$, up to 15% can be developed without much increase or decrease in shear stress, $q = \sigma_1 - \sigma_3$, which is also the characteristics of plastic cohesive soils. The stress–strain curves of the dense samples are shown in Fig. 10.39. Similarly, the axial strain, $\varepsilon_a$, is seen developing continuously while keeping the shear stress at nearly a constant value.

The value of shear strength as read from the results in Figs. 10.37 and 10.39 are plotted in Figs. 10.36a and 10.38a to form the Mohr circles at failure. As well-known, the Mohr circle is drawn by plotting the consolidation pressure, $\sigma_c'$, and the axial stress at failure, $\sigma_{1f}' = \sigma_c' + 2C_u$, as illustrated in Fig. 10.40. From the Mohr circles in Figs. 10.36a and 10.38a, the stress obliquity at failure in terms of total stress is known to be $\tan\beta' = 0.225$ and 0.25 for the loose and dense samples, respectively. The angle of internal friction is read as $\phi' = 30^{\circ}$ for both loose and dense samples from the effective stress plots shown in Figs. 10.36b and 10.38b. As well-known, the shear strength ratio, m, for clays is defined as the ratio of undrained shear strength, $C_u$, to the confining stress, $\sigma_c'$, to which the soil has been consolidated. Thus, with reference to Fig. 10.40, one obtains

$$m = C_u / \sigma_c' = \frac{\sin\phi_{cu}}{1 - \sin\phi_{cu}} \tag{10.1}$$

Introducing the values of $\sin\phi_{cu} = 0.225 - 0.25$ as above, one obtains,

$$m = 0.29 - 0.33 \tag{10.2}$$

As mentioned above, the in-situ deposit of the pumice associated with the sliding is deemed to have been consolidated under a high overburden pressures of the order of $p'_c = 2 - 3\text{MPa}$. Thus, no matter how much the in-situ density might have been, it would be reasonable to assume a unique value of average shear strength ratio, that is, $m \cong 0.3$ in the stability analysis.

#### 10.3.3.2 Mount Ontake Pumice

Undisturbed samples of soils were secured from the exposed failure surface of the deposits which were considered to have been responsible for triggering the slide. In the area of the headwall scarp at Mt. Ontake, samples were obtained from two places of pumice deposits with different altitudes as indicated in Fig. 10.31. The sampling site No. 1 is assumed to have had a depth of about 10 m from the original ground surface prior to the slide, whereas No. 2 site had been at a depth of about 50 m from the original surface. Therefore, the effective overburden pressures at these two sites are assumed to have been approximately 0.02 and 1.0 MPa, respectively. The samples from No. 1 location was yellowish pumice and had a sufficient cohesion for easy handling and transportation. The samples from No. 2 location were also yellowish pumice but had a number of micro-fissures having developed horizontally in the bedding plane, and as such the fissures were easy to open up. Therefore, great precaution was taken during the sampling operation. Undisturbed specimens were tested both under static and dynamic loading conditions using a triaxial test apparatus.

The results of physical tests on samples from No. 2 site indicated the specific gravity of $Gs \cong 2.56$ and non-plastic as listed in Table 10.3. The gradation curve of the pumice is indicated in Fig. 10.41. The material may be identified as sandy silt. Consolidated undrained triaxial tests (CU-test) were performed on undisturbed specimens to determine the static strength of the pumices under monotonic loading conditions. Because samples from different sites were saturated to various degrees, pore water pressure measurement was not made during the undrained phase of load application, and the test results were consistently interpreted on the basis of total stress concept.

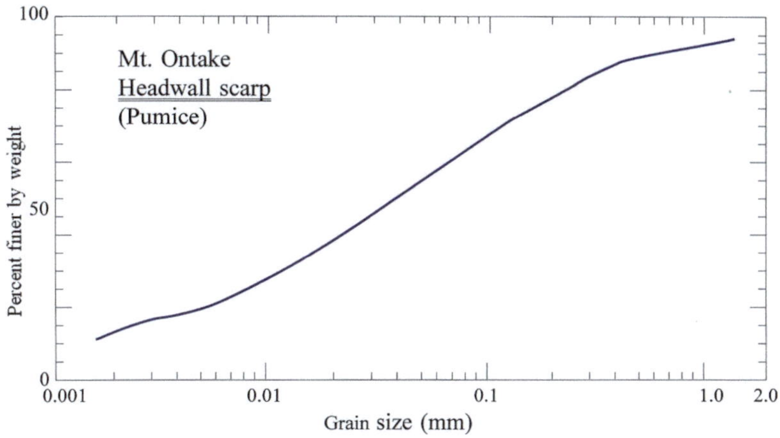


**Fig. 10.41** Typical gradation curve of pumice from the site of landsliding

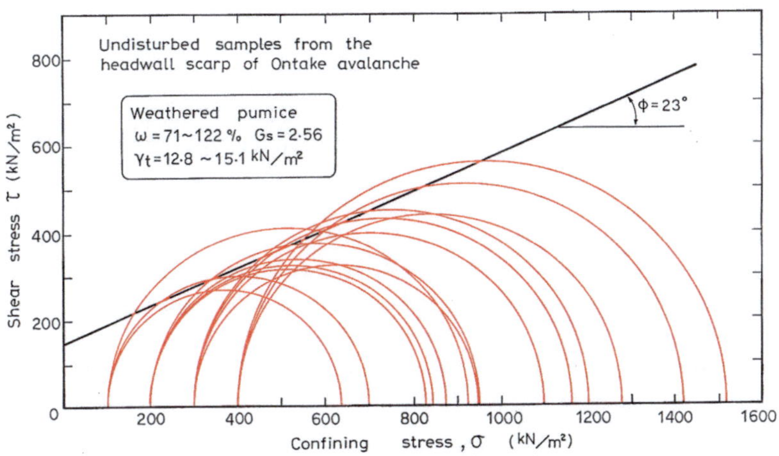


**Fig. 10.42** Mohr circles at failure of undisturbed samples from Mt. Ontake

For the specimens from No. 2 site, it was considered desirable to employ a higher consolidation pressure in the tests in accordance with the high over-burden pressure which had existed before it was removed by the slide. However, because of the limitation in the capacity of cell pressure application, the consolidation pressure used was less than 500 kPa.

The results of tests on the over-consolidated fissured pumice samples from No. 2 site is presented in Fig. 10.42 in terms of Mohr circles at failure stresses, where it may be seen that the angle of internal friction is taken as being $23^{\circ}$ and the cohesion as C = 175kPa, on an average, although there are some scatters in the test results.

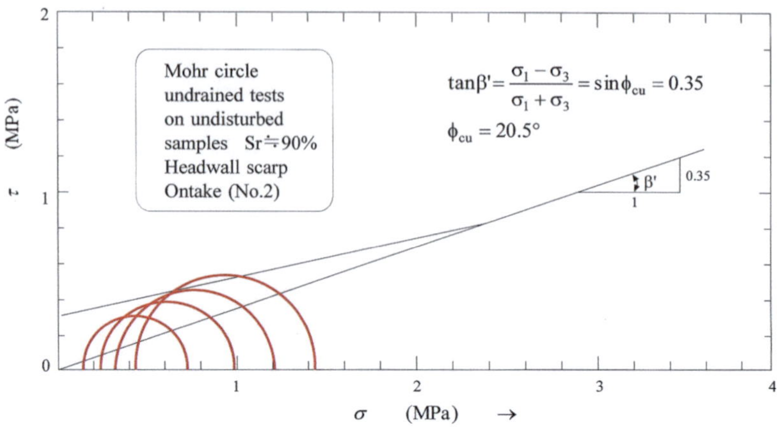


**Fig. 10.43** Failure envelop at high confining stresses inferred from Mor circles at low confining stresses for pumice soil from Mt. Ontake

In view of the fact that the confining stresses at the time of the tests were by far smaller than the in-site overburden pressure to which the pumice had been consolidated, it would be with reasons to assume that the cohesion component at small confining stresses measured in the laboratory tests did emerge from the effects of over-consolidation due to removal of the large overburden pressure by the landslide. With this fact in mind a Mohr circle was redrawn as shown in Fig. 10.43 with extrapolation by judgment to encompass the range of higher confining stresses. In Fig. 10.43, some of the test data quoted from Fig. 10.42 is also displayed. The failure envelop is postulated to branch out into normally and overly consolidated lines around the confining stress of 2.5 MPa. From this extrapolated failure envelop, it may be assumed that the angle of obliquity in total stress was roughly of the order of $\sin\phi_{cu} = 0.35$ . Then, with reference to Eq. 10.1, the shear strength ratio, m, is estimated to have been.

$$m = \frac{C_u}{\sigma_o'} = \frac{\sin\phi_{cu}}{1-\sin\phi_{cu}} = 0.54 \tag{10.3}$$

This value of m will be used in the following stability analysis (Fig. 10.44).

### *10.3.4 Simple Stability Analysis*

To simplify the stability analysis, it would be reasonable to take up major parts of sliding plane which pass through the weakest soil deposit and to assume the slip plane as being represented by a straight line parallel to the ground surface. Then, with reference to Fig. 10.45 the factor of safety during earthquake $F_d$ is given by

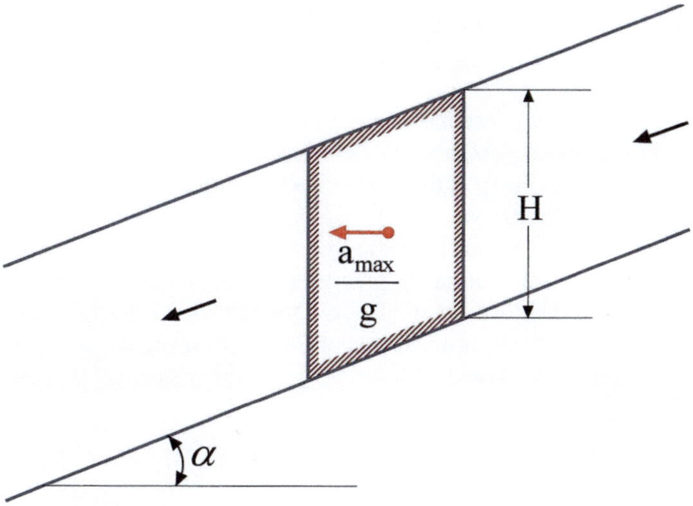


**Fig. 10.44** Scheme of simple method of stability analysis

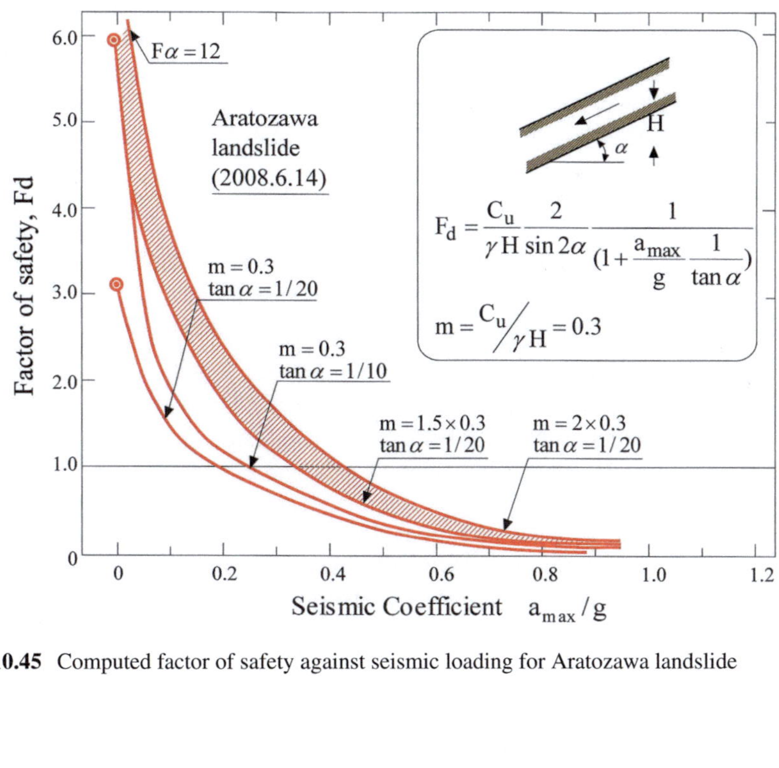


**Fig. 10.45** Computed factor of safety against seismic loading for Aratozawa landslide

$$F_d = \frac{\tan\phi}{\tan\alpha} + \frac{C_u}{\gamma H}\frac{2}{\sin 2\alpha}\frac{1}{(1 + a_{max}/g)} \tag{10.4}$$

where $\phi$ is the angle of internal friction and $C_u$ denotes cohesion of the soil involved in the sliding. $a_{max}$ is the peak or equivalently defined horizontal acceleration during earthquakes and g is the gravity acceleration.

In the case of the Aratozawa and Ontake landslides, the pumice deposits of old era are identified to have induce the slide as described above, and the highly porous pumice subjected to great overburden pressures was shown to exhibit the characteristics which are similar to cohesive clays. Drawing upon this premise, one can ignore the first term on the right-hand side of Eq. 10.4, and make simple analyses by utilizing the formula as follows,

$$F_d = \frac{C_u}{\gamma H}\frac{2}{\sin 2\alpha}\frac{1}{(1 + a_{max}/g)} \tag{10.5}$$

In making the stability analysis considering effects of earthquakes, irregular nature of time history of acceleration needs to be properly processed and taken into account. There are two methods of approach ever adopted for this purpose.

1. The first is to modify given irregular time histories into an equivalent constant-amplitude uniform time variation of acceleration with a certain typical number of cycles, say, 10 or 20 cycles. This has been done based on experienced judgment or on the damage-potential theory. Using the equivalent acceleration thus determined for $a_{max}/g$, the factor of safety can be computed by Eq. 10.5, where the static strength under monotonic loading is used. It is to be noted that this approach is equally applicable to any other construction materials, not material which is specifically relevant to soils.
2. The second approach is to test various soil samples under various irregular loads and to obtain what is termed dynamic strength, $C_{ud}$, which is expressed in terms of a peak value of shear stress in the time histories. The value of $C_{ud}$ thus determined is used in the stability analysis. In this case an actual value of the peak acceleration should be introduced in Eq. 10.5 without any modification.

The author has been engaged in the conduct of laboratory tests in support of the second approach as above (Ishihara and Kasuda 1984. The outcome of these studies showed that (1) effects of irregular loading are manifested on cohesion component in the strength of cohesive soils while keeping the angle of internal friction unchanged and (2) the value of $C_{ud}$ in irregular loading is 1.6–2.4 greater than the cohesion, obtained in the static monotonic loading conditions (Ishihara 1996).

Assuming this conclusion as being applicable to the pumice soils in question, the value of $C_u$ from Aratozawa and Mt. Ontake obtained in the monotonic loading condition may need to be increased by a factor of 1.5–2.0. Thus, the following analyses will be made by using $1.5C_u/\gamma H$ and $2C_u/\gamma H$.

#### 10.3.4.1 Analysis of Aratozawa Landslide

The landslide at Aratozawa is composed of three major blocks of massive rocks and soils. This feature is visualized roughly in the cross sections depicted in Fig. 10.27. It may be assumed, for simplicity, that each block has moved roughly independently without mutual interaction. No matter which cross section is considered, it is apparent that the slip plane passed through the pumice deposit located at depths 100–150 m. With respect to the slide movement, the main stream appears to have been directed along the cross section F-F′. Thus, the inclination of the sliding plane may be assumed to have been $\tan\alpha = 1/20 - 1/10$.

By introducing this value together with the shear strength ratio, $m \cong 1.5\times 0.3$ and $2\times 0.3$, as obtained in Eq. 10.2, the factor of safety during earthquakes was computed and plotted in Fig. 10.45 versus the seismic coefficient, $a_{max}/g$, where $a_{max}$ is the peak horizontal acceleration and g denotes gravity acceleration. As shown in the Fig. 10.45, the static factor of safety in a static condition without seismic excitation is as high as 6–12, but it drops sharply and becomes less than 1.0 when the acceleration becomes greater than (0.35–0.42)g.

As mentioned above, the landslide in Aratozawa took place mainly through the pumice deposit of old geological era, and as such the soil strength can be interpreted in terms of the shear strength ratio of normally consolidated clays as defined by $m = C_u/\gamma H$. This concept renders the consideration of the stability or instability highly simple, as indicated by Eq. 10.5. The analysis results shown in Fig. 10.45 reveal several important features as follows.

1. The stability in the normal quiescent conditions without seismic shaking is sufficiently high due primarily to a low inclination of potential sliding plane. However, once the shaking is applied, the degree of stability is reduced sharply to a factor of safety well below unity with the seismic coefficient of about 0.35–0.42.
2. The actual acceleration at the time of the 2008 earthquake is estimated to have been of the order of 0.8–1.0 times the gravity acceleration. Thus, there are reasons to have the landslide induced at the time of the earthquake.
3. There might have been several other factors which are difficult to quantity. Therefore, actual factor of safety might have been larger than that as computed above. However, the smallness of the seismic coefficient to induce sliding under the simplified condition as above is to be taken as giving a minimum value in the margin of safety factor under seismic conditions.
4. From the observations as above, it is conclusively mentioned that, when the cohesive soils such as highly porous pumice exist in deep-seated deposits, chances are high for them to be in a state of normal consolidation undergoing great overburden pressure, and thus susceptible to sliding when subjected to seismic loading.

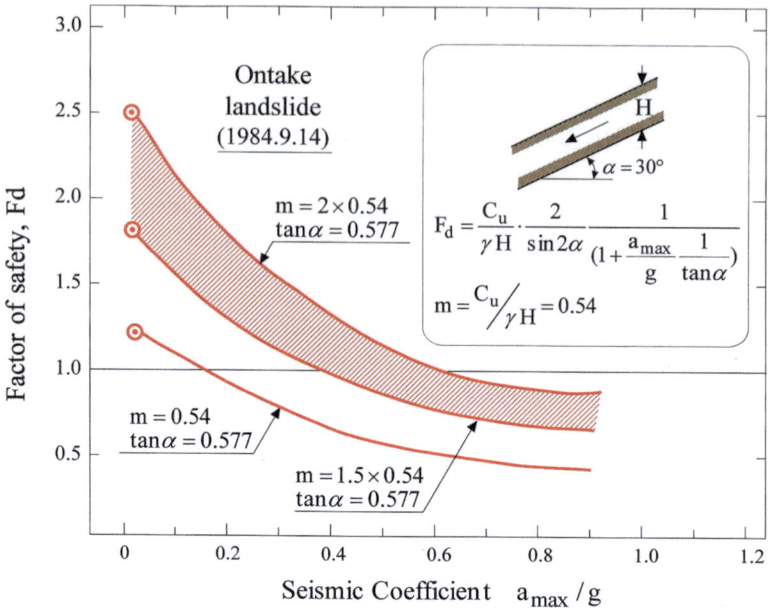


**Fig. 10.46** Computed factor of safety against seismic loading for Ontake landslide

#### 10.3.4.2 Analysis of Mt. Ontake Landslide

The simple analysis was made as well for the Mt. Ontake landslide using the formula in Eq. 10.5 and increased cohesion in the fashion similar to that for the Aratozawa landslide. The results are demonstrated in Fig. 10.46 in terms of the safety factor plotted versus the peak acceleration. The angle of slope in the deep-seated middle portion where the slide is deemed to have occurred through the pumice was about $\alpha = 30^{\circ}$ which is equal to $\tan\alpha = 0.577$. The value of $C_u / \gamma H$ was taken as being 0.54. Using the increased value of $C_u$, the stability analysis was made and the results are shown plotted in Fig. 10.46. The plot indicates that the seismic coefficient to trigger the slide with a safety factor of $F_d = 1.0$ is about 0.4–0.6, a value which is a little smaller than the actual acceleration encountered at the time of the 1984 Naganoken-seibu earthquake. It is also learned in Fig. 10.46 that the factor of safety tends to drop sharply with increasing acceleration particularly in the range of small acceleration.

In the case of the Mt. Ontake landslide being discussed here, there were layers of gravels or lava over the tail and head portion of the sliding area as shown in Fig. 10.32. If the frictional resistance in these layers is properly taken into account, the actual factor of safety in quiescent and seismic loading conditions could be greater than those shown in Fig. 10.46. However, the factor of safety easily becomes less than unity with the acceleration estimated to have actually occurred at the time of the 1984 earthquake. More detailed account of the stability analysis is presented elsewhere (Ishihara et al. 1986).

## 10.4 Concluding Remarks

Performances of high rockfill dams designed and constructed by the up-to-date states of practice are known generally satisfactory even when undergoing high levels of seismic loads with acceleration of the order of 800–1,000 gal. Minor damage not to seriously destroy the integrity of dams could be incurred on the dam body particularly in the vicinity of the crest and also near the abutments. The residual settlements of the crests tend to increase with increasing level of acceleration, and the magnitude in terms of vertical strain, that is, the ratio of crest settlement to the height, has been of the order of 0.1–1.5%.

Large-scale landslides involving 35–70 millions $m^3$ of soils and rocks were triggered in volcanic mountains in Japan by strong shaking during earthquakes. Because of the massive nature of the landslides, the sliding planes were found to lie at depths of 100–150 m where geologically old deposits of pumice exist overlaid by thick layers of tuff rocks and lavas. The overburden pressures are assessed to have been of the order of 2–3 MPa. As a result of the laboratory triaxial tests employing the high confining pressures, it was found that the pumice soils classified as silty sand or sandy silt have a tendency to behave as if they were cohesive soils such as clays. In fact, the compressibility index during consolidation was found to be as large as that of clays of the order of 0.4–0.5. In addition, Mohr circles at failure showed a straight line with the angle of obliquity of $\varphi_{cu} = 13 - 20^{\circ}$ in terms of total stress. Thus, the value of shear strength ratio, defined as the ratio of undrained shear strength to the consolidation pressure, was found to take a value of 0.3–0.54 which is similar to that of clays. On the basis of strength parameter thus determined, simple analysis was made to calculate factor of safety against sliding during earthquakes by assuming the slip plane as being a straight line parallel to the ground surface. The results of the analysis indicated that, while the factor of safety is pretty large in the quiescent state, it tends to drop sharply with increasing acceleration, thereby, bringing about the soil mass to a state of sliding.

Particularly worthy of note is the fact that the massive landslides took place with the slip plane running through pumice deposits with cohesive soil characteristics, existing at a great depth where the overburden pressure is as great as 2–3 MPa. Under such high-pressure environments, even pumice classified as silty sand or sandy silt exhibits characteristics as if it were normally consolidated clay, because of high compressibility due to the crushing or collapse of micro or macro pores existing in the particles. Thus, one of the lessons of engineering significance learned from the above case studies would be to make every possible efforts not to overlook existence of cohesive soil layers at a great depth, particularly when important facilitates such as nuclear power station are to be constructed in the area near the mountains or hills in the back yard. According to the normal practice, soil investigation by boring is stopped at shallow depths of the order of 10–20 m when rocks are encountered. However, the boring should be continued down to depths of 100–150 m in order not to miss the presence of any cohesive soils.

**Acknowledgments** In preparing the draft of this paper, design figures of Ishibuchi dam, Aratozawa dam and Zipingpu dam were offered by the office of respective dam operation office. Information about the failure mode of Aratozawa landslide was offered by Mr. Y. Momikura of Kiso-Jiban Consultants Co., Ltd. in Tokyo. Field investigations on the Aratozawa landslide were performed by the assistance of Mr. I. Morimoto of Kiso-Jiban Consultants. Professor T. Kokusho of Chuo University and Dr. Y.Tsukamoto of Tokyo University of Science helped greatly the in-situ sampling at Aaratozawa. Laboratory tests were performed by Mr. H. Ueshima of Tokyo University of Science. Mr. M. Hayashi of Kiso-Jiban Consultants supervised and assisted the conduct of the laboratory tests using the high pressure triaxial chamber. The author wishes to extend his sincere thanks to the above-mentioned persons.

## References

Ishihara K, Kasuda K (1984) Dynamic strength of cohesive soil. In: Proceeding of the 6th European conference on soil mechanics and foundation engineering, Budapest, pp 91–98

Ishihara K, Hsu Hai-lung, Nakazumi I, Sato D (1986) Analysis of landslides during the 1984 Naganoken-Seibu earthquake. In: Proceeding of the international symposium on engineering geology problems in seismic areas, Bari, vol 2

Ishihara K, Kuwano J, Lee HC (1989) Permanent earthquake deformation of embankment dams. Dam Eng 1(3):221–232

Ishihara K (1996) Soil Behaviour in Earthquake Geotechnics. Oxford Clarendon Press, Oxford, pp 202–206

Momikura M (2009) Data file in Kiso-Jiban Co. Ltd., Tokyo

Sakai T (2008) Status and mechanism of ground deformation of Kashiwazaki-Kariwa NPS by Niigataken Chuetsu-oki earthquake, the international symposium on seismic safety of nuclear power plants and lessons learned from the Niigataken Chuestu-oki earthquake

# Chapter 11
# Bridging the Gap Between Structural and Geotechnical Engineers in SSI for Performance-Based Design

**M.H. El Naggar**

**Abstract** Performance-based design (PBD) involves designing structures to achieve specified performance targets under specified levels of seismic hazard, and requires the analysis of the entire soil-structure system. Dynamic soil-structure interaction (SSI) encompasses linear and nonlinear response features of the structure and foundation soil. Consequently, both structural and geotechnical expertise are needed. Most available computational tools, however, provide either elaborate models for the structure with simplified soil representation, or detailed geotechnical models with simplified structural idealization. This paper discusses SSI pertinent to the analysis of soil-pile-structure interaction (SPSI) problems that are of direct concern to PBD of structures. The paper also presents a model developed with considerations from both geotechnical and structural engineering, thus, facilitating the effective incorporation of SSI in PBD. The model is developed within the framework of Beam-on-Nonlinear-Winkler- Foundation (BNWF) approach and can be easily integrated into commercially-available nonlinear structural analysis software. It accounts for soil nonlinearity and discontinuity conditions at pile/soil interface; cyclic degradation of stiffness and strength due to variable-amplitude loading; and reduction in radiation damping with increased nonlinearity. The model is implemented in a readily available structural package, and hence allows structural engineers to properly account for SSI effects when performing PBD. The model was used to analyze the performance of piled foundations reported in the literature. The results showed that soil cave-in and recompression reduces the pile maximum bending moment and its depth from the ground surface, and increases the hysteretic damping of the soil-pile system, therefore enhancing the performance of the SPSI system. This could be particularly beneficial for situations where plastic hinges develop belowground level. In addition, some important features of pile response to cyclic loading are discussed.

M.H. El Naggar (✉)
Department of Civil and Environmental Engineering,
The University of Western Ontario, London, Canada
e-mail: naggar@uwo.ca

M.A. Sakr and A. Ansal (eds.), *Special Topics in Earthquake Geotechnical Engineering*, Geotechnical, Geological and Earthquake Engineering 16,
DOI 10.1007/978-94-007-2060-2_11, 

## 11.1 Introduction

The computational tools for dynamic structural analysis have improved considerably over the last five decades. Significant advances have been achieved since the initial analysis of a 15-storey building in California was performed assuming fixed-based conditions (Clough 1955). Direct nonlinear soil-structure response analysis is widely used in the seismic design of tall buildings and in many cases mandated by the code (e.g. Nation Building Code of Canada, NBCC 2005). These changes have been brought about by advances in computing technology and the advent of more efficient computational tools.

Soil-structure interaction (SSI) can generally be defined as the combined phenomena that modify the response of a structure due to foundation flexibility, and the response of the supporting foundation due to the presence of the structure. Dynamic SSI encompasses two main components: kinematic interaction, which relates the foundation input motion (FIM) to the free-field (FF) motion due to stiffness contrasts between the foundation and the surrounding soil; and inertial interaction, which results from the transmission of inertial forces from the structure to the soil. The level of sophistication of SSI analysis ranges from a linear structure on linear soil to a nonlinear structure on nonlinear soil. Nonlinearity stems from either material nonlinearity (e.g. yielding of the soil and/or structure), or geometric nonlinearity (e.g., pounding of structures and foundation gapping/separation).

Early advances in the field were motivated by the need for designing safe nuclear containment structures founded on mat foundations (Seed et al. 1977). For building structures, the ATC 3–06 document (ATC 1978) was the first design guideline to account for SSI effects in force-based design (FBD). This was based on the modified fixed-base approach (Veletsos 1977), which involves adjusting the fundamental period and damping ratio of a structure to account for SSI effects. The format of code design spectra (e.g., NBCC 2005) provides, in general, reduced spectral acceleration values. Although true for certain structures, this is an overly simplistic view, and has the effect of crippling design innovation, and blinding the analyst to many important SSI response features (Mylonakis and Gazetas 2000; Allotey and El Naggar 2005a). This, however, to some design engineers, has epitomized SSI and represented all that SSI stands for. On the contrary, dynamic SSI is complex and interdisciplinary, and requires skills in soil mechanics, foundation engineering and structural dynamics (Pecker and Pender 2000).

The current state-of-practice is that programs are developed mainly for the analysis of the superstructure or the substructure. This is evident in the different approaches used by structural and geotechnical engineers in modelling SSI. Structural engineers are inclined to use elaborate models for the structure, with a simplified representation of the soil, while geotechnical engineers are inclined to use the opposite. A stark example of this is the application of FLAC (HCItasca 2002) to study the cyclic response of a foundation (Pender and Ni 2004), but the use of RUAUMOKO (Carr 2001) to study the nonlinear SSI response of a multi-storey building (Wotherspoon et al. 2004) in companion studies. Therefore, there is a need to develop new programs, or enhance existing programs, to incorporate capabilities allowing the performance of both types of analysis, so as to make it appealing for both structural and geotechnical engineers. Hence, for SSI to be better integrated into the design process, robust analysis tools that can account for the important aspects of SSI must be readily available to designers. This is much more so in the context of PBD, where structures must be designed to meet specified performance targets (Celebi and Okawa 1999; Ghobarah 2001).

PBD relies heavily on nonlinear analysis procedures and has renewed the need to revisit simplified modelling approaches. In this regard, the beam-on-a-nonlinear Winkler foundation (BNWF) has received the most attention and has been the focus of several recent studies (e.g., Boulanger et al. 1999; Gerolymos and Gazetas 2005). The main drawback of the approach is its idealization of the soil continuum with discrete soil reactions at different points that are decoupled from each other. For seismic applications, the static BNWF approach also suffers the disadvantage of not being able to account for the cycle-by-cycle SSI response, and an unsatisfactory performance in modelling problems involving significant kinematic interaction and ground motion effects (Finn 2005). These can, however, be accounted for by using dynamic BNWF models. Many dynamic BNWF implementations allow for the reasonable assessment of the effects of SSI on structures, but do not account for important factors such as the effect of cyclic soil degradation that could significantly influence permanent displacements.

The objective of this paper is therefore two-fold. First, to review the importance of SSI in PBD, and then to present a BNWF model that adopts the strengths of existing modelling approaches, and combines these with new ideas, to develop a synthesized model. The developed model has been included in the publicly-available program nonlinear structural analysis program, SeismoStruct (SeismoSoft 2003), as detailed in Allotey and El Naggar (2008a). It can simulate different foundation types and retaining walls under various loading conditions, and its parameters, to a large part, have a physical significance, and can be obtained from standard field and laboratory tests. Two case studies are later presented to highlight the capabilities of the model, and show its potential as a useful tool for PBD.

## 11.2 SSI in Performance-Based Design

The force-based design (FBD) approach in seismic design intends to provide favourable dynamic response and avoid premature collapse. Seismic forces are calculated considering the estimated fundamental period and total mass of the structure with due consideration of the seismic hazard defined in terms of a design spectral acceleration. Lessons learnt from recent earthquakes have shown that although the basic intent of the code to provide life safety was achieved, damage to structures was extensive, leading to large economic losses and high cost of repairs (Eguchi et al. 1998). The performance-based design (PBD) approach, on the other

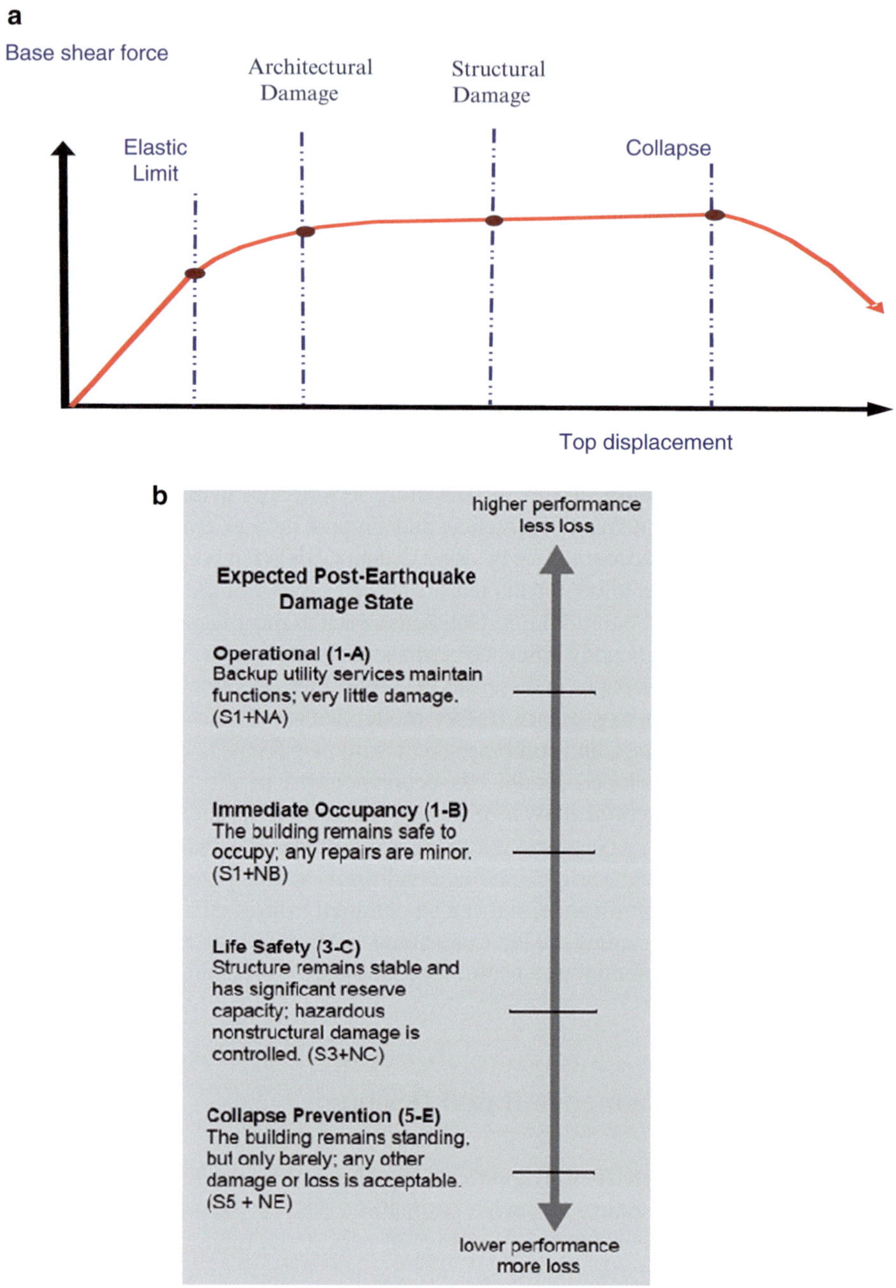


**Fig. 11.1** (**a**) Capacity curve; (**b**) Building target performance levels (After ASCE-7 2010)

hand, provides a more general design philosophy that seeks to achieve specified performance targets under stated levels of seismic hazard as shown in Fig. 11.1. To provide a specified performance at reasonable cost, accurate reliable analysis of the entire structure-foundation-soil system is important. Thus, the design approach relies heavily on nonlinear static and dynamic forms of analysis. Since the response analysis involves the entire system, robust and efficient analysis tools amenable for use by both structural and geotechnical engineers are required (Allotey and El Naggar 2005a).

Dynamic SSI is a complex phenomenon that encompasses different types of response covering a range of sophistication in the analysis from a linear structure/soil to a nonlinear structure/soil. The nonlinearity stems from either material nonlinearity such as yielding of the soil or structure and cyclic strength/stiffness degradation of soil or structure; or geometric nonlinearity such as large displacements and foundation uplift. SSI is often assumed to have beneficial effects on seismic response. This may be attributed to the format of design spectra in most current design codes (Mylonakis and Gazetas 2000). However, observations from recent earthquakes have highlighted the importance of performing realistic SSI analysis (Celebi and Crouse 2001). In addition, PBD requires buildings to be designed to meet specific performance targets, which can only be achieved by ensuring that important factors that affect building response are properly accounted for.

All PBD codes recommend the consideration of SSI in the design process (e.g. SEAOC Vision 2000 (SEAOC 1995), ATC 40 (ATC 1996) and FEMA 356 (ASCE 2000) with its addendum FEMA 440 (FEMA 2004)). The Pacific Earthquake Engineering Research Centre (PEER) has developed a framework methodology for performance-based earthquake engineering (PBEE). The methodology is formulated probabilistically as:

$$\upsilon(DV) = \iiint G(DV / DM)dG(DM / EDP)dG(EDP / IM)d\upsilon(IM) \quad (11.1)$$

Where *DV*, *DM*, *EDP* and *IM* are the decision variable, damage measure, engineering demand parameter and intensity measure, which characterize the important aspects of the problem (Moehle and Deierlein 2004). The geotechnical aspects are accounted for in the term *dG(EDP/IM)*, and involves the analysis of the entire soil-structure system considering the direct approach for SSI. For the substructure approach, *dG(EDP/IM)* can be expressed in terms of its sub-components as (Kramer and Elgamal 2001):

$$dG(EDP / IM) = \iint dG(EDP / FIM)dG(FIM / FF)dG(FF / IM) \quad (11.2)$$

where, *dG(EDP/FIM)*, *dG(FIM/FF)*, *dG(FF/IM)* represent the contributions related to structural, soil-structure and site response analyses, respectively. To obtain reliable estimates of *dG(EDP/FIM)*, all three types of analysis must be treated with the same level of rigor. Thus, proper SSI analysis is as important as structural analysis, and is necessary for reliable designs.

### 11.2.1 Advantages of SSI Modelling

Recent experience obtained from design case histories and observations made during earthquakes have shown that modelling SSI accurately is advantageous in several ways. It results in efficient and reliable designs, and leads to innovative solutions, as illustrated in the following cases:

i. Comartin et al. (1996) reported on the seismic retrofit of an eight-storey shear wall-frame building on spread footings. The foundation was resting on stiff to very stiff clay underlain by sandy clay and dense gravely silty sand. The preliminary analysis showed high inelastic demands on the shear walls due to foundation rocking. This indicated the need for an expensive retrofitting scheme. However, when a more realistic nonlinear SSI analysis was conducted, it was found that two-thirds of the deformation demand was absorbed by the soil, with limited permanent settlement. This resulted in reducing the inelastic demand on the shear walls and maintained it within the life safety performance target of the structure. The retrofitting scheme was changed completely accordingly, resulting in a substantial reduction in the retrofitting cost.
ii. Lubkowski et al. (2004) presented a number of case studies in which the accurate and realistic modelling of the SSI system provided considerable value to the client by meeting the specified performance target at a much lower cost. This was clearly demonstrated in the case of a multi-storey building, with a multi-level basement embedded in interbedded loose to medium sand overlying dense gravel. Using a proper SSI model for the underground part, it was observed that the effective input motion at the top of the basement relative to the free-field motion was amplified for periods less than 1.0 s, but de-amplified for periods more than 1.0 s. The period of the building was about 1.5 s, and thus the seismic load on the superstructure decreased by about 40%. It is worth noting that had the period of the building been less than 1.0 s, the forces experienced by the building would have greatly exceeded those obtained from either the free-field response spectra or the code design spectra. An added advantage of using a realistic SSI model was the ability to identify the predominant load paths for the underground portion, which resulted in the optimization of the foundation design.
iii. Crouse (2002) demonstrated the value of considering the SSI response in the design of five-storey steel concentrically braced structure, founded on potentially liquefiable silty sand overlying fibrous peat. Consideration of SSI facilitated the design of an innovative ductile foundation system. The foundation system was composed of piles with the pile caps connected with grade beams, and possessing a ductile pile cap – pile connection. This connection was needed since the SSI analysis showed that the non-ductile pile to pile-cap connection was a weak link, and could not sustain the estimated curvature demands. This example also underscores one drawback of the FBD approach, in which force-reduction factors are chosen based on the type of the structural system; i.e., a concentrically braced steel structure has a high $R$ value, however, the same structure on a non-ductile foundation, cannot perform in a ductile manner under relatively large foundation movements.

iv. In a study of the measured response of the seven-storey Van Nuys reinforced concrete frame building during the Northridge earthquake, Trifunac et al. (2001a, b) noted that changes in the SSI system period were mainly due to soil nonlinearity and not structural nonlinearity. In addition, a spatial assessment of damaged buildings and breaks in water distribution pipes showed that the nonlinear response of the soil significantly absorbed incident wave energy, and thereby reduced the total energy that would have caused structural damage. This shows that it is possible to harness soil nonlinearity effects in SSI to create an inexpensive base isolation, and energy absorbing system. From simulation analysis, Hayashi et al. (1999) have attributed this nonlinear SSI effect to be the reason why during the Kobe earthquake, out of three buildings located in the same area, two sustained extensive damage but the other survived without much damage.

All these examples together show that accurate modelling of SSI is crucial in PBD where specific performance targets must be met.

## *11.2.2 Important SSI Issues in PBD*

### 11.2.2.1 SSI Computational Tools

Most of existing programs are developed mainly for the analysis of the superstructure or the substructure. Structurally oriented codes incorporate elaborate models for the structure, with a simplified representation of the soil (Fig. 11.2a), while geotechnically focussed programs use sophisticated soil models but over-simplified structural representation (Fig. 11.2b). For example, commercially available structural analysis programs like SAP2000 (CSI, 2011), DRAIN-2DX (Prakash et al. 1993), RUAUMOKO (Carr 2001), CANNY (Li 2002) and SEISMOSTRUCT (SeismoSoft 2003) are capable of modelling the structure well, but do not account for many important aspects of soil behaviour. On the other hand, geotechnical programs like FLAC (HCItasca 2002) and SASSI2000 (SASSI2000 1999) are capable of modelling geotechnical aspects, however, their structural features are relatively overly simplified. A stark example of this is the application of FLAC to study the cyclic response of a foundation (Pender and Ni 2004), but the use of RUAUMOKO to study the nonlinear SSI response of a multi-storey building (Wotherspoon et al. 2004) in companion studies. The OPENSEES computational platform (PEER 2000) is unique in that it offers an integrated environment for complete soil-structure system analysis, as it is developed from both points of view. More programs with this spirit are needed to further promote PBD.

### 11.2.2.2 Effects of SSI on System Period and Ductility

SSI is widely believed to have favourable effects on the seismic response of a structure due to lengthening the period of the structure-foundation system, and enhancing its damping and ductility. This can be easily (and unfortunately sometime deceiving)

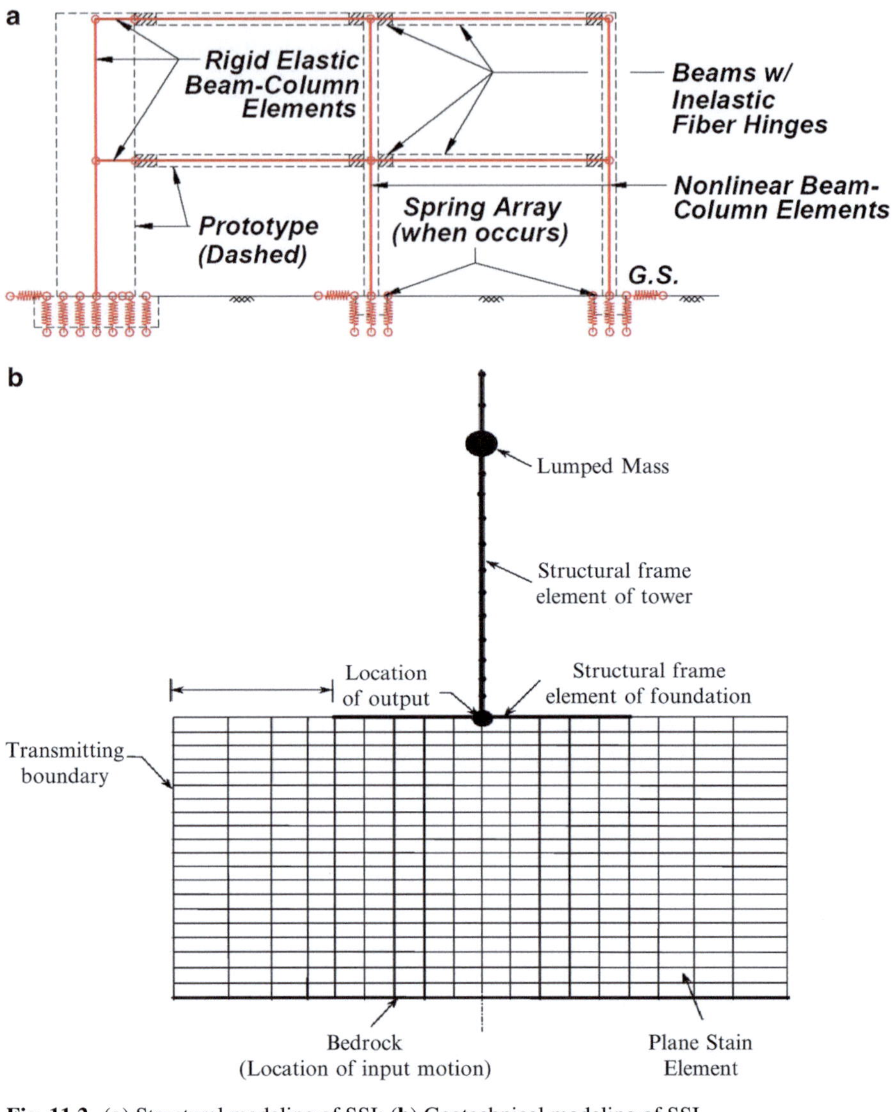


**Fig. 11.2** (**a**) Structural modeling of SSI; (**b**) Geotechnical modeling of SSI

discerned from traditional codes using design response spectrum approaches, which account for SSI effects by using the first mode period-lengthening ratio in combination with an estimate of the system damping (Stewart et al. 2003). With the exception of very short period structures, accounting for SSI in this approach results in reduced base shear forces (e.g. see Fig. 11.3). This reduction contributes to the widely held belief that the effects of SSI on the seismic response of structures are favourable, and hence, there is no need to account for SSI in the case of non-weakening soils.

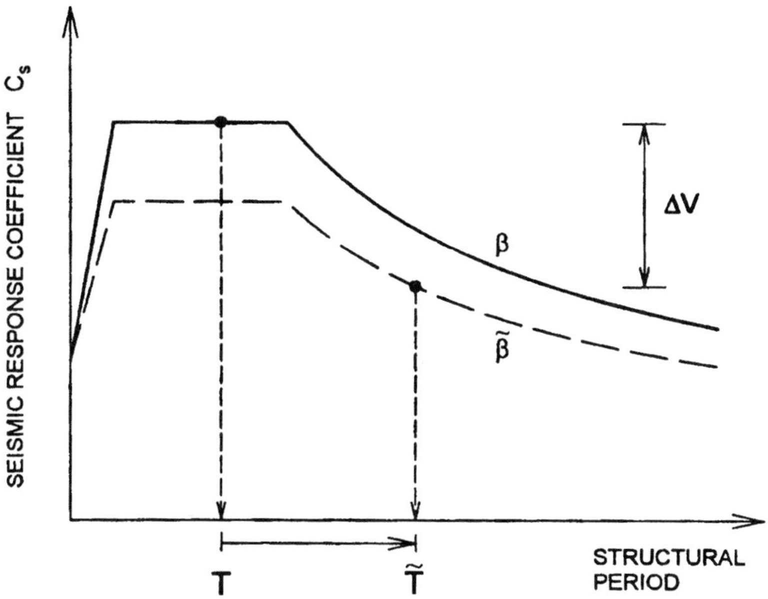


**Fig. 11.3** Reduction in design base shear due to SSI according to NEHRP-97 seismic code

Though this may be true for many structures, it is an overly simplistic view, and as noted by Gazetas and Mylonakis (2001), has the effect of crippling design innovation, and blinding the analyst to important SSI response features.

Gazetas and Mylonakis (2001) and Aviles and Perez-Rocha (2003) studied the response of yielding structure foundations on soft soil. Ground motions at such soft soil sites are usually characterized by long predominant periods. This is also the case for sites in the normal forward fault-rapture direction of near-fault earthquakes (Somerville 1998). For such cases, period-lengthening for structures with fixed based periods less than the predominant period can result in a resonance condition. This can be seen from Fig. 11.4, which shows higher forces and ductility for the flexible base system in comparison with the fixed-base case. Due to the possibility of period-lengthening causing resonance, it is important that every structure with periods shorter than the predominant site period be assessed to check if this condition would be of concern. In addition, site-specific response spectra should be developed to minimize the averaging effects inherent in traditional code design spectra. In the context of PBD, this is important since unsatisfactory input intensity measures (*IM*) directly impact the output decision variable (*DV*).

There are conflicting views about the significance of the SSI system ductility parameter, $\mu_s$, in the literature. Priestley and Park (1987) argued that foundation compliance decreases the ductility capacity of a structure (also Priestley 2000; Calvi 2004). However, Gazetas and Mylonakis (2001) showed that $\mu_s$ is not a measure of the structural distress; rather it is a mathematical parameter that does not have any

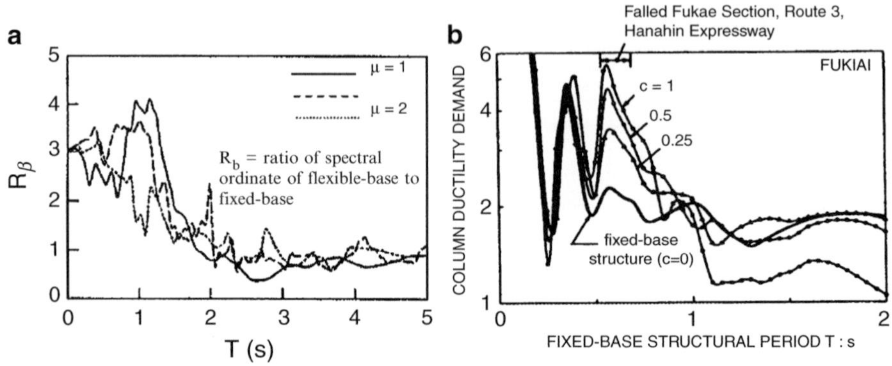


**Fig. 11.4** (**a**) Spectral acceleration ordinate ratios (Aviles and Perez-Rocha 2003); (**b**) ductility demand spectra (Mylonakis and Gazetas 2000)

clear practical significance. Meanwhile, Aviles and Perez-Rocha (2003) derived an expression for the effective ductility of a nonlinear fixed-based oscillator equivalent to the SSI system and showed that $\mu_s$ cannot be linked to the ductility capacity of the structure as suggested by Gazetas and Mylonakis (2001).

#### 11.2.2.3 Foundation Stiffness

The dynamic response of piles is affected by the characteristics of the loading, the structural characteristics of the piles and the interactive pile-soil behaviour. Several investigators have developed numerical methods employing the direct approach in which the soil, pile, and superstructure are modelled and the seismic response is obtained in one step (e.g. Angelides and Roesset 1980; Wu and Finn 1997; and Bentley and El Naggar 2000). The main advantage of the direct approach is the capability of performing the SPSI analysis in a fully coupled manner, without resorting to independent calculations of site or superstructure response. However, it is not commonly used in design offices mainly due to their presumed excessive computational costs and their complexity for common pile dynamic response analysis.

The BNWF method, on the other hand, can account for nonlinear SPSI in a simplified manner and thus is commonly used in professional engineering and research practices. Matlock et al. (1978), Novak et al. (1978), Kagawa and Kraft (1980), Makris and Gazetas (1992), El Naggar and Novak (1995,1996) and El Naggar and Bentley (2000) used BNWF models for piles subjected to lateral dynamic loads. Some BNWF models use the *p-y* curves (unit load transfer curves) approach (Matlok 1970) to represent the soil resistance under static loading because of its simplicity and practical accuracy. In BNWF models that are based on the *p-y* curves approach, the stiffness of the soil is established using *p-y* curves and the damping is established from analytical and/or empirical solutions that account for energy dissipation through wave propagation and soil hysteretic behaviour. Wang et al. (1998) compared several implementations of nonlinear springs based on *p-y* curves and dashpots in parallel and in serial to represent radiation damping in BNWF models of piles subjected to seismic loading. Boulanger et al. (1999) and El Naggar et al. (2005) developed BNWF models utilizing springs in series with dashpots representing radiation damping and used their models to analyze experimental centrifuge tests carried out by Wilson et al. (1997) for seismic loading on piles. El Naggar and Bentley (2000) introduced dynamic *p-y* curves for dynamic lateral response analysis of piles. Programs like DYNA6 (El Naggar et al. 2011) can be used to calculate the stiffness of pile foundations accounting for group effects, and accounting for nonlinearity in an approximate fashion using an annular weak zone around the pile.

#### 11.2.2.4 Cyclic Degradation Effects

Accurate prediction of soil behaviour under cyclic or dynamic loading is of prime importance for predicting the foundation seismic response. Soil degradation occurs mainly due to plastic volumetric strains that lead to changes (increase) in void ratio or pore pressure. This change of state can be explicitly modelled by using effective stress methods, or implicitly modelled using total stress methods. In the latter, this is achieved by using empirical functions based on the number of loading cycles that are derived from constant-amplitude stress or strain laboratory tests. Different models are available in the literature to account for soil degradation, which are based on the number of cycles, including: fatigue contour diagram model (Anderson et al. 2003); degradation index model (Idriss et al. 1978); and pore pressure (liquefaction) model (Seed et al. 1976). The fatigue contour diagram model encompasses three procedures: pore pressure accumulation procedure (suited for sandy soils); permanent shear strain accumulation procedure (useful for soils experiencing significant plastic deformations); and the cyclic shear strain accumulation (gives good predictions for various soil types). The degradation index model is suitable for cohesive soils and simulates cyclic degradation of undrained clay under variable-amplitude strain-controlled loading in total stress cyclic nonlinear analysis. The Seed liquefaction model was proposed to predict pore pressure evolution in effective stress analyses and is obviously suitable for cohesion less soils.

Different models are available to represent the stiffness and/or strength degradation of soil-pile load-transfer curves, i.e., *p-y* and *t-z* curves. These include the models developed by Matlock et al. (1978), Grashuis et al. (1990), and Rajashree and Sundaravadivelu (1996). The Matlock et al. (1978) model is given by:

$$\delta_N = (1-\theta_{mt})(\delta_{N-1} - \delta_f) + \delta_f \tag{11.3}$$

where $\delta_f$ is the maximum amount of degradation, $\theta_{mt}$ is the degradation rate factor and N is the number of cycles. The equation by Matlock is noted to be similar in form to the degradation index curve, and therefore models concave damage evolution trends. This model has been used by Matlock and Foo (1980) and Swane and Poulos (1984) to simulate the cyclic behaviour of piles in stiff clay. The Grashuis et al. (1990) model is similar to the Matlock et al. (1978) model, i.e.

$$\delta_N = 1 - \left(1 - \delta_f\right)\left(1 - e^{-N\theta_{gs}}\right) \tag{11.4}$$

where $\theta_{gs}$ is the degradation rate parameter. Finally, the Rajashree and Sundaravadivelu (1996) model is given by:

$$\delta_N = \quad 1 - \varphi\left(yd\right)\log(N) \quad \varphi\left(yd\right)\log(N) < 1 \tag{11.5}$$

where φ is a function of the normalized displacement ($yd$ = pile displacement/pile diameter) of the pile. This model can only simulate concave degradation trends and was used by Rajashree and Sitharam (2001) to model the cyclic behaviour of piles in soft clay.

#### 11.2.2.5 Modelling Energy Dissipation

The soil damping provides a major source of energy dissipation in soil-foundation systems subjected to dynamic loading. There are two different types of damping that should be considered in seismic SFSI problems, namely radiation and material damping. Usually, radiation (geometric) damping is most important in the far field while material (hysteretic) damping provides most energy dissipation in the near field. Radiation damping is due to wave propagation away from the foundation, and is directly related to the soil compression and shear wave velocities. Hysteretic damping is caused by the plasticity of the soil and possibly discontinuity conditions (uplift and/or sliding) at the foundation-soil interface.

In BNWF models, the soil damping is established from analytical and/or empirical solutions that account for energy dissipation through wave propagation and soil hysteretic behaviour. Wang et al. (1998) have noted that a linear dashpot in parallel with a nonlinear soil spring may result in developing unrealistically large damping forces because forces bypass the hysteretic system. This is particularly important under strong shaking conditions, which effectively reduces the soil shear wave velocity. Nonetheless, El Naggar and Bentley (2000) noted that radiation damping could be modelled using a parallel nonlinear damper; this yielded results similar to a near-field far-field series radiation damping implementation (Wang et al. 1998). Similarly, Badoni and Makris (1996) used a parallel linear dashpot, but limited the value of the damping force that could be developed. Gerolymos and Gazetas (2005) used a viscoplastic approach capable of effectively modelling the reduction of radiation damping due to material/geometric nonlinearity.

#### 11.2.2.6 Foundation Input Motion

Foundation input motion (FIM), i.e., the motion experienced by the foundation due to interaction with the free-field motion, represents another challenge in SSI modelling. Due to lack of data and poor analysis procedures, kinematic interaction effects have mostly been neglected in SSI substructure analysis. However, the measured response of several buildings with basement portions during the Kobe earthquake (Iguchi 2001) showed that there can be a considerable difference between the FIM and free-field motion, which underscores the importance of the accurate prediction of the FIM in PBD.

Ahmad et al. (2007) presented simple calculation schemes to estimate various kinematic response parameters using an artificial neural network (ANN) technique of function approximation, which do not require performing separate site response and soil-pile interaction analyses. The kinematic response parameters include FIM and normalized bending moment at fixed head of pile for homogeneous and two-layer soil profiles. These ANN models represent simple solutions that can be implemented in a simple calculator capable of matrix operation.

## 11.3 Foundation Response Characteristics and Its Modelling

### *11.3.1 General Response Behaviour*

Dynamic soil-pile-structure interaction (SPSI) under extreme loading conditions, such as earthquakes, involves many interacting factors including: soil and structural yielding; pile-soil gap formation with/without soil cave-in and recompression; cyclic degradation of soil stiffness and strength; and radiation damping. These factors directly influence the effective lateral stiffness and strength of the foundation and can govern the design.

### *11.3.2 Stiffness Modelling Technique*

#### 11.3.2.1 Impedance of Single Pile

The pile length, bending and axial stiffness, tip and head conditions, mass, batter and the surrounding soil properties and their variation with depth and layering, affect the dynamic stiffness of a pile. The impedance functions of piles can be described as

$$K_i = k_i(a_0) + i\omega\, c_i(a_0) \tag{11.6}$$

**Table 11.1** Stiffness and damping constants for single piles

| Vertical | Horizontal | Rocking | Coupling | Torsion |
|---|---|---|---|---|
| $k_v = \frac{E_p A}{R} f_{v1}$ | $k_u = \frac{E_p I}{R^3} f_{u1}$ | $k_\varphi = \frac{E_p I}{R} f_{\varphi 1}$ | $k_c = \frac{E_p I}{R^2} f_{c1}$ | $c_\eta = \frac{G_p J}{V_s} f_{\eta 2}$ |
| $c_v = \frac{E_p A}{V_s} f_{v2}$ | $c_u = \frac{E_p I}{R^2 V_s} f_{u1}$ | $c_\varphi = \frac{E_p I}{V_s} f_{\varphi 2}$ | $c_c = \frac{E_p I}{R V_s} f_{c2}$ | $c_\eta = \frac{G_p J}{V_s} f_{\eta 2}$ |

The stiffness constants, $k_i$, and the constants of equivalent viscous damping, $c_i$, for individual motions of the pile head suggested by Novak (1974) are shown in Table 11.1.

These constants are a function of the pile's elastic modulus, $E_p$ cross-sectional area, $A$, and its moment of inertia and torsional stiffness $I$ and $G_pJ$, respectively. R is the radius of circular piles and equivalent radius for non-circular piles. The symbol $f_{1,2}$ in Table 11.1 represents dimensionless stiffness and damping functions whose subscript 1 indicates stiffness and 2 indicates damping. These functions depend on the following parameters: the relative stiffness of the pile and soil, $E_p/G$; dimensionless frequency, $a_0$; the slenderness ratio, L/R, in which L = pile length; material damping of both the soil and pile; the variation of soil and pile properties with depth; and the tip and head conditions. However, $E_p/G$, the soil profile and, for the vertical direction, the tip condition have the strongest effect on the stiffness. The stiffness and damping parameters, $f_{1,2}$, are given in Novak and El Sharnouby (1983).

### 11.3.2.2 Pile-Soil-Pile Interaction

When piles in a group are closely spaced, they interact with each other because the displacement of one pile to contribute to the displacement of others. To obtain an accurate analysis of dynamic behaviour of pile groups it is necessary to use a suitable computer program. However, a simplified approximate analysis can be formulated on the basis of dynamic interaction factors, α, introduced by Kaynia and Kausel (1982) who presented charts for dynamic interaction. For a homogeneous half space, the interaction factors between two piles may be given by (Dobry and Gazetas 1988 and Gazetas and Makris 1991).

$$\alpha_v \approx \frac{1}{\sqrt{2}}\left(\frac{S}{d}\right)^{-0.5} e^{-\beta\omega\frac{S}{V_s}} e^{-i\omega\frac{S}{V_s}} \quad \text{and} \quad \alpha_u(\theta^0) \approx \alpha_u(0^0)\cos^2\theta + \theta_u(90^0)\sin^2\theta \tag{11.7a}$$

$$\alpha_u(0^0) \approx \frac{1}{2}\left(\frac{S}{d}\right)^{-0.5} e^{-\beta\omega\frac{S}{V_{La}}} e^{-i\omega\frac{S}{V_{La}}} \quad \text{and} \quad \alpha_u(90^0) \approx 0.75\alpha_v \tag{11.7b}$$

where $\alpha_v$ and $\alpha_u$ are vertical and horizontal interaction factors, respectively, S/d = pile spacing to diameter ratio, θ is the angle between the direction of load action and the plane in which piles lie, and $V_{La}$ = the so-called Lysmer's analog velocity = $\dfrac{3.4V_s}{\pi(1-\nu)}$.

To calculate the dynamic stiffness of a pile group using the interaction factors approach, the impedance functions of single piles and the interaction factors are calculated first, and then the group impedance functions are computed. The stiffness and damping constants of individual piles are calculated using expressions given in Table 11.1 or formulae due to Gazetas (1991). The interaction factors are calculated using Eq. 11.7 or charts due to Kaynia and Kausel (1982). The impedance functions of a pile group of n piles are then given by (El Naggar and Novak 1995).

$$K_v^G = \bar{k}_v \sum_{i=1}^{n} \sum_{j=1}^{n} \varepsilon_{ij}^v \tag{11.8a}$$

$$K_h^G = \bar{k}_h \sum_{i=1}^{n} \sum_{j=1}^{n} \varepsilon_{2i-1,2j-1}^h \tag{11.8b}$$

$$K_r^G = \bar{k}_h \sum_{i=1}^{n} \sum_{j=1}^{n} \varepsilon_{2i,2j}^h \tag{11.8c}$$

$$K_c^G = \bar{k}_h \sum_{i=1}^{n} \sum_{j=1}^{n} \varepsilon_{2i-1,2j}^h \tag{11.8d}$$

where $K_v^G$, $K_h^G$, $K_r^G$ and $K_c^G$ are the vertical, horizontal, rocking and coupling group stiffness, respectively. In Eq. 11.8, $\bar{k}_v$ is the static vertical stiffness of the single pile, $[\varepsilon^v] = [\alpha]_v^{-1}$ where $\alpha_{ij}^v$ = complex interaction factors between piles *i* and *j*, $\alpha_{ii}^v = \bar{k}_v / K_v$, and $K_v$ is the complex vertical impedance function of the single pile. Similarly, $\bar{k}_h$ is the static horizontal stiffness of the pile $[\varepsilon^h] = [\alpha]_h^{-1}$ where $\alpha_{ij}^h$ = complex interaction coefficients for the horizontal translations and rotations. The formulation of the $[\alpha]_h$ can be found in El Naggar and Novak (1995).

## 11.4 Generalized Dynamic BNWF Model

The dynamic BNWF model by Allotey and El Naggar (2008a) can be classified as a degrading polygonal hysteretic model, with defined rules for loading, reloading and unloading. The model is a compression-dominant element, requiring two elements at each depth, for the modelling of pile-soil interaction. The model is made up of four main parts: the backbone curve, standard reload curve (SRC), general unload curve (GUC) and direct reload curve (DRC). These different parts of the model are briefly described below. The model is capable of accounting for cyclic soil degradation and the reduction of radiation damping with increased soil nonlinearity.

### *11.4.1 Backbone Curve, SRC and GUC*

The backbone curve is a four segment adaptable multi-linear curve (i.e., segments 1, 2, 3 and 4 in Fig. 11.5a) that can either be a monotonically-increasing curve (shown in Fig. 11.5a) or exhibit a post-peak residual behaviour (segments 1 and 2 model the curve up to the peak force, after which segments 3 and 4 model the post-peak behaviour). The various nodes and slopes of the curve are established through curve-fitting to prescribed force-displacement curves. Parameters for different standard *p-y* curves have been derived and can be found in Allotey (2006). The initial horizontal earth pressure is modelled as a pre-straining effect that shifts the backbone curve leftwards; elements on both sides of the pile are therefore pre-loaded at zero pile displacement. A similar approach has been used by Pender and Pranjoto (1996) and Allotey and Foschi (2005).

The standard reload curve (SRC) (i.e., segments 7, 8, 9 and 10 in Fig. 11.5a) and general unload curve (GUC) (i.e., segments 5 and 6 in Fig. 11.5a) are scaled versions of the backbone curve. They are derived using procedures similar to the well-established approach of Extended Masing rules (Vucetic 1990), with the scaling factor developed as a function of the reload or unload point and the current cyclic degradation factor (Allotey and El Naggar 2008c). For monotonic backbone curves, the SRC is comprised of four segments (7–8–9–10). For backbone curves exhibiting post-peak behaviour, the SRC is comprised of three segments (7–8–9, i.e. segments 9 and 10 are merged). The SRC simulates reloading curve reloading is not in the slack zone.

### *11.4.2 Direct Reload Curve (DRC)*

The direct reload curve (DRC) simulates soil reactions when the pile moves in the slack zone. It commences immediately after movement at the minimum force level in the negative direction ends (i.e., after segment 11). The DRC is designed as a convex strain-hardening curve that is controlled by a limiting force parameter $\lambda_f$ $(0 \leq \lambda_f \leq 1)$, which is referenced to the past maximum force. In addition, a curve shape parameter, $\lambda_s$, $(0 \leq \lambda_s \leq 1)$ can also be used to control the shape of the DRC (Fig. 11.5b). For fully unconfined response - pure gap - (e.g., piles in stiff clay), $\lambda_f = 0$, and for fully confined response (e.g., lower portions of piles in dry sand), $\lambda_f = \lambda_s = 1$. Similar parameters are available in the effective stress analysis program, CYCLIC1D (Elgamal et al. 2002; Yang et al. 2003) that efficiently simulates cyclic mobility.

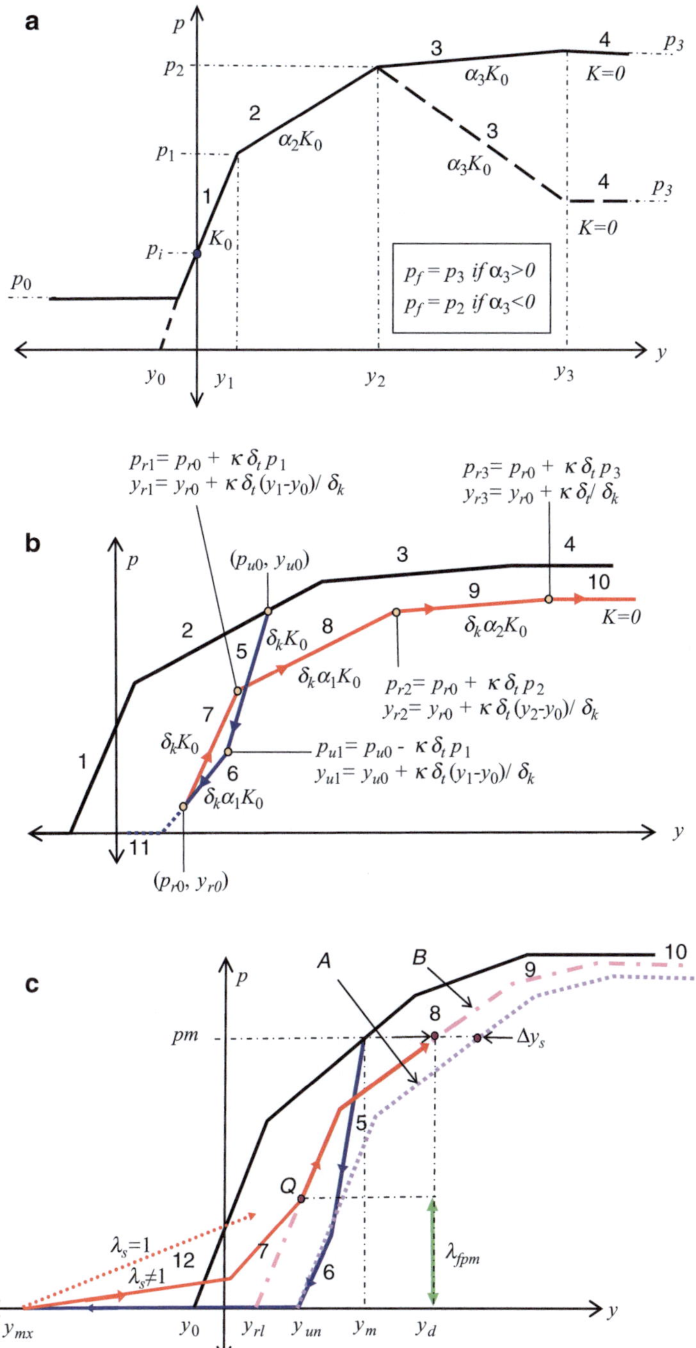


**Fig. 11.5** Schematics of: (**a**) backbone curve forms; (**b**) standard reload and general

Figure 11.5c shows two SRCs: Curve *A* represents the SRC corresponding to stable gap formation (e.g., as occurs in some clays); and Curve *B*, that is offset from Curve *A* to the left, represents a foundation moving back to the point where it last separated from the soil the approach used in PYLAT - El Naggar and Bentley (2000) and FLPIER-D Brown et al. (2001). The offset of Curve *B* (termed the base-SRC) increases with the extent of soil cave-in, and the finite volume that the compressed soil occupies. Allotey (2006) proposed a hyperbolic curve (shown in Fig. 11.5c) to estimate the origin of the base SRC. In Fig. 11.5c, $\varphi_h$, is the cyclic loading ratio ($\varphi_h = -1$ for two-way loading, and $\varphi_h = 0$ for one-way loading) and $\Lambda$ is the soil cave-in parameter. $\Lambda = 5$ gave the best-fit to the mean estimates of Long and Vanneste (1994) and models fully confined behaviour well. For $\Lambda = 0$, no soil cave-in occurs and Curve *B* becomes Curve *A*. The cyclic curve parameters (i.e. $\lambda_f$, $\lambda_s$ and $\Lambda$) vary with soil type and depth. Possible variations with depth include: linear distribution, which represents planar failure; and step-function distribution, which represents caved-in soil partially filling the gap (Allotey 2006).

### *11.4.3 Modelling of Cyclic Degradation*

A modified version of Anthes (1997) rain flow counting technique is developed to calculate cumulative damage, *D,* based on forming "virtual" half-cycle loops and later identifying full-cycle loops accounting for the number of equivalent loading cycles (Fig. 11.6). The incremental damage, $\Delta D$, for the current half-cycle loop is evaluated as (Allotey and El Naggar 2008c):

$$\Delta D_{j,j-1} = \frac{1}{2N_f(S_i)}$$
$$S_i = S_{r_j} - S_{r_{j-1}} \tag{11.9}$$

where, $N_f$ is the number of cycles to failure at a cyclic force (stress) ratio of $S_i$ (taken as a ratio of the soil strength), and $S_{r_j}$ and $S_{r_{j-1}}$ are the beginning and ending force ratios for the current half-cycle loop. $N_f$ is obtained from the failure condition curve (e.g., *S-N* curve from cyclic triaxial or simple shear tests). It is defined by the cyclic force (stress) ratio at $N=1$, $S_1$, and the negative slope of the failure condition line, $\eta_{SN}$. Two possible forms of the failure condition curve are a log-log model (Sharma and Fahey 2003) and a semi-log model (Hyodo et al. 1994). A stress-independent elliptical degradation function (Allotey 2006), is then used to evaluate the stiffness and strength degradation factors, i.e.

$$\delta_\varsigma = 1 + \left(\delta_{m\varsigma} - 1\right)\left[1 - \left(1 - D\right)^{\theta_\varsigma}\right]^{\frac{1}{\theta_\varsigma}} \tag{11.10}$$

where, $\varsigma$ stands for *k* or *t* (for stiffness and strength degradation factors, $\delta_k$ and $\delta_t$), $\delta_{m\varsigma}$ is the minimum/maximum amount of degradation and $\theta_\varsigma$ is the curve shape parameter. These parameters are based on physical quantities and their typical ranges are as follows: for saturated sand, $\eta_{SN}$=0.3–0.4, $S_1$=0.8–1.2, $\theta_k = \theta_t$=0.7–1.1 (De Alba et al. 1976);

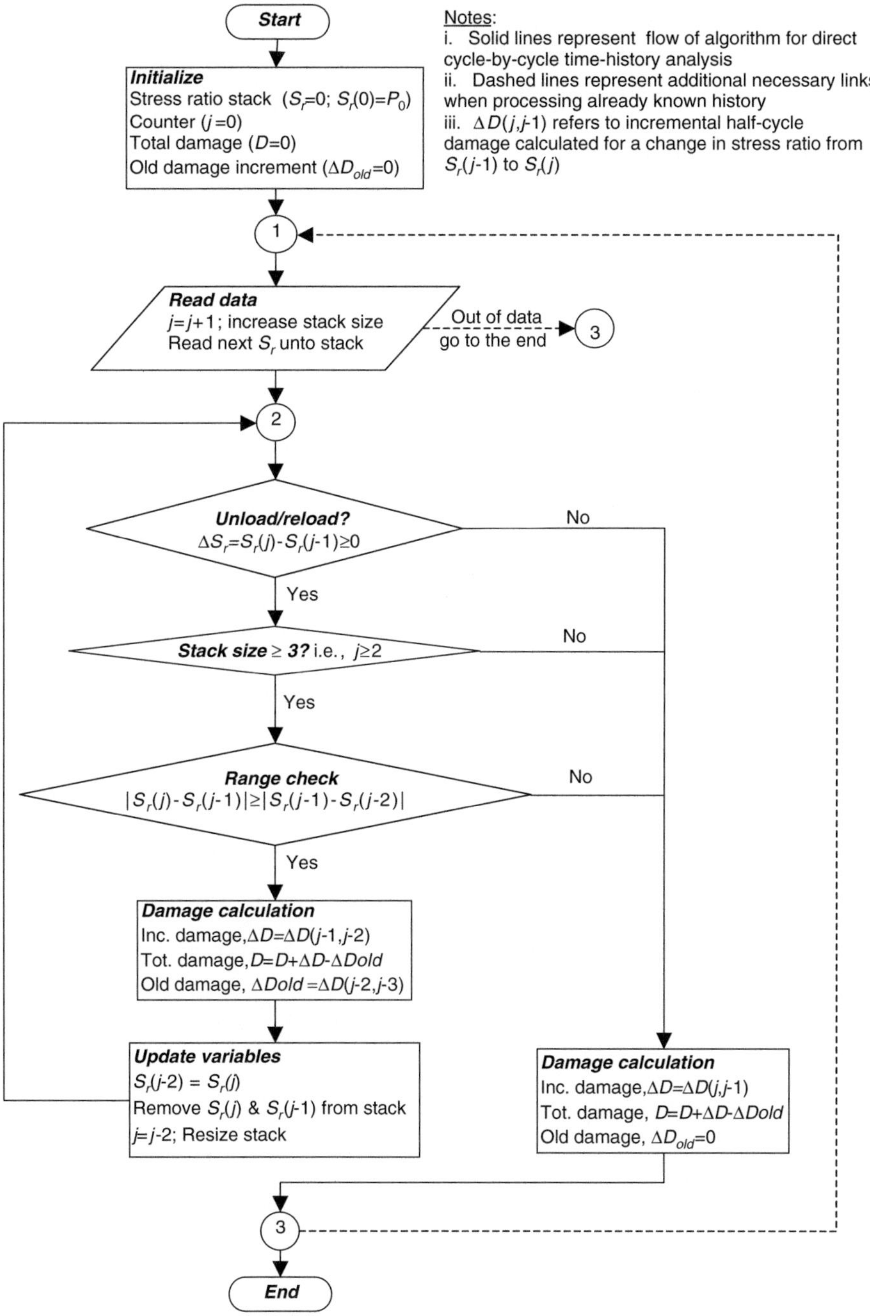


**Fig. 11.6** Flow chart of modified Anthes rain flow counting algorithm (Allotey and El Naggar 2008b)

Popescu and Prevost 1993; and for undrained clay, $\eta_{SN}$=0.07–0.15, $S_1$=1, $\theta_k$=1.5–2.5, $\theta_t$=0.75–0.95 (Hyodo et al. 1994; Andersen et al. 1988; Carter et al. 1982.

## 11.4.4 Model Damping

### 11.4.4.1 Radiation Damping

Radiation damping is modelled using a stiffness-proportional nonlinear damping formulation that comprises a nonlinear dashpot placed in parallel with the nonlinear spring. This is an adaptation and extension of the approach used by Badoni and Makris (1996). The damping constant at each time is related to the current stiffness; with the small-strain initial value estimated using impedance functions available in the literature (e.g., Novak et al. (1978); Gazetas and Dobry (1984)). The proposed damping model is given by:

$$p_d = \left[c(a_o)\alpha_{i\_t}\right]\dot{y} \quad \text{and} \quad c(a_o) = G_{\max} a_o S_{u2}(a_o, \upsilon_s) \tag{11.11}$$

where $S_{u2}$ is a dimensionless constant that is a function of the Poisson ratio, $\upsilon_s$ and the dimensionless frequency, $a_0 = \omega r / V_s$, where, $\omega$ is the circular frequency, $V_s$ is the soil shear wave velocity and $r$ is a characteristic dimension. Also, $G_{max}$ is the small-strain shear modulus, $c$ is the initial damping constant, $\alpha_{i\_t}$ is the current stiffness ratio, and $p_d$ and $\dot{y}$ are the damping force and relative velocity, respectively.

Figure 11.7 compares the damping force-displacement curves for linear damping, stiffness-proportional damping for the hyperbolic curve and a four-segment multi-linear stiffness ($y_o$ in Fig. 11.7 is the maximum displacement). The model performs the same as a linear dashpot under small displacement conditions, but the damping force developed in both stiffness-proportional dampers is reduced compared with the linear damper. With increasing displacement, the response is generally nonlinear, and the total system damping is due mainly to hysteretic damping. The computed radiation damping would be small when movement occurs in the slack zone, and becomes null for the case of a pure gap. A general reduction in radiation damping also occurs for degrading systems, whereas an increase occurs for hardening systems.

### 11.4.4.2 Hysteretic Damping

The hysteretic damping ratio, $\xi_h$, of the model ranges from zero, for a perfectly elastic response, to the largest possible amount of energy dissipation per cycle under two-way cyclic loading, for the case of $\lambda_f = \lambda_s = 1$:

$$\xi_h = \begin{cases} \dfrac{1}{2\pi}\left[2 - \dfrac{K_s}{K_0}\right] & \varphi \le \dfrac{\varphi_1}{1-\varphi_1} \\ \dfrac{1}{2\pi}\left[2 - \dfrac{K_s}{K_0} - \left(\dfrac{K_s}{K_0}\right)\left(1 - \dfrac{(1+\varphi)}{(\varphi/\varphi_1)}\right)^2\left(\dfrac{1}{\varphi_1} - 1\right)\right] & \varphi > \dfrac{\varphi_1}{1-\varphi_1} \end{cases} \tag{11.12}$$

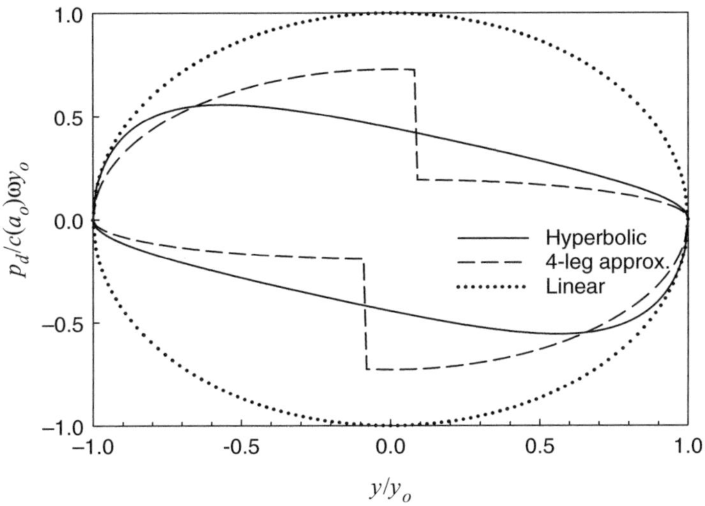


**Fig. 11.7** Comparison between linear and stiffness-proportional radiation damping

In Eq. (19), $\phi_i = p_i/p_f$, $K_s$ is the secant stiffness. For $\phi_1 \geq 0.5$, the first equation controls irrespective of the value of $\phi$. It is noted that the maximum model damping ratio possible is $\xi_h = 1/\pi$. The damping ratios obtained are within the range of damping ratios observed in soil cyclic tests (Allotey and El Naggar 2008a).

## 11.5 Soil-Pile-Structure Interaction (SPSI) Examples

### *11.5.1 Elastic Pile in Medium-Stiff Clay*

The problem involves a 12 m long reinforced concrete pile with diameter, $d = 600$ mm. The pile was embedded in uniform medium-stiff clay with undrained strength, $c_u = 50$ kPa, and unit weight, $\gamma_t = 19$ kN/m$^2$. It was loaded laterally at its head (a distance of one diameter aboveground) with 25 uniform two-way loading cycles with amplitude, 213 kN. The pile yield moment, $M_y = 636$ kNm, and the pile was assumed to stay elastic below the yield moment.

Pender and Pranjoto (1996) and Pranjoto and Pender (2003) modelled the problem using a BNWF model (referred to here as Pender-Pranjoto). The soil at each depth was represented by two springs - one on either side of the pile. Gapping was modelled such that the full gap distance had to be traversed before reloading commenced. Their BNWF formulation is similar to the model used in this study, and their results would serve as a baseline for comparison purposes.

**Table 11.2** *p-y* curve model parameters for Pender-Pranjoto case study

| Backbone curve parameters[a] | |
|---|---|
| $p_1$ | 0.15 |
| $p_2$ | 0.60 |
| $\alpha_2$ | 0.28 |
| $\alpha_3$ | 0.03 |

[a]Curve parameters defined for a unit *p-y* curve

#### 11.5.1.1 Model Description and Input Parameters Estimation

Forty elastic beam-column elements modelled the piles, and the soil modelled using two springs at each pile node. The *p-y* curve used by Pender-Pranjoto was the generalized hyperbolic function:

$$y = \frac{p}{K_0}\frac{{p_f}^{n_{hy}}}{\left({p_f}^{n_{hy}} - p^{n_{hy}}\right)} \tag{11.13}$$

where $p$ is soil reaction, $y$, the pile displacement, $K_0$, the initial stiffness, $p_f$, the ultimate force and $n_{hy}$, the curve shape parameter. For clays, $n_{hy}$ varies between 0.2 and 0.3 ($n_{hy}=0.2$) was used. Table 11.2 lists the model parameters for a 4-segment multi-linear curve fit to the hyperbolic curve. In Table 11.2, $p_i$ and $\alpha_j$ are $i^{\text{th}}$ endpoint and $j^{\text{th}}$ segment of the multi-linear curve.

Pender-Pranjoto assumed $p_f$ to increase linearly from $5c_u$ at the ground surface to $12c_u$ at a depth of $3.5d$, and constant below this depth. The initial *p-y* curve stiffness, $K_0$, was derived from the distributed stiffness, $k_0$, which was estimated by,

$$k_0 = \frac{1.3}{\left(1-\upsilon_s\right)}\sqrt[12]{\frac{E_s d^4}{E_{pl} I_{pl}}}\frac{d}{d_{ref}} \tag{11.14}$$

where $\upsilon_s$ = soil Poisson's ratio, $E_s$ (taken as $600c_u$) and $E_{pl}$ are the Young's modulus of the soil and pile, $I_{pl}$ = pile cross-sectional moment of inertia, and $d_{ref}$ (reference pile diameter) = 1.0.

The six *p-y* model parameter combinations investigated in this study are as follows. Case *A* is representative of the "pure gapping" approach used by Pender-Pranjoto and uses static p-y curves with $\Lambda = \lambda_f = 0$. in Case B, $\Lambda = \lambda_f = 0$ but the *p-y* curves were modified to account for cyclic degradation using the pseudo-static procedure recommended by Matlock (1970), resulting in a negative descending branch for *p-y* curves within the top $3.5d$ of the pile. Cases $C-F$ use cyclic curve shape parameters (i.e., $\Lambda$ and $\lambda_f$, $\lambda_s = 1$) representing soil cave-in and recompression behaviour. Case $C$ ($\lambda_f = 1$ $\Lambda = 5$) represents the condition of no slack zone (i.e. cyclic *p-y* models based on Massing rules in Thavaraj 2001). Case $D$ ($\Lambda = 0$ $\lambda_f = 1$) is an intermediary case between Cases *A* and *C*. Cases *E* and *F* differ in terms of the value $\Lambda$ and $\lambda_f$ within the top-third of the pile, where gapping is assumed to occur to a depth of $6.5d$. However, the caved-in soil in Case *F* is assumed to fill the gap created to a

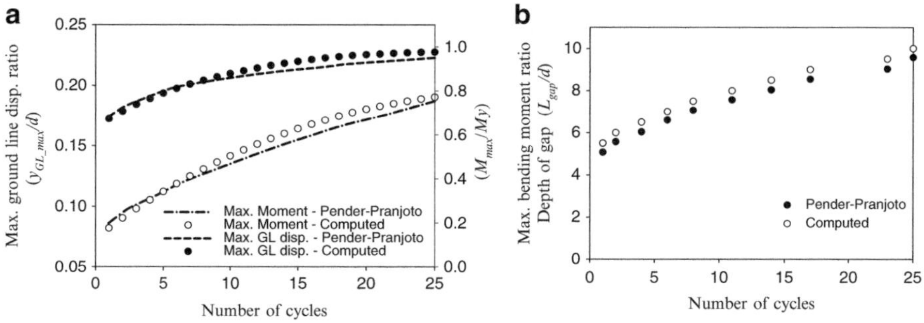


**Fig. 11.8** (**a**) Variation of ground line displacement ratio and maximum moment ratio with number of loading cycles; (**b**) variation of gap depth ratio with number of loading cycles

depth of 1.5*d*. For Case *E*, Λ and $\lambda_f$ vary linearly with depth ($\lambda_f$=0–1: <6.5d, 1: >6.5d and Λ=0–5<6.5*d*, 5: >6.5d). For Case F, Λ and $\lambda_f$ are same as Case E but vary stepwise with depth.

### 11.5.1.2 Results and Discussion

***Base cyclic response behaviour***: Figure 11.8a shows the general agreement of the predicted maximum ground line displacement ratio ($y_{GL\_max}/d$) and maximum bending moment ratio ($M_{max}/M_y$) for Case *A* with those obtained by Pender-Pranjoto. Figure 11.8b shows that the trend of computed gapping depths is similar to (but slightly larger) Pender-Pranjoto's predictions. The differences are attributed to slight differences in the loading curves used in both studies. The maximum ground line displacement ratio, maximum bending moment ratio and gap depth ratio all increase with the number of cycles due to successively increasing permanent displacements, and consequently the tensile forces in springs along the lower portion of the pile, extending the gap to a larger depth which results in larger pile loads and bending moments.

***Effect of cyclic p-y parameter variation***: Figure 11.9 shows the cyclic loops computed for the Cases A-F and Fig. 11.10 shows the corresponding maximum moment ratio profiles. Case *C* displayed an oval shape, with slight increase in displacement with cycling. Figure 11.10 shows that the bending moment profile for this case was close to the *N*=1 profile, i.e., load cycling had a limited effect on response. The loops for Case *D* were similar to Case *C*, but with larger displacement and bending moment, with the maximum moment occurring at a lower depth. The two cases differed only in the value of parameter Λ, which affected the gapping depth and bending moment. Similar observations are made for Case *E* (Λ increases linearly with depth), where the displacements and moments were smaller compared with Case *D*, and maximum bending moment occurred at a higher depth. For Case F, accounting

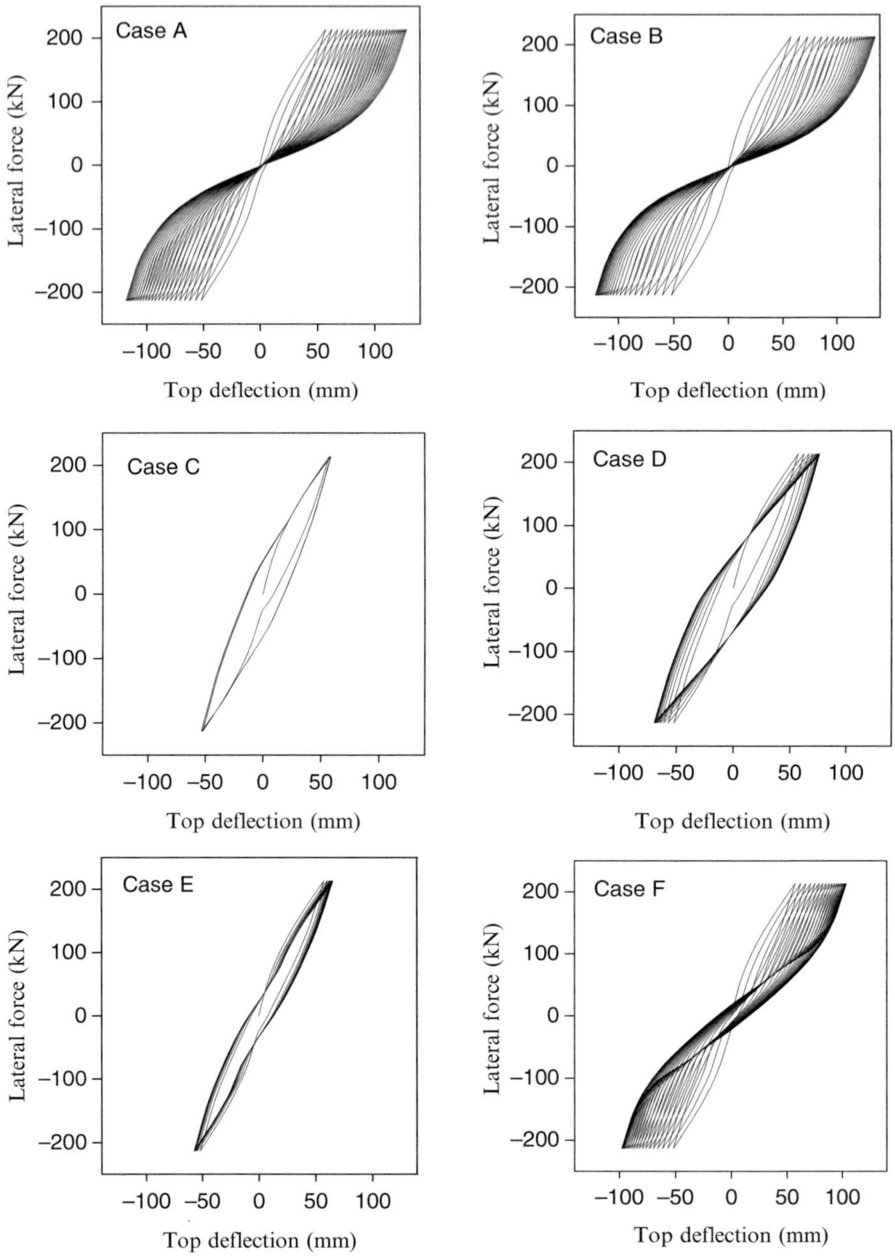


**Fig. 11.9** Computed cyclic *p-y* loops for six curve shape parameter cases

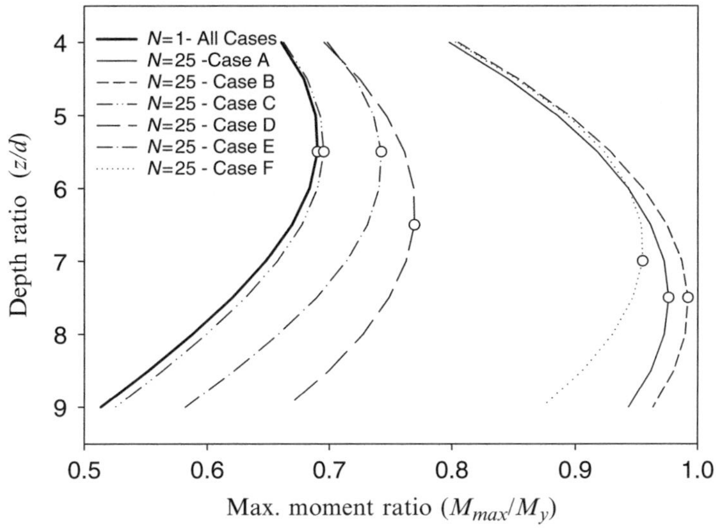


**Fig. 11.10** Maximum moment ratio depth profile for six curve shape parameter cases

for cyclic degradation resulted in an increase in the pile maximum displacement and bending moment. The response curve, however, remained similar to Case *A*. Also, the results show that $\lambda_f$ increases hysteretic damping but has a limited influence on the pile bending moment.

The results showed that soil cave-in and recompression affect the response of SPSI systems in three main ways: reduces the pile maximum bending moment; moves the point of maximum bending moment closer to the ground surface; and increases the hysteretic damping of the soil-pile system. These effects were more pronounced when soil cave-in occurred within the pile topmost part, which is attributed to increases in the effective confinement of the pile due to soil cave-in, therefore enhancing the performance of the SPSI system. This could be particularly beneficial for situations where plastic hinges develop belowground level. In addition, the calculated responses suggest the need for caution in using BNWF models that are based only on either a pure gap formation (e.g., FLPIER-D) or on Masing rules (e.g., Kagawa and Kraft 1980) to model the cyclic response of soil-pile systems.

## 11.5.2 Reinforced Concrete Piles in Sand

### 11.5.2.1 Experimental Setup

Full-scale SPSI experiments were conducted at the University of California at Davis involving reversed cyclic quasi-static testing of four RC piles of diameter, $d = 406$ mm, embedded in loose and dense sand beds to a depth of 13.5*d*. The soil

**Table 11.3** Relevant soil information

| Soil and pile properties | | |
|---|---|---|
| | $2d$ | $6d$ |
| Pile diameter, $d$ (mm) | 406 | 406 |
| Embedded pile length | $13.5d$ | $13.5d$ |
| Soil relative density, $D_r$ (%) | 53 | 84 |
| Soil friction angle, $\phi'$ | 37° | 42° |
| Soil bulk density, $\gamma_t$ (kN/m$^3$) | 17 | 18.2 |
| Reference shear wave velocity, $V_{sr}$ (m/s) | 171 | 261 |
| Reference depth, $z_{Vs}$ | $3d$ | $7.3d$ |

was contained in a cylindrical container 6.7 m in diameter and 5.5 m deep. The distance from the pile to container wall was greater than $6d$ to minimize the boundary effects on the measured pile response (Park et al. 1987). The pile head was $2d$ above ground surface for two piles, and $6d$ for the other two. The sand used was classified as clean, poorly graded sand with 3% fines, $C_u = 4.4$ and $D_{50} \approx 0.5–0.6$. Two piles were considered herein: one with $2d$ free length in loose sand and another with $6d$ free length in dense sand; these two cases will be referred to as $2d$ and $6d$ experiments, respectively.

The relative density, $D_r$, (estimated from CPT tests) was 53% and 84%, and the soil unit weight was 17 and 18.2 kN/m$^3$ and the piles concrete compressive strength was 41 MPa and 47.5 MPa for the $2d$ and $6d$ experiments. Table 11.3 summarizes the relevant soil information. The piles were reinforced with seven longitudinal bars. The piles were instrumented to measure longitudinal bar strain, pile curvatures, axial load, lateral load and displacement. Both piles were subjected to an initial axial compression of 445 kN. The lateral loading started with load-controlled cycles, followed by displacement-control after the soil-pile system softened. Further test information including pile concrete and reinforcement details can be found in Chai and Hutchinson (1999).

#### 11.5.2.2 Model Description and Parameter Estimation

The pile was modelled using 31 beam-column elements, each $0.5d$ in length. The beam-column elements were modelled in SeismoStruct using the fibre element approach. Two types of $p$-$y$ curves were used in the analysis: the API recommended $p$-$y$ curve for sand (API 1993), and the Yan and Byrne (1992) $p$-$y$ curve, i.e.

API (1993):

$$p = Ap_u \tanh\left(\frac{khy}{Ap_u}\right) \tag{11.15a}$$

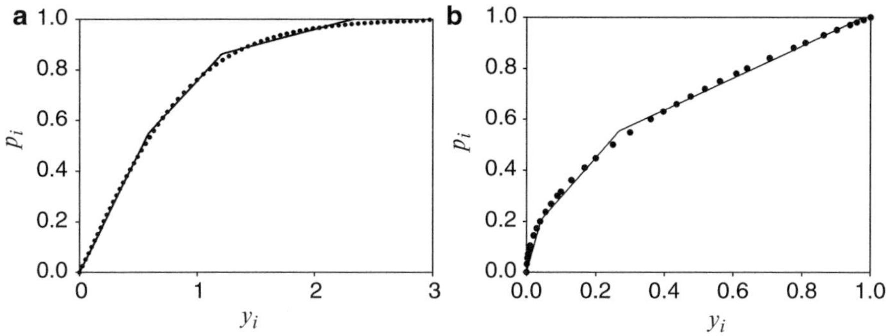


**Fig. 11.11** Four-segment multi-linear fit to: (**a**) unit API *p*-*y* curve for sand; (**b**) Yan and Byrne curve for 2*d* example

$$p_u = \begin{cases} p_{us} = C_1 h + C_2 d)\gamma' h & \text{if } p_{us} \le p_{ud} \\ p_{ud} = C_3 d\gamma' h & \text{if } p_{ud} < p_{us} \end{cases} \tag{11.15b}$$

where, $p_u$ = ultimate resistance, $\gamma'$ = effective soil unit weight, $h$ = depth, $A$ = 0.9 for cyclic loading, $k$ = sub grade modulus and $C_1 - C_3$, constants that can be found in the reference.

Yan and Byrne (1992):

$$p = \begin{cases} E_{\max}\ y & \text{if } y \le \alpha^2 d \\ E_{\max}\ \alpha\ d\left(\dfrac{y}{d}\right)^{0.5} & \text{if } y > \alpha^2 d \end{cases} \tag{11.16}$$

where $\alpha$ depends on the soil relative density and is given as $\alpha = 0.5(D_r)^{0.8}$.

Figure 11.11 shows the curve fitting for the Yan and Byrne *p*-*y* curve. The first segment of the backbone curve was the same as the initial portion of Yan and Byrne's curve. Two soil shear wave velocity profiles were considered: linear variation of $V_s$ from 0 to 1.5d, and constant $V_s$ = 171 m/s afterwards; and parabolic variation of $V_s$ with depth, $V_s$ = 231 m/s at pile toe. Table 11.4 presents the model parameters: Case I corresponded to pure gapping; Cases II and III correspond to a linear variation of $\Lambda$ and $\lambda_f$ with depth up to one-third and two-thirds of the pile length, representing significant and moderate soil cave-in and recompression. The lateral cyclic loading was applied as a displacement history comprising three-cycles of stepwise increasing displacement with a maximum of 350 and 400 mm for the 2*d* and 6*d* examples.

#### 11.5.2.3 Results and Discussion

The ATC-32 (ATC 1996) bridge seismic design guideline rates extended pile shafts as limited ductility structures, and assigns them a displacement ductility factor of

**Table 11.4** *p-y* curve model parameters for 2*d* and 6*d* aboveground bridge bent experiments

| Backbone curve parameters[a] | | | |
|---|---|---|---|
| | API | Yan & Byrne (2*d*) $D_r$=53% | Yan & Byrne (6*d*) $D_r$=84% |
| $p_1$ | 0.55 | 0.20 | 0.15 |
| $p_2$ | 0.86 | 0.55 | 0.51 |
| $\alpha_2$ | 0.54 | 0.32 | 0.28 |
| $\alpha_3$ | 0.13 | 0.13 | 0.095 |
| Cyclic curve shape parameters | | | |
| | Case I | Case II | Case III |
| $\Lambda$ | 0 | 0–5<4.5*d*<br>5: >4.5*d* | 0–5<9*d*<br>5: >9*d* |
| $\lambda_f\ (\lambda_s=1)$ | 0 | 0–1: <4.5*d*<br>1: >4.5*d* | 0–1: <9*d*<br>1: >9*d* |

[a]Curve parameters defined for a unit *p-y* curve

$\mu_\Delta=3$. Global and local responses of the SPSI system play important roles in their design, and are discussed below. The results obtained for the 2*d* and 6*d* tests are presented in Tables 11.5, 11.6 and 11.7 and Figs. 11.12–11.15.

### Base Cyclic Response Behaviour

Limited soil cave-in was observed and gaps were noted to extend to significant depths. For example, a soil gap of 75 mm extending to a depth = 2.03 m was observed at a lateral force of 66.7 kN (i.e., about half the peak force, $P_u$) in the 2*d* experiment. Based on this, Case I (with no soil cave-in) should best represent the condition of both tests, and therefore they are compared with the measured cyclic response.

Figure 11.12 shows that the initial stiffness was reasonably predicted, but the capacity, $P_u$, of the hysteresis loops computed using the API *p-y* curves was under-predicted for both experiments. This can be attributed to the ability of the soil to sustain higher loads at larger displacements than predicted by the API curves. Figure 11.13 shows that the predictions obtained with Yan and Byrne's *p-y* curves were generally satisfactory, with the parabolic profile giving better results, with total energy dissipated (given by the area enclosed by the loops) matched closely the measured results for both the 2*d* and 6*d* examples (Table 11.5).

Figure 11.14 shows a comparison between the predicted longitudinal strains for bar # L1 based on the parabolic profile and the measured longitudinal strains. For both tests, the predicted depth to maximum damage compared favourably with the measured value, i.e., for the 2*d* test: measured = 3.3*d*, predicted = 3.0*d*; for the 6*d* test: measured = 1.25*d*, predicted = 1.5*d*. From Table 11.6, the predicted depth to maximum curvature for both tests was also found to be in good agreement with the measured values, i.e., for the 2*d* test: measured = 3.4*d*, predicted = 3–3.5*d*; for the 6*d* test: measured = 2.1*d*, predicted = 1.5–2*d*.

**Table 11.5** Summary of global results for 2*d* and 6*d* aboveground bridge bent experiments

| | $P_{yi}$ [a] (kN) | $\Delta_{yi}$ [a] (mm) | $P_u$ [b] (kN) | $\Delta_u$ [b] (mm) | $P_{res}$ [c] (kN) | $K_{re}$ [c] (kN/m) | $\mu_\Delta$ [d] |
|---|---|---|---|---|---|---|---|
| ***2d*** | | | | | | | |
| Measured | 97.0 | 68.2 | 124.1 | 144.1 | 94.9 | 142.6 | 4.1 |
| API (Base case) | 67.6 | 51.5 | 85.2 | 109.3 | 36.4 | 201.9 | 5.3 |
| Yan – linear (Base case) | 99.8 | 49.8 | 124.3 | 89.4 | 72.5 | 198.6 | 5.5 |
| Yan – parab.(Base case) | 97.2 | 57.5 | 116.4 | 98.8 | 70.1 | 184.3 | 5.1 |
| Yan – parab.(Case II) | 97.4 | 57.9 | 116.5 | 99.8 | 75.4 | 164.3 | 4.85 |
| Yan – parab(Case III) | 97.2 | 57.5 | 116.4 | 99.2 | 75.2 | 164.2 | 4.88 |
| ***6d*** | | | | | | | |
| Measured | 43.3 | 78.3 | 52.5 | 137.5 | 5.5 | 166.4 | 4.6 |
| API (Base case) | 38.7 | 80.3 | 45.3 | 140.6 | 9.1 | 130.6 | 4.5 |
| Yan – linear(Base case) | 44.5 | 81.2 | 53.8 | 139.8 | 14.5 | 140.4 | 4.3 |
| Yan – parab.(Base case) | 44.1 | 81.4 | 54.1 | 140.1 | 13.8 | 143.6 | 4.3 |
| Yan – parab.(Case II) | 44.7 | 82.3 | 54.0 | 140.1 | 14.0 | 142.8 | 4.3 |
| Yan – parab (Case III) | 44.5 | 81.8 | 54.1 | 140.1 | 13.8 | 143.6 | 4.3 |

[a] $P_{yi}$ and $\Delta_{yi}$ are the first yield force and yield displacement
[b] $P_u$ and $\Delta_u$ are the peak force and peak displacement
[c] $P_{res}$ and $K_{res}$ are the residual force and post-peak stiffness
[d] $\mu_\Delta$ is the displacement ductility

**Table 11.6** Summary of local results for 2*d* and 6*d* aboveground bridge bent experiments

| | *2d* | | | *6d* | | |
|---|---|---|---|---|---|---|
| | $\varphi_{max}$ [a] (mrad/m) | *z/d* | $\mu_\varphi$ [b] | $\varphi_{max}$ (mrad/m) | *z/d* | $\mu_\varphi$ |
| Measured | 167 | 3.4 | 12.7 | 114 | 2.1 | 8.6 |
| API (Base case) | 172 | 4–4.5 | 13.1 | 170 | 2.5–3 | 12.8 |
| Yan – linear(Base case) | 155 | 2–2.5 | 11.8 | 140 | 1.5–2 | 10.5 |
| Yan – parab.(Base case) | 162 | 3–3.5 | 12.4 | 130 | 1.5–2 | 9.7 |
| Yan – parab.(Case II) | 110 | 1–2 | 8.4 | 95 | 1–1.5 | 7.1 |
| Yan – parab (Case III) | 130 | 1.5–2 | 9.9 | 115 | 1–1.5 | 8.6 |

[a] Curvature at a displacement ductility, $\mu_\Delta = 2.8$
[b] $\mu_\varphi$ = curvature ductility (for elasto-plastic yield)

**Table 11.7** Total energy dissipated in 2*d* and 6*d* aboveground bridge bent experiments

| | Energy dissipated (kNm) | |
|---|---|---|
| | *2d* | *6d* |
| Measured | 325.5 | 244.5 |
| API (Base case) | 237.5 | 187.9 |
| Yan – linear (Base case) | 413.7 | 257.9 |
| Yan – parab.(Base case) | 362.6 | 252.8 |
| Yan – parab.(Case II) | 550.7 | 305.6 |
| Yan – parab (Case III) | 447.1 | 268.5 |

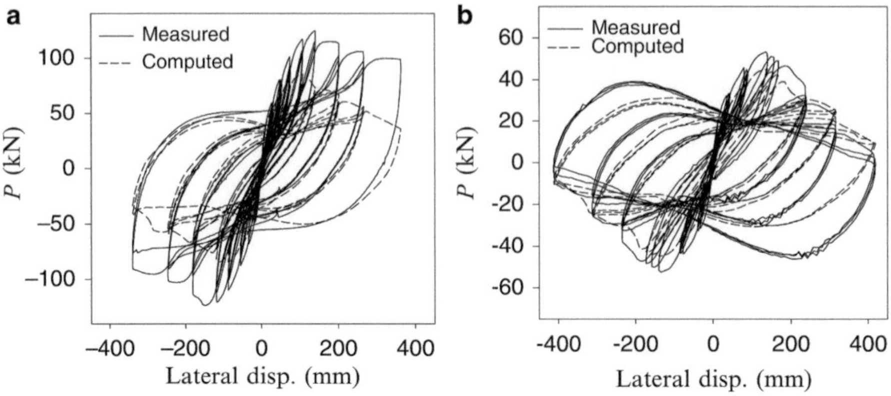


**Fig. 11.12** Comparison of measured and base case of API *p-y* curve lateral load – displacement hysteresis loops for: (**a**) 2*d* example; (**b**) 6*d* example

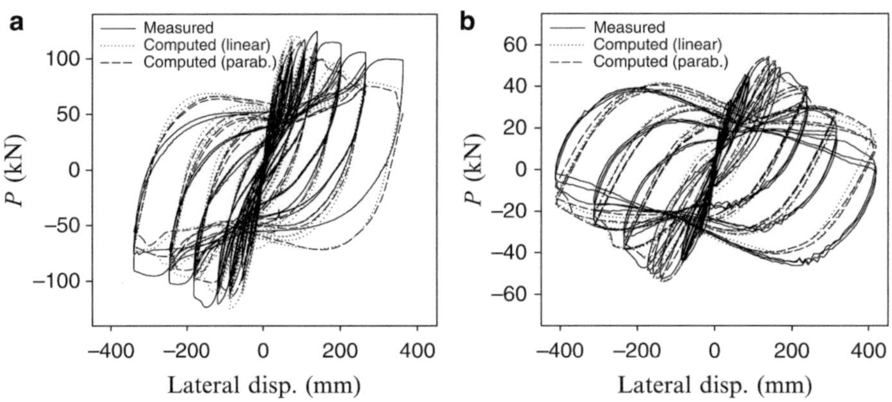


**Fig. 11.13** Comparison of measured and base case of Yan and Byrne *p-y* curve lateral load – displacement hysteresis loops for: (**a**) 2*d* example; (**b**) 6*d* example

## Effect of Cyclic p-y Parameter Variation

Based on the good agreement obtained between the measured and predicted results for the parabolic profile, the influence of soil cave-in and compression on the computed response was assessed by comparing the results of the three cases for this profile. From Fig. 11.15, a considerable increase in the energy dissipation is noted for Cases II and III although the results of the global response show only a slight

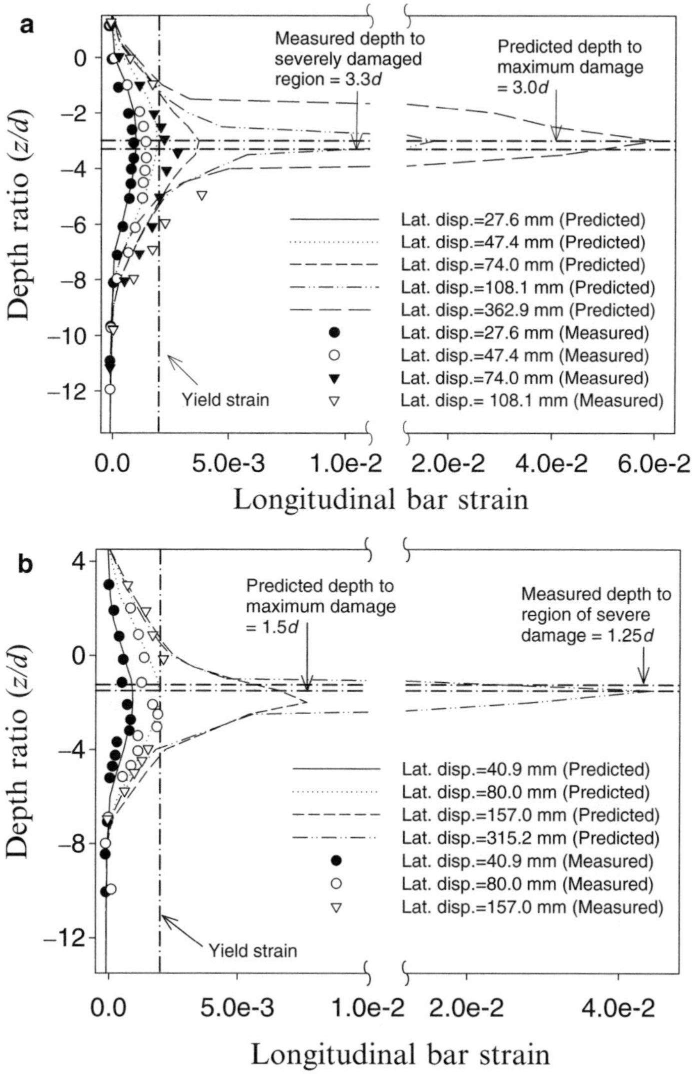


**Fig. 11.14** Comparison of measured and predicted longitudinal strain – depth profiles for bar # L1 for the parabolic base case: (**a**) 2*d* example; (**b**) 6*d* example

increase in the post-peak residual load (Table 11.7). Also, the results of the local response show a significant reduction in the maximum curvature, and corresponding curvature ductility. A study of the curvature profile showed curvatures to be spread over a larger length of the pile, thus, effectively decreasing the maximum curvature at a given displacement ductility. In other words, soil cave-in helps increase the effective length of the plastic hinge.

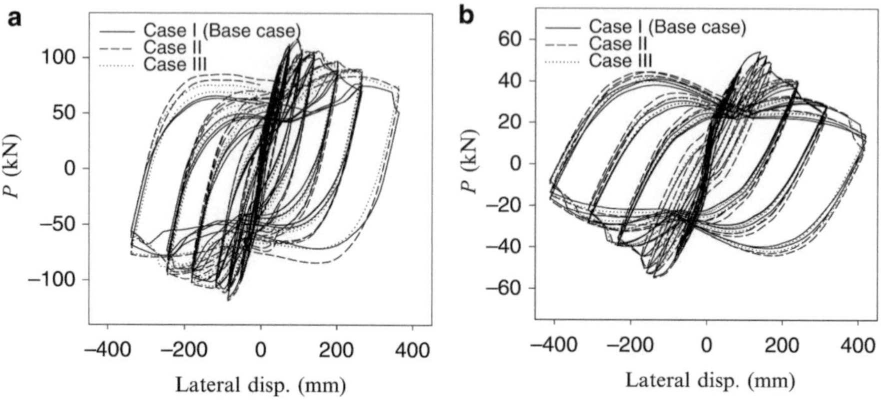


**Fig. 11.15** Comparison of lateral load – displacement hysteresis loops for cases I, II and III for parabolic profile based on Yan and Byrne's *p-y* curve: (**a**) 2*d* example; (**b**) 6*d* example

### 11.5.2.4 Discussion

The results demonstrated the beneficial effect of soil cave-in and recompression. It increases the effective confinement of the pile and spreads the curvature demand, thereby minimizing localized curvature and increasing the effective length of the plastic hinge. This is partly corroborated by the results of full-scale drilled shaft experiments by Wallace et al. (2001), who noted that spalled-off concrete increased the effective soil-pile frictional resistance.

For damaged piles, the extra confinement provided to the damaged zone by the cave-in soil could enhance the performance of the SPSI system. Two recent sets of large-scale cyclic lateral load tests of RC piles with diameters ranging from 0.4 to 1.8 m supported these findings (Juirnarongrit and Ashford 2003; and Budek et al. 2004). Both studies reported that the soil surrounding the damaged portion of the pile was observed to provide significant external confinement. In fact, Budek et al. (2004) concluded that soil confinement can play a very significant role in pile shaft response, as the increase in section effective confinement retards development of localized plastic rotation, which can increase the ductility capacity. This conclusion validates the observations made from the current study, and further verifies the ability of the proposed approach to simulate these phenomena. Currently, no rational approach exists to account for the confining effect of soil in design, and more efforts are required to adequately quantify it. This is particularly relevant considering recent changes in design philosophy to allow for foundation yielding (Martin and Lam 2000).

**Acknowledgement** The author would like to acknowledge the significant contributions of his former student: Dr. Nii Allotey. The presented work cannot be completed without his intellectual contributions.

## References

Ahmad I, El Naggar MH, Khan AN (2007) Artificial neural network application to estimate kinematic soil pile interaction response parameters. Soil Dyn Earthquake Eng 27:892–905

Allotey N, El Naggar MH (2005) Soil-structure interaction in performance-based design – a review. In: Proceedings of the 11th IACMAG, 19–24 June , vol 3. Torino, pp 595–602

Allotey N, El Naggar MH (2006) Cyclic degradation/hardening models in total stress analysis: new equations. First European conference on earthquake engineering and seismology (a joint event of the 13th ECEE and 30th General Assembly of ESC), Paper no 1015, Geneva

Allotey N, El Naggar MH (2008a) Generalized dynamic Winkler model for nonlinear soil-structure interaction analysis. Can Geotech J 45(4):560–573

Allotey N, El Naggar MH (2008b) An investigation into the Winkler modelling of the cyclic response of rigid footings. Soil Dyn Earthquake Eng 28(1):44–57

Allotey N, El Naggar MH (2008c) A consistent soil fatigue framework based on the number of equivalent cycles. J Geotech Geol Eng 26(1):65–77

Allotey NK, Foschi RO (2005) Coupled *p-y t-z* analysis of single piles in cohesion less soil under vertical and/or horizontal ground motion. J Earthquake Eng 9(6):755–776

American Petroleum Institute (API) (1993) Recommended practice for planning designing and constructing fixed offshore platforms, API-RP2A-WSD. American Petroleum Institute, Washington, DC

American Society of Civil Engineers (ASCE) (2000) FEMA-356-Pre-standard and commentary for the seismic rehabilitation of buildings, Washington, DC

Andersen KH, Kleven A, Heine D (1988) Cyclic soil data for design of gravity structures. J Geotech Eng, ASCE 114(5):517–539

Anderson DL (2003) Effect of foundation rocking on the seismic response of shear walls. Can J Civil Eng 30:360–365

Angelides DC, Roesset JM (1980) Non-linear dynamic stiffness of piles. Research report R80-13, Department of civil engineering, MIT, Cambridge, MA

Anthes RJ (1997) Modified rain flow counting keeping the load sequence. Int J Fatigue 19(7):529–535

Applied Technology Council (ATC) (1978) Tentative provisions for the development of seismic regulations for buildings: ATC 3–06. Applied Technology Council, Washington, DC

Applied Technology Council (ATC) (1996) ATC 40 – The seismic evaluation and retrofit of concrete buildings, vol I & II. ATC, Redwood

ASCE-7 (2010) ASCE 7–10 minimum design loads for buildings and other structures. American Society of Civil Engineers, 12 May 2010, 658 pp, ISBN: 9780784410851

Aviles J, Perez-Rocha LE (2003) Soil-structure interaction in yielding systems. Earthquake Eng Struct Dyn 32:1749–1771

Badoni D, Makris N (1996) Nonlinear response of single piles under lateral inertial and seismic loads. Soil Dyn Earthquake Eng 15:29–43

Bentley KJ, El Naggar MH (2000) Numerical analysis of kinematic response of single piles. Can Geotech J 37:1368–1382

Boulanger RW, Curras JC, Kutter BL, Wilson DW, Abghari A (1999) Seismic soil-pile-structure interaction experiments and analysis. J Geotech Geoenviron Eng, ASCE 125(9):750–759

Budek AM, Priestley MJN, Benzoni G (2004) The effect of external confinement on the flexural hinging in drilled pile shafts. Earthquake Spectra 20(1):1–24

Calvi GM (2004) Recent experience and innovative approaches in design and assessment of bridges. In: Proceedings of the 13th World conference on earthquake engineering, Paper no 5009, Vancouver, BC

Carr AJ (2001) Ruaumoko 3D – user's manual. University of Canterbury, Christchurch, New Zealand

Carter JP, Booker JR, Wroth CP (1982) A critical state soil model for cyclic loading. In: Pande GN, Zienkiewicz OC (eds) Soil mechanics – transient and cyclic loads. Wiley, Chichester, pp 219–252

Celebi M, Crouse CB (2001) Recommendations for soil structure interaction (SSI) instrumentation, report for consortium of organizations for strong-motion observation systems, Emeryville, CA

Chai YH, Hutchinson TC (1999) Flexural strength and ductility of reinforced concrete bridge piles. Report no UCD/STR-99/02, University of California, Davis, CA

Clough RW (1955) On the importance of higher modes of vibration in the earthquake response of a tall building. Bull Seismological Soc Am 45:289–301

Comartin CD, Keaton JR, Grant PW, Martin GR, Power MS (1996) Transitions in seismic analysis and design procedures for buildings and their foundations. In: Proceedings of the 6th workshop on the improvement of structural design and construction practice in the US and Japan, ATC Report no ATC 15–5, Victoria, BC

Crouse CB (2002) Commentary on soil-structure interaction in US seismic provisions. In: Proceedings of the 7th US national conference on earthquake engineering, Boston, Massachusetts

CSI (2011) SAP2000 version 15.0 nonlinear – user's manual. Computers and Structures Inc., Berkeley, CA

De Alba P, Seed HB, Chan CK (1976) Sand liquefaction in large-scale simple shear tests. J Geotech Eng Div, ASCE, 102(GT9):909–927

Dobry R, Gazetas G (1988) Simple method for dynamic stiffness and damping of floating pile groups. Geotechnique 38(4):557–574

Eguchi RT, Goltz JD, Taylor CE, Chang SE, Flores PJ, Johnson LA, Seligson HA, Blais NC (1998) Direct economic losses in the Northridge earthquake: a three-year post-event perspective. Earthquake Spectra 14:245–264

Elgamal A, Yang Z, Parra E (2002) Computational modeling of cyclic mobility and post-liquefaction site response. Soil Dyn Earthq Eng 22:259–271

El Naggar MH, Bentley KJ (2000) Dynamic analysis for laterally loaded piles and dynamic p-y curves. Can Geotech J 37(6):1166–1183

El Naggar MH, Novak M (1995) Nonlinear lateral interaction in pile dynamics. J Soil Dyn Earthquake Eng 14(3):141–157

El Naggar MH, Novak M (1996) Nonlinear analysis for dynamic lateral pile response. J Soil Dyn Earthquake Eng 15(4):233–244

El Naggar MH, Novak M, Sheta M, El-Hifnawy L, El-Marsafawi H (2011) DYNA6, version 6.0 a computer program for calculation of foundation response to dynamic loads. Geotechnical Research Centre, The University of Western Ontario, London

El Naggar MH, Shayanfar MA, Kimiaei M, Aghakouchak AA (2005) Simplified BNWF model for nonlinear seismic response analysis of offshore piles with nonlinear input ground motion analysis. Can Geotechn J 42:365–380

Federal Emergency Management Agency (FEMA) (2004) FEMA 440 – Improvements to inelastic seismic analysis procedures, Washington, DC

Finn WDL (2005) A study of piles during earthquakes: issues of design and analysis. Bull Earthquake Eng 3:141–234

Gazetas G (1991) Formulas and charts for impedance of surface and embedded foundations. J Geotech Eng, ASCE 117(9):1363–1381

Gazetas G (2001) SSI issues in two European projects and a recent earthquake. In: Proceedings of the 2nd UJNR workshop on soil-structure interaction, Tsukuba, Japan

Gazetas G, Dobry R (1984) Simple radiation damping model for piles and footings. J Eng Mech, ASCE 110(6):937–956

Gazetas G, Makris N (1991) Dynamic pile-soil-pile interaction, part I: analysis of axial vibration. Earthquake Eng Struc 20:115–132

Gazetas G, Mylonakis G (2001) Soil structure interaction effects on elastic and inelastic structures. In: Proceedings of the 4th international conference on recent advances in geotechnical earthquake engineering and soil dynamics, Paper no. SOAP-2, San Diego, CA